To Barbara

MERRILL'S SERIES IN MECHANICAL AND CIVIL TECHNOLOGY

Bateson, *INTRODUCTION TO CONTROL SYSTEM TECHNOLOGY*, 2e   (8255-2)

Hopeman, *PRODUCTION AND OPERATIONS MANAGEMENT:*   (8140-8)
    Planning, Analysis, and Control, 4e

Humphries, *MOTORS AND CONTROLS*   (20235-3)

Keyser, *MATERIALS SCIENCE IN ENGINEERING*, 4e   (20401-1)

Lamit/Paige, *COMPUTER-AIDED DESIGN AND DRAFTING*   (20475-5)

Maruggi, *TECHNICAL GRAPHICS: Electronics Worktext*   (20311-2)

Mott, *APPLIED FLUID MECHANICS*, 2e   (08305-2)

Mott, *MACHINE ELEMENTS IN MECHANICAL DESIGN*   (20326-0)

Rolle, *INTRODUCTION TO THERMODYNAMICS*, 2e   (8268-4)

Rosenblatt/Friedman, *DIRECT AND ALTERNATING CURRENT*   (20160-8)
    MACHINERY, 2e

Webb, *PROGRAMMABLE CONTROLLERS: Principles and*   (20452-6)
    Applications

Wolf, *STATICS AND STRENGTH OF MATERIALS*   (20622-7)

Preface

This book, which is intended for both civil and mechanical engineering technology programs, uses a parallel approach to cover the topics of statics and strength of materials. In this approach the topics are carefully integrated, bringing forward topics of strength of materials as soon as the student understands the prerequisite statics topics.

Why such an approach? I have undertaken this project in this way so that the student can be more highly motivated by seeing the applications of statics topics that are found in strength of materials. Technology students are very pragmatic; this makes them valuable in an engineering team after they graduate. Yet, this basic pragmatism also makes the typical course in statics seem very abstract. The problems appear to have no practical resolution; they seem to start and end with abstract forces without leading to conclusions about the configuration or suitability of a structure or member for a particular application. Furthermore, very few interesting demonstrations or lab experiments are available with a typical statics course. However, in the parallel approach, students study the mechanical properties of materials along with statics. Exciting labs and demonstrations that involve tensile tests can be introduced early in the course to capture the students' interest and to reinforce their understanding. In addition, the students can begin immediately to size simple members or to estimate load ratings. Thus, statics becomes more interesting because its topics are applied immediately.

In the parallel approach, the level of mathematics increases gradually, with the topics divided approximately into thirds. The first third requires an understanding of college algebra, the second third requires a proficiency in trigonometry, and the final third presumes a familiarity with polynomial calculus (used only where natural and appropriate). Thus, only a background in algebra and trigonometry is necessary for the majority of the text, and the latter part of this text can be covered concurrently with the student's first course in calculus.

Table 1 shows a comparison between the parallel approach (on the left) and a topical outline of the serial approach (on the right). From this comparison, an instructor can see that all of the topics are covered in both approaches; however, the pedagogical advantages of the parallel approach are found inherently in a different organization, which this text uses.

Table 1

The Parallel Approach	*The Serial Approach*

Part I

Structures, Forces, and Equilibrium
    Force, weight, and mass
    Unidirectional equilibrium
	Properties of Forces
	    Force, weight, and mass
	    Resultant forces
	    Components of forces
	    Summation of components
	    Equilibrium
	    Moments and couples

Straight-Line Tension Structures
    Tension stresses and strains
    Modulus of elasticity
    The tension test	

Compression and Bearing
    The compression test
    Thermal expansion
    Members of two materials
	Planar Force Systems
	    Concurrent force equilibrium
	    Parallel force systems
	    General force systems

Shear Elements
    Shear stress and strain
    Fasteners and welds
    Friction
	Distributed Forces
	    Distributed loads
	    Areas and centroids

Torsion
    Moments and couples
    Torsion stress
    Torsion deformation
    The torsion test
	Trusses and Frames
	    Method of joints
	    Method of sections
	    Frames

Angled Structures
    Resultants of forces
    Components of forces
    Summation of components
    Equilibrium of concurrent forces
    Equilibrium of general forces
    Frames
    Inclined planes
    Angle of friction
	Cables and Arches
	    Parabolic
	    Catenary

	Friction
	    Friction
	    Angle of friction
	    Inclined planes

Trusses
    Method of joints
    Method of sections	

Part II

Shaped Structures
    Parabolic cables and arches
    Catenary cables and arches
    Pressure vessels
	Stress and Strain
	    Uniaxial stress and strain
	    Modulus of elasticity
	    The tension test
	    The compression test
	    Thermal expansion
	    Members of two materials

Bending Members
    Equilibrium of parallel forces
    Distributed loads
    Bending moment and shear force
    Stress in rectangular beams
    Shear in rectangular beams
	Shear Stress
	    Direct shear
	    Fasteners and welds
	    Pressure vessels
	    Torsion stress
	    Torsion deformation
	    The torsion test

(End for single non-calculus course)	

Beams of Non-Rectangular Cross Section
    Areas and centroids
    Moment of inertia
    Bending stresses
    Shear stresses
	Properties of Areas
	    Areas and centroids
	    Moment of inertia

Shear Force and Bending Moment Diagrams
    Shear force diagrams
    Bending moment diagrams
	Shearing Force and Bending Moment
	    Shear force and bending moment
	    Shear force diagrams
	    Bending moment diagrams

Deformation of Beams
    Bending moment and curvature
    Slope and deflection diagrams
    Variable EI (optional)
	Bending
	    Bending stress
	    Shear stress

Statically Indeterminate Beams (optional)
    Beam on three supports
    Beam built-in at both ends
    Other statically indeterminate beams
	Deformation of Beams
	    Bending moment and curvature
	    Slope and deflection diagrams
	    Variable EI (optional)

Columns
    Euler's relationship
    Steel columns
    Wood columns
	Statically Indeterminate Beams (optional)
	    Beam on three supports
	    Beam built-in at both ends
	    Other statically indeterminate beams

Compound Stresses (optional)
    Mohr's Circle
    Principal stresses
    Failure theories
	Columns
	    Euler's relationship
	    Steel columns
	    Wood columns

	Compound Stresses (optional)
	    Mohr's Circle
	    Principal stresses
	    Failure theories

This text is well-suited for two- or three-term courses in Statics and Strength of Materials; however, it can also be used for a one-term noncalculus Survey of Structural Fundamentals course by covering Chapters 1–9.

"Never overestimate the knowledge of your audience nor underestimate their intelligence." These words of Eric Severied have been taken to heart in my writing. Since all persons using this text will not have the same preparation, a great deal of information, which may be familiar to some but unfamiliar to others, is included at every possible juncture. Illustrative problems include considerable details, such as algebraic manipulations and the treatment of units. Even though such details are provided, students are encouraged to ponder the practical implications and results of the problems and to work through a sophisticated and rigorous presentation. Although derivations are included and computers are introduced, math and computers do not obscure the primary thrust of this book, which is an understanding of structures and their wonderful technology.

Acknowledgments

I thank the following reviewers for their helpful comments and suggestions: Ray Dickerson (Guilford Technical Community College), C. Quentin Ford (New Mexico State University), Edward Garcia (SUNY/Farmingdale), James Knowlton (Wentworth Institute of Technology), Jerry McDonald (Columbus Technical Institute), James Moore (Tarrant County Junior College), Hal Roach (Purdue University), and Harold Wirtz (Vermont Technical College). I would also like to thank Dr. K. I. Jacob for his thorough working of each problem in this text to ensure its accuracy, my colleagues at the University of Houston for allowing me to experiment with our students, and two classes of engineering technology students who patiently and fondly endured the development of these problems. Most of all I wish to thank Barbara, not only for her word processing but especially for her technical editing and encouragement.

Contents

Chapter Nine   Bending Members   261

Chapter Ten   Beams of Non-Rectangular Cross Section   293

Chapter Eleven   Shear Force and Bending Moment Diagrams   337

CHAPTER ONE

$$\sum F \mathbf{\ddagger} = 0$$

Structures, Forces, and Equilibrium

Beginning with a discussion of what constitutes a structure, this chapter defines *loads* and *forces*. Kinds of forces are briefly reviewed. Both the English Gravitational System (EGS) of units and the Système International (SI) are examined in terms of their differing philosophies concerning force and mass. This examination requires some consideration of the concepts of velocity and acceleration before the EGS unit of mass and the SI unit of force can be derived in terms of Newton's Second Law. Conversion from one system of units to the other is accomplished by considering freely falling objects.

Two basic analytical skills, the cancellation of units and power-of-ten notation, are exercised. Unit prefixes and abbreviations are developed in detail.

The chapter closes by introducing the concept of equilibrium of forces. The use of the free-body diagram and static equilibrium equations is demonstrated on some pulley systems.

1.1 STRUCTURES

A structure is anything that transmits forces or supports loads. Structures may occur in nature, or they can be constructed or manufactured. A bridge is the purest example of a structure. Practically every part of a bridge contributes to its ability to carry loads. Buildings also are structures. But a building is much more than a load-carrying device. A building's primary purpose is to provide an environment for its occupants or the things contained within. Much of the building is devoted to appearance, lighting, heating, and other functions. Nevertheless, when we think of buildings, we usually think of structure, because the first step in constructing a building is the erection of its structure, which announces the arrival of the building on the scene.

1

Though often disguised or covered from view, vehicles and machines also have structure. Although we give more thought to the function of machines when we use them, structure is of prime importance in their design and construction. While buildings and bridges must simply resist the forces and loads applied to them, machines generate additional forces, due to motion, that may be enormous in comparison to the applied loads. All manufactured machines have their structural system, just as the human body (a biological machine) has its skeletal system to carry the forces and loads that it will encounter.

Any structure, be it a bridge, building, or machine, that is not strong enough to perform its tasks can kill or maim people and destroy property. It is also true that whereas a structure must be strong enough to do the job, excess strength above the appropriate margin for safety is a waste of money, energy, and resources. Since structures must be strong enough, but not too strong, an understanding of structures always involves the question "How much?" We describe this kind of understanding as quantitative. An understanding of structures is of little value unless it is quantitative.

Structures, or components of structures, can be classified into basic categories such as tension structures, compression structures, shear structures, torsion structures, angle structures, trusses, shaped structures, beams, and levers. Each of these categories offers certain advantages, limitations, and characteristics in their responses to forces. Later on in this book these classifications are identified by their characteristics. They are then investigated as to the forces prevailing, the possible ways in which failure may happen, and appropriate design criteria.

Structures can be fascinating. In addition, the study of structures has many practical rewards because the applications are many. Everything in nature has its structural aspects. Those who have invested some of their time and energies to an understanding of structures usually reflect that they have enjoyed it, and that it has proved to be useful professionally. But, most of all, they have attained a better appreciation and understanding of the nature of things.

1.2 FORCES

A force is a push or a pull. Forces upon structures may be caused by several things: gravity, aerodynamic and hydrodynamic forces, electrostatic forces, magnetism, dynamic forces, and contact forces.

Gravitational Attraction

Gravity is the force that all matter in space exerts upon other matter. Gravitational attraction holds together the earth and all heavenly bodies and confines them to their regular paths or orbits. On earth, gravitational attraction causes what we call *weight,* which is the pull of gravity. Structures must withstand their own weight and the weight of any objects or loads they must support. Weight is a force acting downward, or, more precisely, in the direction of the center of the earth.

Aerodynamic and Hydrodynamic Forces

When air or any moving gas or liquid impinges upon an object, forces result. These forces are due to the push upon the object by direct collision as well as to any pull upon the object resulting from the flow of the fluid around it. *Dynamic* means moving. *Aero* and *hydro* are prefixes meaning air and water. Aerodynamics and hydrodynamics are sciences devoted to the study of the forces due to the movement of gases and liquids. Such forces occur whether the body is stationary and the fluid is moving (wind) or whether the fluid is still and the object is moving through it (a vehicle). Aerodynamic forces produce what is known as lift and drag upon structures and vehicles. *Lift forces* tend to raise an aircraft or, in the negative sense, help to hold a wheeled vehicle on the road. *Drag forces* tend to oppose the motion of the vehicle.

Aerodynamic forces cause *wind loads* upon stationary structures. Such loads are mainly horizontal. They are discussed more fully in chapter 9.

Electrostatic Forces

Electrostatic forces are caused by the attraction or repulsion of electrical charges. An example we have all experienced is the attraction between dry hair and a comb. Electrostatic forces may be significant in the structure of electrical equipment.

Magnetic Forces

Magnetic forces are caused by magnetic attraction. They may result from an electromagnet (generated intentionally, such as in motors, generators, or solenoids) or a permanent magnet (used as holding and locking devices). However, magnetic forces may also be an unavoidable side-effect in some types of electrical equipment. Magnetic forces are unusual in that they are material selective. They act upon only iron and low-alloy steels. They also act upon other magnets, whether permanent or electrically induced.

Dynamic Forces

Dynamic forces are pushes and pulls caused by the motions of an object or parts of a machine. In building structures and machinery, some dynamic forces are caused by vibrations. If due to earthquake motions, dynamic forces are referred to as *seismic* forces. Sometimes confusion exists as to whether the motion causes forces or forces cause the motion. To avoid that ''chicken and egg'' argument, let us say that when motion is present there are likely to be dynamic forces that may be very large. To determine the severity of dynamic forces in machinery, some calculation by an engineering specialist known as a *structural dynamicist* may be required.

Contact Forces

Contact forces occur when one object touches or is connected to another object. The objects might be forced together by weight or electrostatic, magnetic, or dynamic forces. Friction is one form of contact force.

Physicists argue that all forces finally can be attributed to gravity, electromagnetism, and two other forces internal to the atom. From the viewpoint of physics, contact forces and dynamic forces are only by-products or manifestations of gravity, electricity, or magnetism. However, when we try to understand structures, it is best to add dynamic and contact forces to the list, as we have done here.

like to build sme Thing put sme Thing to Together

1.3 LOADS

In the world of structures the terms *load* and *force* are sometimes used interchangeably. But, in more precise technical usage, *load* is a force that a vehicle, building, or machine was intended to carry. Load is thus deliberately applied. The term *force,* on the other hand, includes both the applied loads and any reactions to applied loads, even though these reactions may be unintended or unwanted.

1.4 UNITS OF FORCE

The units of force commonly used in structures are pounds (lb), newtons (N), tons (T), kilograms (kg), and kilopounds (kips). In structural practice all of these units are used. Considerable confusion exists regarding proper use of units, some of which is attributable to conversion to the metric system. However, in fairness to the metric system, confusion was common even before the United States decided to apply the metric system to its technology.

Lack of clarity regarding units greatly complicates the work of engineers, technicians, and technologists. Responsibility for translation often falls upon those who find themselves in the position of interpreting between engineering and the crafts. Persons with a thorough knowledge of the two systems of units and the rationales upon which they are based can make many contributions in construction and manufacturing.

The two systems of units are the *English Gravitational System (EGS)* and the *Système International (SI)* or International System, also called the metric system. One uses pounds and feet and the other uses kilograms and meters; but differences go further than that. The most fundamental yet subtle difference is that the EGS is a force-based system, and the SI is a mass-based system. Let us explore this difference.

Telling Them apart like discribe sme Thing

1.5 FORCE AND WEIGHT

The first step in distinguishing between EGS and SI units is understanding the difference between force and weight.

> *Force* is a push or a pull.
> *Weight* is the force of gravity.

A 1-lb object produces a force of 1 lb due to the gravitational pull of the earth at the equator at sea level. For common structural purposes the difference in the pull of gravity from the lowest to the highest point on the earth's surface and from the equator to the poles is negligible. So, think of the weight of an object as the push or pull caused by gravity on that object at the surface of the earth.

Beyond the earth, the weight of an object varies considerably with gravitational pull. For example, an object weighing 60 lb on the earth's surface will have a weight of only

10 lb on the moon's surface, because the gravitational pull on the surface of the moon is only one-sixth that of the earth.

1.6 MASS

Mass must also be distinguished from weight.

> *Mass* is the measure of the quantity of matter.

The SI system of units is mass-based. There is a standard kilogram at the International Bureau of Weights and Measures in Paris. Using balances, masses of objects are compared to this international standard. The mass of an object does not vary with its location, as does weight. A kilogram mass is 1 kg on the surface of the earth and also 1 kg on the surface of the moon. Although an object's weight would be reduced on the moon to one-sixth of its value on earth, its mass remains constant.

In the original metric system, which was the precursor of the SI, 1 g (gram) was defined as the mass of 1 cm^3 (cubic centimeter) of pure water at a temperature of 0° C. One kg was, therefore, equal in mass to 1 000 cm^3 or 1 L (liter) of water. However, the water-based standard has been abandoned in favor of the standard kilogram in Paris and is today only an approximation.

> **Important Concepts**
> The EGS is a force-based system.
> Weight is the force due to gravity.
> The SI (metric) is a mass-based system.
> Mass is a measure of the amount of matter in an object.
> The pound in the EGS is the unit of force.
> The kilogram in the SI is the unit of mass.

Since by definition pounds and kilograms are units describing two different quantities, theoretically there can be no conversion factor between the kilogram and the pound. But, as a practical matter, pounds and kilograms are used for both force and mass. In order to rationalize this seeming inconsistency, we will go further into the theory upon which these unit systems are based.

1.7 RELATIONSHIP BETWEEN FORCE AND MASS

When a force (push or pull) acts upon a mass (quantity of matter), it causes the mass to accelerate (change its velocity). The relationship between force, mass, and acceleration is known as *Newton's Second Law of Motion*, which is expressed as follows:

$$F = ma \tag{1.1}$$

In this relationship, F is force, m is mass, and a is acceleration. In the English Gravitational System, force is measured in pounds and acceleration in feet per second per second (or, per second squared).

Acceleration is a change in velocity or speed. When you drive a car you experience acceleration whenever you step on the gas pedal. The change in velocity can be observed by watching the speedometer. When the speed increases, you undergo an acceleration.

Velocities may be expressed in any distance unit over any time unit, such as kilometers per hour or meters per second. But in the English Gravitational System, velocity is usually expressed in feet per second; 1 mile per hour (mph) is equal to 1.457 feet per second (ft/s). Because acceleration is the change in feet per second per second, the units are feet per second squared (ft/s^2).

An acceleration of 1 ft/s^2 means that each second, the velocity of the object is increasing by one foot per second. If the object is started from rest, after 2 s it would have a velocity of 2 ft/s, as shown in figure 1.1a. After 2 s, it would have a velocity of 2 ft/s. After 100 s, its velocity would be 100 ft/s.

Mass in the English Gravitational System is expressed in units of slugs.

> One *slug* is the amount of mass which, when acted upon by a force of 1 lb, will accelerate at a rate of 1 ft/s^2.

The slug is a derived, rather than a defined, unit. By this we mean that the slug is derived from other units. There is no standard for a slug.

It is doubtful that you will ever encounter the unit of slugs in the field or shop. It wasn't popular when the EGS units were the standard in the United States. And, it is likely to enjoy a well-deserved oblivion as technology moves to SI units.

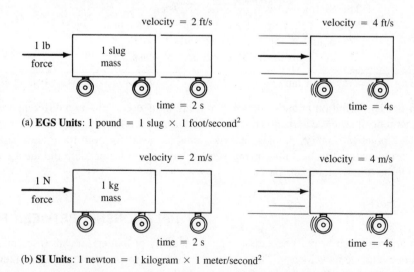

(a) **EGS Units**: 1 pound = 1 slug × 1 foot/second^2

(b) **SI Units**: 1 newton = 1 kilogram × 1 meter/second^2

Figure 1.1 Newton's Second Law of Motion in two different systems of units.

In SI the kilogram is the unit of mass based upon the standard kilogram kept in Paris. The SI units of acceleration are meters per second per second. These are based upon a standard meter and a standard second. The second in SI units is identical to that in EGS units.

In SI units, the unit of force is the newton.

One *newton* is the force required to accelerate a mass of 1 kg at a rate of 1 m/s².

If a mass of 1 kg has an acceleration of 1 m/s², as shown in figure 1.1b, it is being propelled with a force of 1 N (newton). The newton is defined in terms of acceleration and mass. The newton (SI), like the slug (EGS), is a derived unit and rarely used among shop and construction personnel. However, as SI gains momentum, considerable pressure is building to establish the newton as the preferred unit of force in all technological work. Therefore, engineers, technicians, and technologists (upon whom the burden of translation will fall) must be familiar with its use.

Table 1.1   Units in SI and EGS

Quantity	SI Units	EGS Units
force	newton (N)*	pound (lb)
mass	kilogram (kg)	slug*
length	meter (m)	foot (ft)
time	second (s)	second (s)

*Derived unit from Newton's Second Law

In SI the first letter of the unit is *never* capitalized even though it may be the name of a person, such as the newton. However, an abbreviation for a unit is capitalized if it is the name of a person, (e.g., N).

The abbreviations in table 1.1 for the EGS are most common in technical usage.

ILLUSTRATIVE PROBLEMS

1.1 How much force is required to give a mass of 5 slugs an acceleration of 450 ft/s²?

Solution:

$$F = ma$$
$$m = 5 \text{ slugs}$$
$$a = 450 \text{ ft/s}^2$$

Therefore,

$$F = 5 \times 450 = 2\,250 \text{ lb}$$

1.2 What will be the acceleration of a 10-kg mass when acted upon by a force of 0.990 N?

Solution:

$$F = ma$$
$$F = 0.990 \text{ N}$$
$$m = 10 \text{ kg}$$

Solving the relationship algebraically for *a*.

$$a = \frac{F}{m}$$

$$a = \frac{0.990}{10} = 0.099\ 0 \text{ m/s}^2$$

PROBLEM SET 1.1

1.1 A drag racing car accelerates from zero to 200 mph in 10 s. What is its average acceleration?

1.2 A car skids from a velocity of 20 m/s to zero in one second. What is its average acceleration?

1.3 A 2 000-kg spacecraft fires its retrorockets, having a constant force of 1 000 N for 100 s. If its forward velocity was 20 000 m/s before, what is its velocity after the burn? What was the acceleration?

1.4 A barge has a mass of 20 000 slugs. A tow boat exerts a force of 1 000 lb upon the barge. What acceleration is produced?

1.8 CONVERSION FACTORS

The following conversion factors have been reduced from ASTM (American Society for Testing Materials) Specification No. 380.[1]

$$
\begin{array}{l}
1 \text{ lb} = 4.448\ 222 \text{ N} \\
1 \text{ slug} = 14.593\ 90 \text{ kg} \\
1 \text{ ft} = 0.304\ 8 \text{ m}
\end{array}
$$

These equivalencies may be expressed as fractions such as

$$\frac{1 \text{ lb}}{4.448\ 222 \text{ N}} \quad \text{and} \quad \frac{1 \text{ ft}}{0.304\ 8 \text{ m}}$$

or

$$\frac{4.448\ 222 \text{ N}}{1 \text{ lb}} \quad \text{and} \quad \frac{0.304\ 8 \text{ m}}{1 \text{ ft}}$$

Since the fractions have numerators equivalent to their denominators, when taken together with their units, the fractions are equal to one. Any quantity may be multiplied by the number one without changing its value. This principle is employed in the proper use of conversion factors as we will now discover.

Suppose that you wished to express 304 ft in meters. Write

$$304 \text{ ft} \times \frac{0.304 \text{ 8 m}}{1 \text{ ft}}$$

The conversion factor is placed as a fraction with the unit (ft) in the denominator. Then cancel the unwanted unit, leaving only the desired unit of meters as follows:

$$304 \; \cancel{\text{ft}} \times \frac{0.304 \text{ 8 m}}{1 \; \cancel{\text{ft}}} = 92.659 \text{ 2 m}$$

Similarly, to convert 0.063 5 N to pounds:

$$0.063 \text{ 5} \; \cancel{\text{N}} \times \frac{1 \text{ lb}}{4.448 \text{ 222} \; \cancel{\text{N}}} = 0.014 \text{ 275 366 65 lb}$$

1.9 SIGNIFICANT FIGURES

In the previous calculations the number 0.014 275 366 65 is far too precise for most structural applications. It would be regarded as absurd to use so many decimal places, because it implies that the weight was measured with an accuracy of one hundred millionth of a pound! In technology the figures in a number imply the degree of physical measurement or ability to measure all the figures indicated. These are called *significant figures*. Three or four significant figures is normal structural practice. For structural applications it is good to use the following rule:

> When the first significant digit is a one, round the number to four figures. If the first significant digit is two through nine, round it to three figures.

The two unit conversions used in the examples of section 1.8 would be rounded as follows:

92.659 2 m would be rounded to three places, 92.7 m

0.014 275 366 65 lb would be rounded to four places, 0.014 28 lb

Note that in the second case, the zero to the right of the decimal point didn't count when rounding to four figures. The zero is merely a spacer for placing the decimal point. It is not used in the count of four.

Do not give more significant places in an answer than you had in the input information. This rule may mean using fewer than three or four significant places. With modern electronic calculators and computers it is best to carry as many decimal places as possible while performing a calculation. But when the calculation is complete, round it off to the proper number of significant figures.

For example, the quantity 325.643 lb implies the ability to measure over 300 lb to an accuracy of 0.001 lb. An extremely sensitive balance, operating in a precisely controlled environment, would be required to make such a measurement. An accurate platform scale would give a reading of 326 lb under normal shop or field conditions, because it can be read only to the nearest pound. So, unless highly refined measurements are actually used or intended, the number should be rounded to 326 lb.

In another example, the length 1.643 92 m, with six significant figures, implies the ability to measure something approximately the height of a person with an accuracy of 0.000 01 m. This distance is about one-tenth the thickness of a human hair, measurable only with precise equipment in a temperature-controlled environment. That length should be rounded to 1.644 m, which implies quite accurate measurement for structural purposes.

In a third example, precision may also be expressed by zeros. In the measurement 2.000 in. (inches), the three zeros are significant figures. They mean that the length is neither 1.999 or 2.001 in., but rather 2.000 ± 0.0005 in. Implied is a measurement precision of one-thousandth of an inch, requiring a micrometer.

1.10 POWER-OF-TEN NOTATION

In structures, loads and forces can be very large. Millions of pounds of force are not uncommon. Conversely, deformations can be very small. Rigid steel structures may deform under load only millionths of a meter. *Power-of-ten notation* is a convenient means of handling calculations involving these outsized and undersized numbers.

Table 1.2 Power-of-ten notation

$$1\ 000\ 000 = 10^6$$
$$100\ 000 = 10^5$$
$$10\ 000 = 10^4$$
$$1\ 000 = 10^3$$
$$100 = 10^2$$
$$10 = 10^1$$
$$1 = 10^0$$
$$0.1 = 10^{-1}$$
$$0.01 = 10^{-2}$$
$$0.001 = 10^{-3}$$
$$0.000\ 1 = 10^{-4}$$
$$0.000\ 01 = 10^{-5}$$
$$0.000\ 001 = 10^{-6}$$

To convert to power-of-ten notation, count the places from the decimal point to the right of the 1. The resulting number becomes the power of ten. For example:

$$10\,000. = 10^4 \leftarrow \text{exponent}$$

When you count to the right, the exponent is positive. To the left, it is negative.

$$0.\underset{\longleftarrow}{000\ 01} = 10^{-5 \leftarrow \text{ exponent}}$$

Numbers can be expressed in power-of-ten notation as follows:

$$59\ 376\ 000. = 5.937\ 6 \times 10\ 000\ 000 = 5.937\ 6 \times 10^7$$
$$0.001\ 735 = 1.735 \times 0.001 = 1.735 \times 10^{-3}$$

Or, more directly:

$$\underset{7\ 6\ 5\ 4\ 3\ 2\ 1}{59\ 376\ 000.} = 5.937\ 6 \times 10^7$$

$$\underset{1\ 2\ 3}{0.001\ 735} = 1.735 \times 10^{-3}$$

1.11 E FORMAT

To avoid complications in printing and typing numbers in power-of-ten notation, E format was developed. With E format it is not necessary to type or print exponents halfway above the line. In E format the capital letter E means exponent of ten. The letter E is then followed by the exponent. For example, the number 2.537×10^6 is represented in E format as 2.537E6. More examples follow:

$$1\ 809. = 1.809 \times 10^3 = 1.809E3$$
$$37\ 200\ 000. = 3.72 \times 10^7 = 3.72E7$$
$$0.070\ 3 = 7.03 \times 10^{-2} = 7.03E-2$$
$$0.000\ 001\ 906 = 1.906 \times 10^{-6} = 1.906E-6$$
$$4.97 = 4.97 \times 10^0 = 4.97E0$$

Computations in E format involve only addition or subtraction. When multiplying numbers in E format, add the exponents. For example:

$$1.376E3 \times 2.058E7 = 2.83E10$$
$$3.78E2 \times 9.85E-3 = 37.2E-1 = 3.72E0$$
$$7.80E-3 \times 2.37E-4 = 18.49E-7 = 1.849E-6$$

When dividing numbers in E format, subtract the exponent of the denominator from that of the numerator. For example:

$$\frac{4.03E3}{1.623E1} = 2.48E2 \qquad\qquad \frac{4.73E-3}{3.06E4} = 1.546E-7$$

$$\frac{1.032E5}{2.43E7} = 0.425E-2 = 4.25E-3 \qquad\qquad \frac{7.76E8}{3.80E-2} = 2.04E10$$

$$\frac{9.81E-3}{7.31E-1} = 1.342E-2$$

1.12 UNIT PREFIXES

Unit prefixes, like power-of-ten notation, are used to handle very large or very small numbers. Kilopounds, millimeters, kilonewtons, microinches, and centimeters are examples.

In structural practice, the most commonly used unit prefixes are as shown in table 1.3.

Table 1.3 Unit prefixes

Prefix	Symbol	Means	Amount	E Format
mega-	M	one million times	1 000 000	E6
kilo-	k	one thousand times	1 000	E3
centi-	c	one hundredth of	0.01	E−2
milli-	m	one thousandth of	0.001	E−3
micro-	μ	one millionth of	0.000 001	E−6

When using unit prefixes, remember:

☐ Avoid the *centi-* prefix, except in the case of centimeters.

☐ No prefixes should be used on kilogram or kilopound units. In other words, do not form compounds such as *microkilograms* or *megakilopounds*.

Examples:

$$150 \text{ kN} = 150 \text{ kilonewtons} = 150E3N = 150\ 000 \text{ N}$$
$$3\ 000\ \mu \text{ in.} = 3\ 000 \text{ microinch} = 3\ 000E{-}6 \text{ in.} = 0.003 \text{ in.}$$
$$7.52 \text{ kips} = 7.52 \text{ kilopounds} = 7.52E3 \text{ lb} = 7\ 520 \text{ lb}$$

(The kip is a special abbreviation for the kilopound, used in structural practice.)

$$0.532 \text{ cm} = 0.532 \text{ centimeters} = 0.532E{-}2 \text{ m} = 0.005\ 32 \text{ m}$$
$$20.1 \text{ mm} = 20.1 \text{ millimeters} = 20.1E{-}3 \text{ m} = 0.020\ 1 \text{ m}$$

1.13 THE ACCELERATION OF GRAVITY

Galileo discovered that all objects fall to the earth at the same rate of acceleration. As an experiment, he simultaneously dropped a heavy cannonball and a lighter cannonball from the Leaning Tower of Pisa and observed that they struck the ground together. He concluded that when wind resistance is not a factor, as he felt was the case with the streamlined cannonballs, the acceleration of freely falling objects is independent of their weight. Years later, freely falling bodies were found to accelerate in a vacuum at a rate of about 32.2 ft/s^2 or 9.80 m/s^2. This acceleration is known as g (the acceleration of gravity). As a standard, g is 32.174 0 ft/s^2 or 9.806 65 m/s^2.

1.14 UNITS DERIVED FROM NEWTON'S LAW APPLIED TO A FALLING BODY

In section 1.7, the slug and the newton were said to be derived units. This derivation is done from Newton's Second Law applied to a freely falling body as follows.

When an object is released to fall freely, the only force upon it will be its weight, which is the pull of gravity (figure 1.2). At the time of release, wind resistance is not yet a factor. Therefore, the object will experience 1 g of acceleration. Applying Newton's Second Law, relationship 1.1, to the freely falling object:

$$F = ma$$

Here $F = W$ (weight), because the force applied to the object in this case is the weight of the object, and $a = g$, because the object is in free fall. Therefore:

$$W = mg \qquad \qquad \textbf{(1.2)}$$

Solving relationship 1.2 for m,

$$m = \frac{W}{g} \qquad \qquad \textbf{(1.3)}$$

In EGS units:

$$m = \frac{1 \text{ lb}}{32.174\ 0 \text{ ft/s}^2}$$

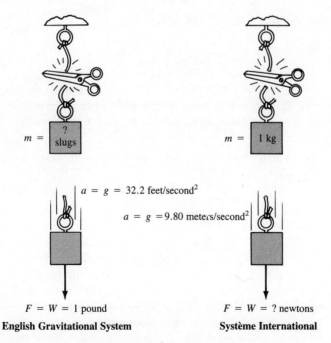

Figure 1.2 The freely falling object in EGS and SI units.

Table 1.4 Basic units and conversions

Quantity	Unit	Abbreviation	Prefixes	Conversion Factors
force	pound	lb	none	1 lb = 4.45 N***
	ton	T	none	1 T = 2 000 lb**
	kip	none	none	1 kg* = 2.20 lb
	newton	N	kilo (kN)	
	kilogram*	kg	none	
mass	kilogram	kg	none	1 kg = 2.20 lb*
	pound*	lb	none	
length	foot	ft	none	1 ft = 12 in.
	inch	in.	micro (μ in.)	1 ft = 0.305 m
	mile	none, except in	none	1 in. = 2.54 cm
		mph (miles per hour)		1 mile = 5 280 ft
	meter	m	kilo (km)	1 km = 0.621 miles
time	second	s	none	
	hour	hr	none	

*Commercial usage only.

**Numbers without decimal point are exact.

***Numbers with decimal point are rounded to the usual number of significant figures expected in this book.

Because the slug is defined as 1 lb per ft/s^2,

$$m = \frac{1}{32.174\ 0}\text{slugs}$$

Therefore:

An object weighing one pound on earth has a mass of approximately 1/32.2 slugs.

In SI units, which are mass-based, the force unit must be derived using relationship 1.2:

$$W = mg$$
$$W = 1 \text{ kg} \times 9.806\ 65 \text{ m/s}^2$$

Because 1 N is defined as 1 kg · m/s^2,

$$W = 9.806\ 65 \text{ N}$$

Therefore:

An object weighing 9.80 N on earth has a mass of approximately 1 kg.

Here, the conversion factor was truncated to three significant places, rather than rounded. This cutting is done so that the conversion factor from kilograms to pounds (derived in the next section) works out to the most widely accepted value of 9.80 in three significant figures.

1.15 CONVERTING KILOGRAMS TO POUNDS

It is *not* scientifically accurate to convert from kilograms to pounds, because the pound is a unit of force and the kilogram is a unit of mass.

However, it *is* precise to say that 1 kg of mass on the surface of the earth will exert a force of approximately 2.20 lb. This fact is shown in figure 1.3 and can be reasoned in the following way. The 1 kg of mass weighs 9.80 N. Therefore,

$$W = 9.80 \text{ N}$$

Using the conversion factor 1 lb = 4.448 222 N,

$$W = 9.80 \text{ N} \times \frac{1 \text{ lb}}{4.448\ 222 \text{ N}} = 2.204\ 623 \text{ lb}$$

Therefore, 1 kg of mass will exert a force of approximately 2.20 lb on earth.

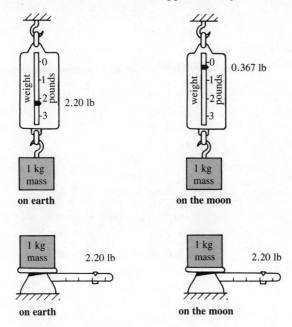

Figure 1.3 Mass and weight on earth and on the moon. Spring scales measure force; platform balances compare masses. The 1-kg mass would weigh less on the moon if measured on the spring scale, because weight is the force due to gravity. The platform balance would show no difference on earth from on the moon, because mass does not change in going from one place to the other.

Commercially, the conversion between pounds and kilograms is done all the time. They are used both as weight (force) and as mass. Commerce does not make a distinction between force and mass; only science does. Commerce uses the basic units of kilograms and pounds and equates 2.20 lb to 1 kg. Like shop and construction personnel, commerce does not use units of slugs or newtons (table 1.4).

PROBLEM SET 1.2

1.5 Express the following quantities in three significant figures, if the first digit is from 2 to 9; and four significant figures, if the first digit is 1.

a.	35 927 kN	i.	0.000 009 285 N
b.	0.001 893 8 lb	j.	7.536 T
c.	4.792 7 ft	k.	0.057 34 kg
d.	1.9 m	l.	1 753 810 km
e.	33.76 T	m.	0.031 257 lb
f.	0.987 69 m	n.	75 kg
g.	805.793 ft	o.	0.14 kips
h.	452.9 T	p.	0.009 08 m

1.6 Convert the following quantities from EGS units to corresponding SI units.

a.	580 lb	i.	83 ft
b.	9.80 ft	j.	0.010 43 lb
c.	0.138 2 lb	k.	65 T
d.	6 950 000 s	l.	0.750 s
e.	0.005 28 ft	m.	1.809 lb
f.	1 375 000 lb	n.	95.0 T
g.	15.83 ft	o.	803 miles
h.	0.037 5 miles	p.	5.50 kips

1.7 Convert the following SI units to corresponding EGS units.

a.	1.785 N	i.	0.043 s
b.	0.035 6 m	j.	0.000 932 km
c.	80.3 km	k.	17 m
d.	75.1 N	l.	0.095 5 N
e.	0.000 147 3 m	m.	38 200 m
f.	5 090 N	n.	0.043 8 s
g.	753 m	o.	27.3 N
h.	8.52 s	p.	0.802 m

1.8 Convert the following numbers to E format with the decimal point after the first digit.

a.	10 340	g.	14.32
b.	98.5	h.	0.000 426
c.	3 500 000	i.	178.3
d.	0.278	j.	9 050 000 000
e.	0.000 173 2	k.	105.2
f.	0.000 000 000 724	l.	385

1.9 Convert the following numbers from E format to standard format.

   a. 7.00E2   **g.** 2.80E5

   b. 3.55E−2   **h.** 1.302E3

   c. 1.473E1   **i.** 5.55E−6

   d. 9.80E−5   **j.** 7.38E8

   e. 1.023E−3   **k.** 6.89E−1

   f. 4.78E6   **l.** 8.02E−4

1.10 Convert the following quantities from prefixed units to basic units.

   a. 5.03 km   **g.** 1.430 km

   b. 0.043 1 mm   **h.** 8.07 kg

   c. 425 μ in.   **i.** 9.53 Mm

   d. 0.000 483 MN   **j.** 0.356 mm

   e. 83.0 cm   **k.** 19.82 kN

   f. 47 500 kN   **l.** 105 372 kN

1.11 Perform the following multiplications or divisions and express the answers in E format.

   a. 1.532E7 × 4.83E5   **g.** $\dfrac{1.043E-5}{2.46E-7}$

   b. 9.80E−2 × 8.37E4

   c. 1.003E−4 × 3.96E−2   **h.** $\dfrac{5.92E-3}{1.732E-1}$

   d. 6.08E8 × 3.42E−4

   e. $\dfrac{1.980E7}{4.53E4}$ **i.** $\dfrac{4.52E-6}{7.03E3}$

   f. $\dfrac{7.873E1}{7.05E-1}$

1.12 Convert the following quantities to the units indicated.

   a. 1.302 lb = _____ N

   b. 3.75E2 N = _____ lb

   c. 8.32 T = _____ kN

   d. 0.174 2 m = _____ ft

   e. 5.28E3 ft = _____ km

   f. 7.32E−3 mm = _____ μ in.

   g. 750 miles = _____ km

   h. 2.86E6 kg = _____ lb

   i. 9.32E−4 kg = _____ N

   j. 5.76 kips = _____ N

   k. 9.05E3 kN = _____ kips

   l. 1.052E3 m = _____ in.

   m. 2.05E3 kN = _____ T

   n. 4.56 km = _____ ft

   o. 1.07E4 lb = _____ kg

1.13 Astronaut Jack Schmitt of the *Apollo 17* mission collected a lunar rock of "mare basalt" that had a mass of 8.110 kg. Later analysis showed that the rock was older than 99.9 percent of the rock on earth. How many pounds does the rock weigh on earth? What is its mass in kilograms on earth? How many pounds of force did Astronaut Schmitt exert when he lifted the rock on the moon?

1.16 EQUILIBRIUM OF FORCES IN A SINGLE DIRECTION

According to Newton's Second Law, an object acted upon by an unbalanced force will accelerate. In this book we will investigate only structures that are *not* accelerating. These are called *static structures*. (*Static* means still or unmoving.)

If a structure is static, all the forces acting upon it must be in balance or "equilibrium." Equilibrium in a straight-line structure means that all the forces acting up must be balanced by those acting down. Or, if the line of action is horizontal, all the forces acting to the right must be equal to those acting to the left. In other words, the forces along the

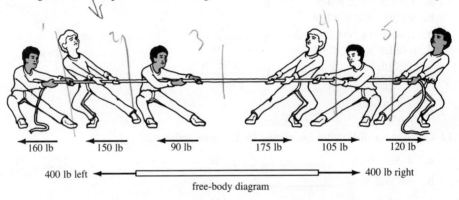

| 160 lb | 150 lb | 90 lb | 175 lb | 105 lb | 120 lb |

400 lb left ⟷ 400 lb right

free-body diagram

(a) In equilibrium, all pulls to the right balance those to the left.

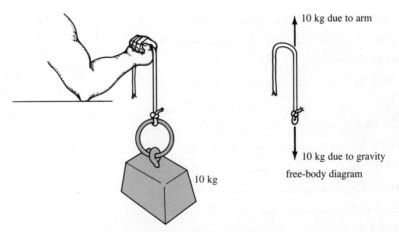

10 kg due to arm

10 kg due to gravity

free-body diagram

10 kg

(b) In equilibrium, all pulls up must balance all those acting down.

Figure 1.4 Equilibrium in horizontal (a) and vertical (b) directions.

line of action of the structure, no matter which direction that line may take, must be in balance.

Normally, a positive value will be assigned to the forces acting to the right and a negative value to those acting to the left. When the forces are vertical, a positive value will be assigned to those acting up and a negative value to those acting down.

This convention is merely a bookkeeping system. The reverse sign convention is equally valid and is used by some engineers and technologists. The important thing is to choose a sign convention and use it consistently through the entire problem. We use this convention throughout this book: Vertical forces will be positive when acting upward; horizontal forces will be positive when acting to the right. Using a sign convention, the law of static equilibrium can then be written

$$\sum V \updownarrow = 0 \tag{1.4}$$

or

$$\sum H \leftrightarrow = 0 \tag{1.5}$$

The symbol $\sum$ is the Greek letter sigma and means "the sum of."

The shorthand, $\sum V \updownarrow = 0$, means "The sum of all vertical forces taken positive upward is equal to zero."

Similarly, $\sum H \leftrightarrow = 0$ means "The sum of all horizontal forces taken positive to the right is equal to zero." The expression $\sum H \leftrightarrow = 0$ can be expanded to represent the tug of war depicted in figure 1.4a as follows:

$$\sum H \leftrightarrow = -160 \text{ lb} - 150 \text{ lb} - 90 \text{ lb} + 175 \text{ lb} + 105 \text{ lb} + 120 \text{ lb} = 0$$

$\sum V \updownarrow = 0$ can be expanded to represent the weight lifting of figure 1.4b as follows:

$$\sum V \updownarrow = 10 \text{ kg} - 10 \text{ kg} = 0$$

1.17 THE FREE-BODY DIAGRAM

To clearly illustrate the forces acting on a structure, the free-body diagram technique was developed.

A free-body diagram is a diagram of a structure or a part of a structure removed from its supports. The supports are replaced with arrows representing the forces that the supports provide.

As an example, a free-body diagram of the weight shown in figure 1.4b is constructed in figure 1.5. The body in this case is the weight. It is drawn free from its supports. The only thing supporting the weight is a rope. The rope was removed and replaced with an arrow representing the tension force, T, with which the rope would pull on the weight. The only other force on the body is the gravitational force, which on a 10-kg weight would be 98.0 N.

Using the principle of equilibrium, relationship 1.4, one can write:

$$\sum V \updownarrow = 0 \quad \text{or} \quad \sum V \updownarrow = T - 98.0 \text{ N} = 0$$

Therefore, it is possible to solve for T,

$$T = 98.0 \text{ N}$$

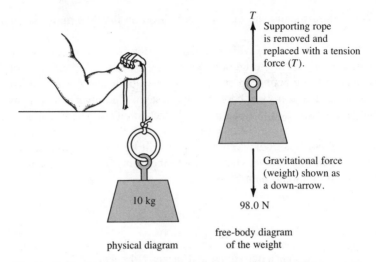

Figure 1.5 Construction of the free-body diagram of the weight.

"But," you say, "it is obvious in this case that the tension in the rope has to be equal to the weight." However, sometimes it isn't obvious. When it isn't obvious, the method of the free-body diagram and equilibrium equation can help you understand the load paths represented by the structure.

1.18 PULLEY SYSTEMS: MECHANICAL ADVANTAGE

A pulley system is a machine consisting of one or more cables wound over pulleys and configured to multiply the force applied when moving loads. In fact, the cable may be anything that holds its length under pull but is very flexible laterally. It may be a wire or fiber rope, a string, a rubber or leather belt, or even a chain. The pulleys are shaped to accommodate the type of cable used. Pulleys may be replaced with sprocket wheels in the case of chains.

The static equilibrium relationships 1.4 and 1.5 can be used to analyze pulleys. Although the pulleys are not truly static or motionless, they are assumed to be *quasistatic*, meaning that they are essentially at static equilibrium at all points over the course of their motion. While laypersons may not think of pulleys as structures, pulley systems can be regarded as structures of variable configuration. The use of free-body diagrams and static equilibrium is easier to demonstrate than to explain in advance. Illustrative problems 1.3, 1.4, and 1.5 demonstrate this use on pulley systems and constitute an introduction to the more general concept of equilibrium that will be used throughout this book.

It is assumed that the pulleys are frictionless on their axles and that the cables are completely flexible. Since this cannot be true, the answers received will be imperfect, like all answers. But they do constitute either an upper or lower boundary to reality and can be interpreted in that light. The extent to which the answers conform to reality depends upon how nearly the real system is frictionless and flexible. If frictionless and flexible, the tension in a cable is identical on both sides of a pulley.

Mechanical advantage is a measure of the effectiveness of a force-multiplying machine such as a pulley system. It is the ratio of the output force to the input force.

ILLUSTRATIVE PROBLEMS

1.3 What is the tension in the cable rigged as shown in figure 1.6, supporting the concrete placement bucket weighing 1 500 lb? What is the *mechanical advantage* of the pulley mechanism?

Solution:

First, construct a free-body diagram. Select a body that, when cut free, will expose the desired force. Here the desired force is the cable tension. The bucket is thus selected as the free body. It is cut free by separating the four cable lengths supporting the buckets and replacing them with unknown tensions.

Since there is only one cable rigged over freely turning pulleys, the tension will be the same throughout the cable. It will be the same at any cut. The tension is given the name T. Each cut in the cable exposes a tension having the unknown value T.

The equilibrium condition is $\Sigma V \updownarrow = 0$. Expanding this into an equation, we get

$$4T - 1\,500 = 0$$

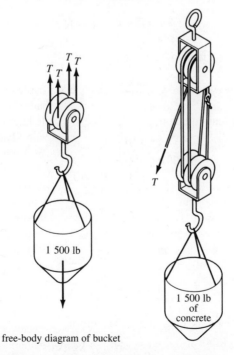

free-body diagram of bucket

Figure 1.6 Illustrative problem 1.3. A free-body diagram of the lower pulley and bucket is "cut" from the system.

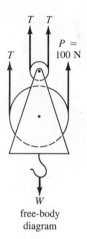

T T

$P =$ 100 N

T

W

free-body
diagram

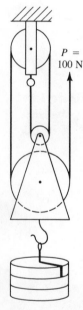

$P =$ 100 N

Figure 1.7
Illustrative
problem 1.4.
A free-body
diagram of
lower pulley
can be used
to develop an
equilibrium
equation for
vertical
forces.

Solving the equation, we get

$$T = \frac{1\ 500}{4}\ \text{lb} = 375\ \text{lb}$$

The cable tension is 375 lb, which would be required to hold the load. Slightly more than 375 lb would raise the load. Less then 375 lb would lower the load.

Here, the *mechanical advantage* is 4 to 1. In other words, only 1 lb of force is required by the machine in order to hold a load of 4 lb.

1.4 How many kilograms of mass can the 100-N force P raise when acting through the machine shown in figure 1.7?

Solution:

First, construct a free-body diagram of the lower pulley assembly as shown. The mass on the hook causes an unknown force, W. The tension in the cable is indicated by the forces T.

In this problem we assume that the weight of the lower pulley assembly is negligible compared to the weight carried by the pulley. This assumption is usually a good one and should be made unless the weight of the pulley system is given in the problem.

Because there is only one cable, the tension is uniform through all the parts of the cable. For this reason,

$$T = P = 100\ \text{N}$$

For the free-body diagram to be in equilibrium,

$$\Sigma\ V \updownarrow\ = 0$$

Expanding this into an equation,

$$3T + P - W = 0$$
$$3(100) + 100 - W = 0$$
$$400 - W = 0$$
$$W = 400\ \text{N}$$

Since $W = mg$,

$$m = \frac{W}{g} = \frac{400\ \text{N}}{9.80\ \text{m/s}^2}$$

$$m = 40.8\frac{\text{N}}{\text{m/s}^2}$$

A newton is defined as $1\ \text{kg} \cdot \text{m/s}^2$. Therefore,

$$m = 40.8\frac{\cancel{\text{N}}}{\cancel{\text{m/s}^2}} \times \frac{\text{kg} \cdot \cancel{\text{m/s}^2}}{1\ \cancel{\text{N}}}$$

$$m = 40.8\ \text{kg}$$

1.5 What amount of force does the pulley assembly shown in figure 1.7 exert upon the overhead support?

Solution:

A free-body diagram of the entire pulley assembly is shown in figure 1.8. The force that the overhead support must supply is shown by the unknown, F. For equilibrium,

$$\Sigma V \updownarrow = 0$$
$$F + P - W = 0$$

P is given as 100 N. W was computed in illustrative problem 1.4 to be 400 N.

$$F + 100 - 400 \text{ N} = 0$$

Finally,

$$F = 300 \text{ N}$$

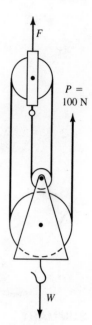

Figure 1.8
Illustrative problem 1.5.

PROBLEM SET 1.3

1.14 What is the mechanical advantage of the pulley system shown in figure 1.7 (illustrative problem 1.4)?

1.15 A pulley rigging used to draw in the boom of a sailboat is shown in figure 1.9. If the tension on the sheet (rope) is 42 lb, what force does the rigging exert on the boom?

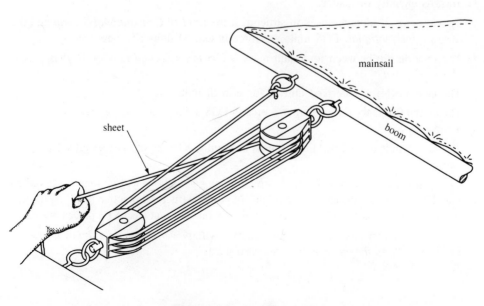

Figure 1.9 Problem 1.15.

1.16 In each of the pulley arrangements shown in figure 1.10 (a through d), determine the weight W required to balance the 968-kg load, and the mechanical advantage.

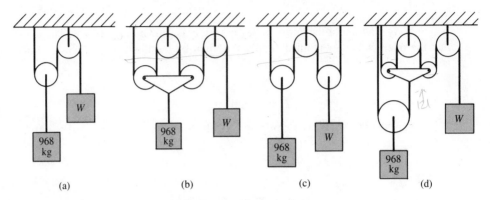

Figure 1.10 Problem 1.16.

SUMMARY

☐ A *structure* is anything that transmits forces or supports loads.

☐ A *force* is a push or a pull. A force can be caused by gravitational attraction, aerodynamics, hydrodynamics, electrostatics, magnetics, motion, or contact between solid objects.

☐ *Weight* is force due to gravity.

☐ *Mass* is quantity of matter.

☐ Two systems of units are used in structures, the *English Gravitational System* and the *Système International*. EGS units are force-based. SI units are mass-based.

☐ The relationship between force and mass is *Newton's Second Law of Motion:*
   (1.1) $F = ma$

☐ The newton (N) is a derived unit of force in SI units.

☐ The pound (lb) is the defined unit of force in EGS units. 1 lb = 4.448 222 N. *Weight* is a force.

☐ The kilogram is the defined unit of mass in SI units. One kg of mass weighs 2.20 lb on earth.

☐ *Conversion factors* can be entered into complex calculations as fractions contrived so that the unwanted units can be cancelled. In normal structural practice *three significant places* are used except when the first digit is a one. In that case four places are used.

☐ Very large or very small numbers are handled with *power-of-ten notation* or *E format*. Prefixes such as *mega-* or *micro-* are also used.

□ All forces in any direction acting upon a static structure must be balanced. This concept is called *static equilibrium*.

□ A *free-body diagram* is a diagram of an object in which the supports have been removed and replaced with arrows indicating the possible forces.

□ An *equilibrium equation* results when all the forces in a particular direction are set mathematically equal to zero.

REFERENCES

1. American Society for Testing Materials. *Standards for Metric Practice,* E 380–76. ASTM, 1976.

CHAPTER TWO

Straight-Line Tension Structures

Tension forces are the predominant forces in many structures and structural elements. Static equilibrium in such structures need only consider forces along a single direction along the line of the axis of the structures. Pressure pistons and oil drilling rigs are important examples of static equilibrium in a single direction.

From equilibrium, stress can be defined for the case of tension. Cross-sectional area is discussed in detail, including the reduction of area due to threads and stress concentrations due to abrupt changes of cross section.

The elongation of a tension member is considered as the concept of strain is introduced. Methods of applying and measuring stress and strain in a tensile test are discussed from a historical perspective as well as from the technologist's point of view. Determination from a tensile test of the four fundamental material properties of yield strength, ultimate strength, elastic modulus, and percent elongation is explained.

The chapter closes with the variation of material properties with cold working and the estimation of the deformation of tension members and structures under load.

2.1 SOME EXAMPLES OF TENSION STRUCTURES

Tension structures are the simplest of all structures. They carry a load in the direction of their axes. They are also said to be *stable* because the action of the tension load tends to pull them into shape. Therefore, a tension structure may be made either of rigid rods, pipes, structural shapes, or flexible components such as cables, chains, and ropes. We all know that you cannot push on a rope. But you can pull on it. Tension means ''to pull.''

Tension structures are very efficient in their use of material. The strength per pound of material is always greatest when a material having good tensile strength is fashioned

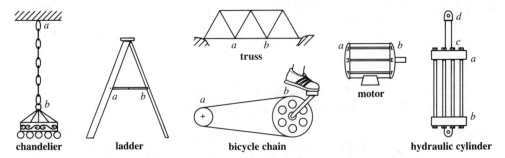

Figure 2.1 Examples of tension members.

into some type of a tension structure. This chapter only addresses straight-line tension structures. However, it is also possible to suspend cables and chains to form curved tension structures. Such spans, which derive considerable strength from their shape, are considered in chapter 9 on "shaped structures." Straight-line tension structures, as considered in this chapter, rarely form a complete structure. Usually they are components of a larger structure. The word *member* is used to imply part of a larger structure in the same sense that an individual may be a member of a team. Straight-line tension structures are usually structural members.

In figure 2.1 the chain supporting the chandelier is a tension member. The bottom members of the truss, such as *ab*, are tension members. The tie rod, *ab*, in the ladder is

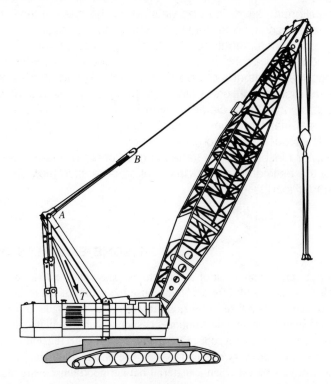

Figure 2.2 Some machines are essentially structures of variable geometry.

a tension member. The portion of the bicycle chain between the pedal sprocket and the wheel sprocket, *ab,* is also a tension member. The tie rods on the motor holding the end plates together are tension members. The tie rods, *ab,* on the hydraulic cylinder are tension members. They resist the hydraulic pressure trying to separate the end plates on the cylinder. The piston rod, *cd,* may also be a tension member if or when the mechanism is contracting.

The pulley system *A-B* of the crane shown in figure 2.2 can be regarded as a structural tension element rather than as a machine element. When the crane is holding a load rather than raising or lowering a load, there is no distinction between a structure and a machine. Furthermore, while slowly raising or lowering the load in a controlled manner, such machines are essentially structures that have variable configurations.

ILLUSTRATIVE PROBLEM

2.1 The crane shown in figure 2.2 has a pulley system *A-B* to raise and lower the boom. What tension must be exerted by the drawworks upon the pulley cable in order for the pulley system to apply 1.134E6 N of pull to the boom? The details of the pulley system are shown in figure 2.3.

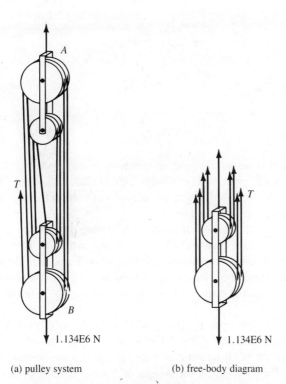

(a) pulley system   (b) free-body diagram

Figure 2.3 Illustrative problem 2.1. Pulley system for raising the boom of the crane in figure 2.2.

Solution:

The pulley system *A-B* is used to provide a mechanical advantage so that the draw-works need not absorb the full 1.134E6 N of force. The more lengths of cable used in the rigging, the greater will be the mechanical advantage. This pulley system uses two pulley axles top and bottom and also has two pulleys on each axle. The system is shown oriented vertically in figure 2.3a.

The structure can be understood by making a free-body diagram of the lower pulley assembly as shown in figure 2.3b. At point *B*, a total of nine lengths of the same cable are attached to the lower pulleys. Neglecting the small differences due to friction, each of these lengths will exert the same tension, *T*. One of these lengths goes to the drawworks.

In the free-body diagram of the lower pulleys, the cables are "cut" and replaced with the tension forces, *T*. For vertical equilibrium of the free body, the sum of all vertical forces must be zero.

$$\Sigma \, V \updownarrow \, = 0$$

Expanding, we get

$$9T - 1.134\text{E}6 \text{ N} = 0$$

Therefore,

$$T = \frac{1.134\text{E}6 \text{ N}}{9}$$

$$T = 0.126 \, 0\text{E}6 \text{ N} = 126 \, 000 \text{ N, or } 126.0 \text{ kN}$$

The drawworks must pull with a force of 126.0 kN in order to apply 1.134 MN to the boom.

2.2 OIL WELL DRILLING

The longest tension structures in modern technology are oil well drill "strings." Using currently available drill pipe, it is possible to extend such strings ten miles down into the earth; however, at the time of this writing, record wells are in the 6-mile range. Economics will compel progressively deeper wells.

Since the process of drilling is to push while turning on a bit, the fact that drilling rigs must constantly pull rather than push is certainly not intuitive. Therefore, people don't think of oil well drill strings as tension structures of such imposing length. The length of the drill string is disguised because it is assembled as it is put in the hole. And, it is always hidden in the hole. Figure 2.4 shows oil workers "making up" a threaded joint in a string of drill pipe.

Well drilling depends upon the string being a tension member that tends to pull itself into a straight line. In such long lengths, drill pipe is quite flexible. And, like a rope, if pushed down from above, it would go off to the side. So the downward drilling force is provided not from above, but by the weight of several lengths of heavy wall drill pipe called *collar* near the bottom of the hole immediately behind the rock bit. The collar is followed by pipe in approximately 40-ft lengths. The pipe has special threaded joints,

Figure 2.4 The longest straight-line tension structures are oil well drilling strings. Workers assemble drill pipe into strings extending as much as 10 miles into the earth. (Source: Hughes Tool Co.)

shown in figure 2.4, which may have a larger outside diameter than the pipe. The drill collar is about five times heavier than the same length of drill pipe.

Instead of pushing down from above, the rig (figure 2.5) pulls upward on the drill string with a force greater than the total weight of the drill pipe without the collar, thereby keeping every foot of the drill pipe in tension. Only the smaller part of the drill string, which is composed of collar, is in compression. It is this compression provided by gravity that pushes the bit into the earth. As long as the drawwork's pull does not exceed the total weight of the pipe and collar, there will be compression at the bit. The driller is provided with a scale which continuously measures the load on the hook. The driller operates the controls to slowly lower the string during drilling, keeping the hook load great enough to keep the drill pipe in tension but small enough to permit compression only in the collar behind the bit.

The oil well drill string is not only an imposing tension structure; it is also an excellent example of equilibrium along a straight line.

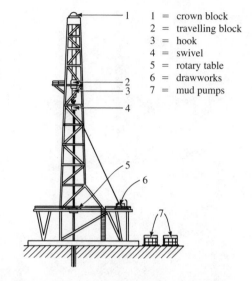

Figure 2.5 Main components of a deep well drilling rig. (Source: Fag Kugelfischer, Schweinfurt, West Germany, Stamford, Connecticut 06904)

Figure 2.6 Illustrative problem 2.2. Free-body diagram of oil well drill string.

ILLUSTRATIVE PROBLEM

2.2 A rig is drilling an oil well at a depth of 12 000 ft. It is using a $7\frac{7}{8}$-in.-diameter rock bit with $4\frac{1}{2}$-in. drill pipe. The drill pipe weighs 20 lb/ft. The drill string has thirty-two 20-ft lengths of collar weighing 105 lb/ft. (The buoyant effect of the drilling fluids is neglected for simplicity.) The driller is trying to maintain a force of 45 000 lb on the bit. What tension force must the driller maintain at the drawworks to obtain the desired bit force?

Solution:

The purpose of this problem is to calculate the correct hook load. The free-body diagram in this problem is the entire drill string shown in figure 2.6.

The length of the drill pipe is

12 000 ft total depth − 640 ft collar = 11 360 ft

The weight of the drill pipe is

11 360 ft × 20 lb/ft = 227 200 lb

The weight of the drill collar is

640 ft × 105 lb/ft = 67 200 lb

For equilibrium,

$$\Sum V \updownarrow = 0$$

Expanding,

$$\Sigma\ V\ \updownarrow\ :\ T\ -\ 227\ 200\ -\ 67\ 200\ +\ 45\ 000\ =\ 0$$
$$T\ -\ 249\ 400\ =\ 0$$

We see that $T = 249\ 000$ lb, or 249 kips when rounding to appropriate accuracy. The driller must maintain a force of 249 kips at the drawworks.

PROBLEM SET 2.1

2.1 A drill string is composed of 84 lengths of drill pipe weighing 2 000 kg each. If the desired drilling force is 200 000 N and no more than 80 percent of the collar must be in compression, how many lengths of collar weighing 5 060 kg each will be required? What must be the drawworks force?

2.2 The driller on an oil rig draws up the drill string and reads the drawworks scale at 524 000 lb. The load remains constant as it travels upward. The crew adds a section of pipe weighing 1 040 lb to the string. Then the driller lowers the string until a force reading of 463 500 lb appears on the scale. What is the bit force?

2.3 STRESS AND STRAIN

The words *stress* and *strain* in our everyday vocabulary mean approximately the same thing. We hear people say, "I'm working under a lot of stress today," or, "He is really straining to keep up." The common usages of stress and strain are so similar that the words confuse students learning their precise technological definitions. In technology, stress and strain have two entirely different meanings.

Stress is the internal distribution of forces within an object due to externally applied forces or loads.

like busiunss gose To deffent places

Strain is the deformation throughout the structure due to the loads.

These definitions will be developed more precisely throughout this book on a case-by-case basis. For now it is sufficient to understand that stress has to do with internal forces and that strain has to do with deformation.

Stress is internal force divided by area. In the tension member shown in figure 2.7a the 1 000-lb load is assumed to be distributed uniformly across a cross-sectional area perpendicular to the axis of force. This is shown in the free-body diagram of a portion of the member in figure 2.7b. The stress distributes itself equally, or uniformly, across the cross section in tension except near the ends of the members or near any change or discontinuity in the cross-sectional area. Since the stress is uniformly distributed through the cross section, as shown in figure 2.7b, the value of stress can be computed by dividing the tension load at the cross section by the cross-sectional area. To hold the free body in

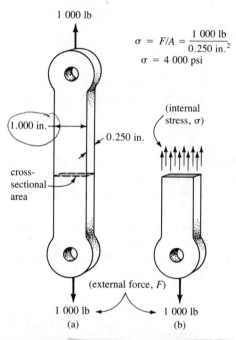

$$\sigma = F/A = \frac{1\ 000\ \text{lb}}{0.250\ \text{in.}^2}$$
$$\sigma = 4\ 000\ \text{psi}$$

Figure 2.7 Definition of stress: The stress is the internal force per unit area.

equilibrium, the stress, σ, times the cross-sectional area, A, must be equal to the external force, F:

$$F = \sigma \times A$$

Therefore, we see that

$$\boxed{\sigma = F/A} \tag{2.1}$$

In the example shown in figure 2.7, the area, A, is equal to 1 in. $\times$ 0.25 in., or 0.25 in.^{2} Therefore,

$$\sigma = \frac{1\ 000\ \text{lb}}{0.25\ \text{in.}^2} = 4\ 000\ \text{lb/in.}^2$$

In computing tension stress, area is always cross-sectional area. This is the area which would be exposed by sawing the member perpendicular to the line of action of the tension force at the point where one wishes to compute stress. These cross-sectional areas may have many shapes. Most common are the square, rectangle, circle, an annular area such as in a pipe, and structural shapes like angles, channels, and "I's." Areas for common geometric shapes can be computed using the relationships given in Appendix III. Areas of cross sections of common structural shapes can be obtained for the various structural sizes directly from the tables of Appendix V.

Circular Areas

Many areas used in structures are circular. Therefore, one must be proficient at calculating circular area. From geometry,

$$A = \pi R^2$$

Since $R = D/2$, and diameters rather than radii are more commonly specified,

$$A = \pi(D/2)^2 = (\pi/4)D^2$$

$$\boxed{A = 0.785D^2}$$

For hollow circular sections, where D is the outside diameter and d is the inside diameter,

$$\boxed{A = 0.785(D^2 - d^2)}$$

Although it is obvious in the preceding formula that the diameters must be squared before subtracting, an error made all too frequently is subtracting before squaring. So, be careful.

Units of Stress

The units of area used in stress are length units squared. In EGS units, square inches are most common. Square feet are occasionally used for low-strength materials such as soils. In SI units, square centimeters or square millimeters are used, with square millimeters preferred. Square meters are rarely used, because a square meter is an enormous cross-sectional area for most structures.

Keep in mind the following conversion factors:

$$
\begin{array}{l}
1 \text{ ft}^2 = 144 \text{ in.}^2 \\
1 \text{ m}^2 = 1E4 \text{ cm}^2 \\
1 \text{ m}^2 = 1E6 \text{ mm}^2 \\
1 \text{ cm}^2 = 1E2 \text{ mm}^2
\end{array}
$$

In EGS units the abbreviation lb/in.^{2} for pounds per square inch is cumbersome. Therefore, the abbreviation psi is commonly used. When carrying units through your calculations, you will need to remember that the abbreviation psi can be replaced with lb/in.^{2}

An SI unit, the pascal, is now widely used for stress. The pascal is abbreviated Pa. The abbreviation is capitalized because the unit is named after a person, Blaise Pascal. The pascal is defined as 1 newton per square meter.

$$\boxed{1 \text{ Pa} = 1 \text{ N/m}^2}$$

Because of the size of the values of stress commonly found in structures, the units ksi and MPa are frequently used. The first, ksi, is an EGS unit denoting kips per square inch. MPa means megapascals and is a common SI unit of stress.

$$1 \text{ ksi} = 1E3 \text{ psi}$$
$$1 \text{ MPa} = 1E6 \text{ Pa}$$

To convert from one system of units to the other, the following conversion factors are useful:

$$1 \text{ in.}^2 = 6.45 \text{ cm}^2$$
$$1 \text{ in.}^2 = 6.45E2 \text{ mm}^2$$
$$1 \text{ psi} = 6.89E3 \text{ Pa}$$
$$1 \text{ psi} = 6.89E{-}3 \text{ MPa}$$
$$1 \text{ ksi} = 6.89 \text{ MPa}$$

These conversion factors and others are summarized on the inside front cover for your convenience.

ILLUSTRATIVE PROBLEMS

2.3 What is the stress in the draw bar shown in figure 2.8? It has a diameter of 22.0 mm and pulls a load of 250 000 N.

250 000 N

22 mm diameter

250 000 N

Figure 2.8
Illustrative
problem 2.3.

Solution:
The cross-sectional area is a circle with a diameter of 22 mm.

$$A = 0.785D^2 = 0.785(22.0 \text{ mm})^2$$
$$A = 380 \text{ mm}^2$$

and

$$\sigma = F/A$$
$$\sigma = \frac{250\ 000 \text{ N}}{380 \text{ mm}^2} = 657 \text{ N/mm}^2$$

Since 1 mm = 1E−3 m,

$$1 \text{ mm}^2 = 1E{-}6 \text{ m}^2$$

Therefore,

$$\sigma = 657 \text{ N/E}{-}6 \text{ m}^2$$
$$\sigma = 657E6 \text{ N/m}^2$$

Because 1 Pa is defined as 1 N/m²,

$$\sigma = 657E6 \text{ Pa} \quad \text{or} \quad \sigma = 657 \text{ MPa}$$

The stress is 657 MPa. Megapascals are common units for stress.

Conclusion:
When force is in newtons and area in square millimeters, stress is in megapascals.

$$1 \text{ N/mm}^2 = 1 \text{ MPa}$$

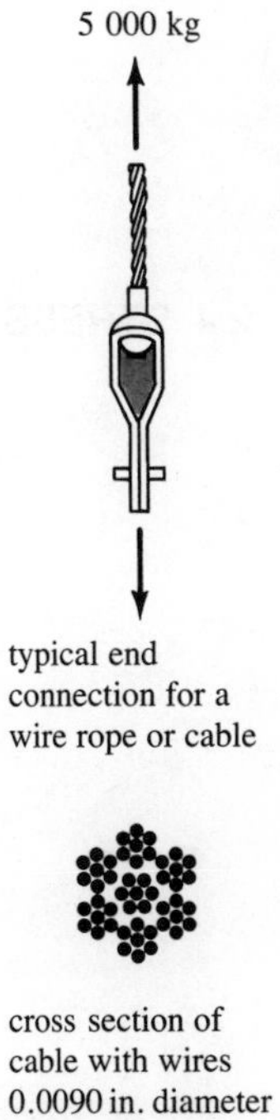

2.4 A pulley system in an aircraft is rigged with a wire rope (or cable) shown in figure 2.9. Each wire is 0.0090 in. in diameter. If the force on the cable is 5 000 kg, what is the average stress in the wire expressed in MPa?

Solution:
The wire rope has seven strands of seven wires each, called a 7×7. Each wire has a diameter of 0.090 in. The total cross-sectional area, *A,* therefore, is

$$A = 7 \times 7 \times 0.785 \times (0.090 \text{ in.})^2 = 0.312 \text{ in.}^2$$

The force is expressed in kilograms. It is the same as the force in newtons that would be produced by 5 000 kg of mass on earth.

$$F = 5\ 000 \text{ kg} \times \frac{9.80 \text{ N}}{\text{kg}} = 49\ 000 \text{ N}$$

$$\sigma = F/A = \frac{49\ 000 \text{ N}}{0.312 \text{ in.}^2} \times \frac{1 \text{ in.}^2}{645 \text{ mm}^2} = 243 \text{ MPa}$$

Or, if EGS units are desired,

$$\sigma = 243 \text{ MPa} \times \frac{1 \text{ ksi}}{6.89 \text{ MPa}} = 35.3 \text{ ksi}$$

The stress is 243 MPa or 35.3 ksi.

5 000 kg

typical end
connection for a
wire rope or cable

cross section of
cable with wires
0.0090 in. diameter

Figure 2.9
Illustrative
problem 2.4.

PROBLEM SET 2.2

2.3 A flat bar of cold-rolled steel 1.000 in. wide by 0.188 in. thick is made into a tension member. It carries a load of 4 T. What is the tensile stress in the bar?

2.4 A tension member is to be made of an aluminum rod 0.785 in. in diameter. What load can the rod support if the stress is not to exceed 11 500 psi?

2.5 What level of stress is produced in a titanium tube that has an outside diameter of 2.50 cm and a wall thickness of 4 mm by a tension load of 4 150 N?

2.6 A plastic strap for binding packages for shipment is tested in tension. It has a breaking strength of 1 276 lb. The strap is 2 mm thick and 16 mm wide. What is the breaking stress in both ksi and MPa?

2.7 A tension brace on a water tower is expected to carry a load of 41 kips. It is proposed that the brace be made from a $3 \times 4 \times \frac{3}{8}$-in. standard structural angle. What would be the stress in the brace?

2.8 What is the highest tensile stress in the $4\frac{1}{2}$-in. oil well drill pipe of illustrative problem 2.2? The inside diameter of the drill pipe is 3.640 in.

2.9 A pulley system for a sailboat uses a single nylon rope. The nylon rope is wound from monofilament line 0.03 in. in diameter into a 7×7 pattern. The pulley system has a mechanical advantage of 6 and is used to apply a force of 1 250 lb. What is the stress in the rope?

2.4 STRESS CONCENTRATION

Changes in cross section due to threads, shoulders, holes, and notches cause stress to concentrate. In the region of such discontinuities, the stress will be higher than that calculated from F/A. This is true even after considering that the cross-sectional area, A, is reduced by the material removed to form the threads, shoulders, holes, or notches. Stress is increased because its distribution is not uniform across the section. Instead, stress tends to be concentrated in intensity toward where the material was removed, as in figure 2.10.

The peak value of stress can be calculated by using a *stress concentration* factor, k, as follows:

$$\sigma = kF/A \qquad\qquad (2.2)$$

Stress concentration factors have been determined by advanced stress analysis and by experimental techniques such as shown in figure 2.11 for many kinds of discontinuities in

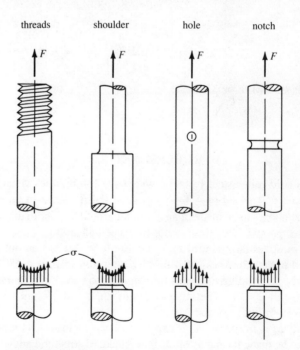

Figure 2.10 Discontinuities cause the stress to concentrate toward where the material was removed.

Figure 2.11 The determination of a stress concentration factor using a technique called photoelasticity. The machine shown is called a polariscope because it uses polarized light to make stress concentrations visible in scale models of the structural part made of clear plastic. The phenomenon used is exactly the same that is observed when looking through polarized sunglasses at stressed glass or clear plastic. However, when precise technology is employed, highly accurate quantitative results are obtained. (Source: Courtesy Measurements Group, Inc., Raleigh, North Carolina, USA)

round and rectangular shapes for tension, bending, and torsion loadings. An excellent reference work on stress concentration factors is *Stress Concentration Factors,* by R. E. Peterson.[1] Figure 2.12 gives two graphs from this source. The following illustrative problem shows how to use such graphs of stress concentration factors.

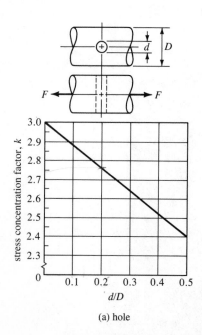

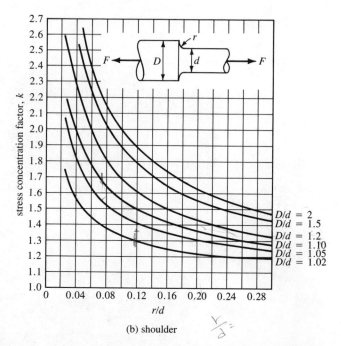

Figure 2.12 A sample of stress concentration factor graphs from Peterson's *Stress Concentration Factors*.

ILLUSTRATIVE PROBLEM

2.5 A tension rod has a diameter of 1.000 in. with shoulders increasing the diameter to 1.500 in. on each end. The inside radius at the shoulder is 0.100 in. There is also a 0.125-in.-diameter hole drilled perpendicular to the center line of the rod on the 1.000-in. diameter. The rod carries a tension load of 20 800 lb. What is the maximum stress in the rod?

Solution:
The highest stress will occur either at the shoulder or at the hole. Both places must be investigated.

The Shoulder:
In figure 2.12b,

$$D = 1.500 \text{ in.}$$
$$d = 1.000 \text{ in.}$$
$$r = 0.100 \text{ in.}$$
$$D/d = 1.500/1.000 = 1.500$$
$$r/d = 0.100/1.000 = 0.100$$

It is necessary to follow in figure 2.12b the curve for $D/d = 1.500$. The imaginary curve is extended to where it crosses the vertical line at $r/d = 0.100$. The stress concentration is read to the left.

$$k = 1.88$$
$$A = 0.785d^2 = 0.785(1.000 \text{ in.})^2 = 0.785 \text{ in.}^2$$

Using relationship 2.2, we get

$$\sigma = kF/A$$
$$\sigma = 1.88 \times 20\ 800 \text{ lb}/0.785 \text{ in.}^2$$
$$\sigma = 49\ 800 \text{ psi at the shoulder}$$

The Hole:
From figure 2.12a,

$$D = 1.000 \text{ in.}$$
$$d = 0.125 \text{ in.}$$
$$d/D = 0.125/1.000 = 0.125$$

The stress concentration factor, k, is read to the left of the intersection of the line and the vertical at $d/D = 0.125$. Thus, $k = 2.85$.

$$A = 0.785D^2 - d \times D \text{ (approximately)}$$
$$A = 0.785 \text{ in.}^2 - 0.125 \times 1.000 \text{ in.}^2$$
$$A = 0.660 \text{ in.}^2$$
$$A = kF/A$$

and

$$\sigma = 2.85 \times 20\ 800 \text{ lb}/0.660 \text{ in.}^2$$
$$\sigma = 89\ 800 \text{ psi at the hole}$$

The maximum stress occurs at the hole. Stress concentrations disappear within one diameter, D, from the shoulders or the hole. The stress in the rod away from the ends or the hole where the stress is not concentrated is given by relationship 2.1:

$$\sigma = F/A = 20\ 800 \text{ lb}/0.785 \text{ in.}^2 = 26\ 500 \text{ psi}$$

The stress is increased to 49 000 psi at the shoulder due to stress concentration. It is increased to 89 800 psi at the hole due to a combination of material removal and stress concentration.

Usually, stress concentration can be reduced by increasing the radius, r. Sharp corners can cause very large stress concentrations. Area reductions should be gradual. Examples of ways to reduce the stress concentrations are shown in figure 2.13.

2.5 THREADS

Four kinds of standard threads are commonly encountered in structural practice. These are the coarse threads, the fine threads, the 8-thread series, and metric threads. See Appendix II for thread sizes and dimensions.

Coarse Threads

Coarse threads, designated NC, have a greater pitch (figure 2.14) for a given thread diameter than do fine threads. As an example, $\frac{1}{4}$-20NC means $\frac{1}{4}$-in. diameter, 20 threads per in. They are general-purpose threads.

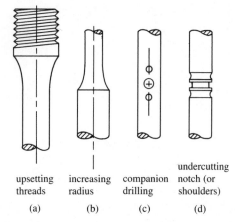

upsetting threads
(a)

increasing radius
(b)

companion drilling
(c)

undercutting notch (or shoulders)
(d)

Figure 2.13 Stress concentrations can be reduced by the addition of material (a) and (b) or the removal of material (c) and (d).

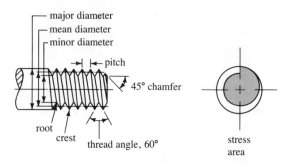

Figure 2.14 Thread terminology and dimensions.

Fine Threads

Fine threads, designated NF, have a smaller pitch and larger root area. They are used in high-strength applications. For example, $\frac{1}{4}$-28NF means $\frac{1}{4}$-in. diameter, 28 threads per in.

8-Thread Series

The 8-thread series, designated 8N, is used for very large threads up to 6 in. in diameter. They have 8 threads per in. regardless of the diameter. In other words, the pitch is always $\frac{1}{8}$ in.

Metric Threads

Metric threads, designated M, are becoming more common in all applications. They are in SI unit sizes.

Stresses in threads are calculated either on the basis of root area (area of a circle equal to the minor diameter) or the stress area, depending on the design code or standard used. Root area is conservative. Stress area is the area that would be seen if the thread were cut perpendicular to its centerline (figure 2.14). Stress area is approximately equal to the area based on major diameter plus that based upon minor diameter, divided by 2.

Sometimes stress concentration due to threads is an important consideration. When it is, threads on a tension member are best cut on an upset end. Upsets can be rolled or swagged on threaded parts in a mass-production setting. Upset ends can also be welded on to tension members, as in the oil well drill pipe shown in figure 2.4. Only in rare cases is lathe work used to reduce the diameter to form upsets, since it costs several times more to remove metal on a lathe than it does to purchase an equal volume of the metal. In severe applications, where brittle failure, shock loads, or hundreds of thousands of load cycles are anticipated, it is best to upset threads at all costs. Stress concentrations can also be reduced substantially if the threads are formed by rolling rather than by using a die or a lathe to cut them into the shaft.

PROBLEM SET 2.3

2.10 The bolts on a high-pressure steam turbine casing have $3\frac{1}{2}$-8N threads. When pretensioned to a load of 192 000 lb, what is the stress (a) in the shank of the bolt and (b) in the threads?

2.11 By what percentage is the strength of a $\frac{5}{8}$-in. bolt increased by changing from NC to NF threads?

2.12 A 1.25-cm-diameter tension rod carries a load of 4 320 kg. The rod has a 0.25-cm-diameter hole drilled perpendicular to the centerline. What is the stress in the rod near the hole? Give the answer in MPa.

2.13 A tension specimen has a shape similar to the round specimen shown in figure 2.25. If d, r, and D equal 0.750, 0.150, and 1.00 in., respectively, what is the stress concentration factor? What is the maximum stress in the gage length under a load of 21 kips?

2.6 THE DISTINCTION BETWEEN STRESS AND PRESSURE

Pressure is the force per unit area in a fluid. It is discussed here for two reasons. First, many solid pieces of machines and structures interface with fluids under pressure. Second, pressure and stress are both defined as force per unit area and have the same units. Yet they are two entirely different things and must be understood separately.

The difference is explained by *Pascal's Principle,* which applies to fluids but not to solids. According to Pascal's Principle, *a fluid has the ability to transfer pressure to all parts of its container equally in all directions.* Pascal's Principle applies for static conditions, neglecting the weight of the fluid itself. When the fluid is moving or when the weight of the fluid is not negligible, Pascal's Principle must be modified using other principles from hydraulics, pneumatics, and fluid mechanics, which are each separate areas of study. The unit *pascal* is named in honor of the man who formulated much of fluid mechanics.

Pressure, from Pascal's Principle, is a *scalar quantity,* meaning that it has equal magnitude in all directions. Solids do not conform to this principle. Stress applied in the axial direction of a round metal bar, for example, is not transferred in the traverse direction. Stress has complex directional properties, which are discussed in chapter 15. Even though they are defined similarly and have identical units, pressure must be used only with fluids and stress only with solids. Sometimes the line between a solid and a liquid is not clear, such as in gelatinous or thickly viscous materials. But such materials are uncommon.

Illustrative problem 2.6 demonstrates how stress and pressure can interface in the same problem.

ILLUSTRATIVE PROBLEM

2.6 What bore diameter and piston rod diameter of a hydraulic cylinder mechanism must be used for the lifting mechanism of a fork-lift truck if it must produce at least 1 500 lb of force on contraction? The hydraulic system pressure is to be 1 000 psi. The allowable stress of the piston rod is 20 000 psi (figure 2.15). Neglect seal and piston friction.

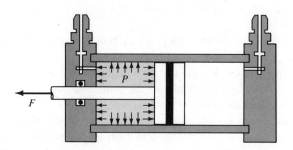

Figure 2.15   Illustrative problem 2.6.

Solution:

To contract the cylinder mechanism, the pressure, *P,* must be applied to the chamber shown in figure 2.15. If a tension force, *F,* of 1 500 lb must be carried by a rod not stressed more than 20 000 psi, the diameter of the rod can be computed from the definition of stress,

(a) $$\sigma = F/A$$

where

(b) $$A = 0.785d^2 \text{ for a circular rod}$$

Therefore,

(c) $$d = \sqrt{\frac{F}{0.785\sigma}}$$

$$d = \sqrt{\frac{1\ 500\ \text{lb}}{0.785 \times 20\ 000\ \text{psi}}}$$

$$d = 0.309 \text{ in.}$$

A standard $\frac{5}{16}$-in.-diameter rod, which has a decimal equivalent of 0.3125 in., probably should be specified for the piston rod. A free-body diagram of the piston can be drawn as shown in figure 2.16.

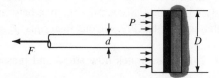

Figure 2.16   Illustrative problem 2.6.

The equilibrium relationship using the definition of pressure can be written as:

(a)
$$\Sigma H \leftrightarrow : -F + PA_p = 0$$

where A_p is the face area of the piston over which the pressurized fluid acts.

(b)
$$A_p = 0.785(D^2 - d^2)$$

Substituting (b) into (a) and solving for D,

$$-F + 0.785P\,(D^2 - d^2) = 0$$

$$-\frac{F}{0.785P} + D^2 - d^2 = 0$$

$$D = \sqrt{\frac{F}{0.785P} + d^2}$$

$$D = \sqrt{\frac{1\ 500\ \text{lb}}{0.785 \times 1\ 000\ \text{psi}} + 0.3125^2\ \text{in.}^2}$$

$$D = 1.417\ \text{in.}$$

The bore of the hydraulic cylinder must be at least 1.417 in. Probably 1.420 should be specified, but more would need to be known about the details of construction.

PROBLEM SET 2.4

2.14 A hydraulic cylinder shown in figure 2.17 is actuated by a pressure of 1.53 N/m^2 in chamber A. What is the tension force in the cylinder rod? Neglect the seal drag.

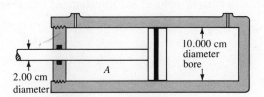

10.000 cm diameter bore

A

2.00 cm diameter

Figure 2.17   Problem 2.14.

2.15 The piston rod on a reciprocating water pump is exerting a tension of 100 lb. What is the suction pressure (or negative pressure) produced in chamber B of the pump shown in figure 2.18?

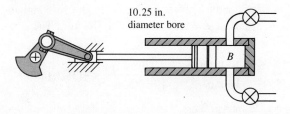

10.25 in. diameter bore

B

Figure 2.18   Problem 2.15.

2.16 Calculate the tension force on the 8 tie rods of the pneumatic cylinder shown in figure 2.19 when a pressure of 110.4 psi is acting in chamber A. An excess force of 50.0 lb, over that required to restrain the pressure, is present to seat the gasket.

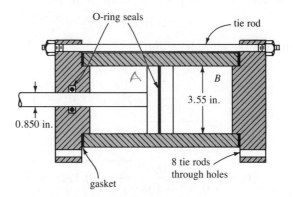

Figure 2.19 Problem 2.16.

2.17 A hydraulic cylinder has a bore of 90 mm. The end plates are retained by twelve threaded tie rods 4 mm in diameter. The piston rod is 10 mm in diameter and capable of providing a pull of 10 kN. What is the tension stress in the tie rod threads at maximum pull, ignoring the gasket forces?

2.7 STRAIN IN A TENSION MEMBER

Strain in a tension member can be defined as elongation per unit length. Strain, ϵ, can be expressed by

$$\epsilon = \delta/L \qquad (2.3)$$

where δ is elongation and L is the length of any portion of the tension member having a uniform cross-sectional area.

This relationship is demonstrated in figure 2.20, which is a tension specimen for determining the strength of a material. The ends are upset for gripping in a tension-testing machine. A large radius reduces the cross section to 1 cm in diameter. The large radius serves to minimize the effects of stress concentration. The ''gage length'' of the specimen is the portion 1 cm in diameter between the radii. In the case shown, it is 10 cm long. When a load is applied, the gage length elongates 0.100 mm. The strain can be computed from relationship 2.3 as follows:

$$\epsilon = \frac{\delta}{L} = \frac{0.100 \text{ mm}}{10 \text{ cm}} = \frac{0.100E-3 \text{ m}}{10E-2 \text{ m}} = 0.001 \text{ m/m} = 0.001$$

Note that strain is a ''dimensionless'' quantity. Because units of length are divided by units of length, the length units are cancelled, as shown in this example. Millimeters and centimeters were reduced to meters, and the meters were cancelled.

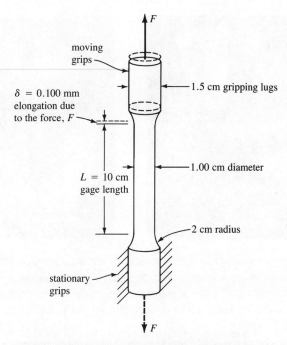

Figure 2.20 A specimen being tested in tension to determine the structural properties of the material.

Even though dimensionless, strain is often referred to in terms of inches per inch or, because of the very small values of strain normally encountered in structural practice, microinches per inch. Inches per inch can always be cancelled and are therefore superfluous. The word *micro* is sometimes used to indicate microinches per inch or microcentimeters per centimeter. Because strain is a dimensionless quantity, one inch per inch is identical to one centimeter per centimeter, or one meter per meter. A value of strain reduced to dimensionless form in SI units is identical to that in EGS units.

All structures deform under load. When a small bird lands on a mighty bridge, the bridge deforms. There is no way to prevent a structure from deforming when loads are applied. The deformations are normally very small and undetectable to the unaided eye. However, instruments are available that can measure these small deformations. The intent of structural design is to make the structure strong enough so that it restores itself to its original shape when the loads are removed.

Strains in a tensile test can be determined by measuring elongation and dividing it by gage length. For a structural steel specimen, a strain of 0.001 is very large. Each inch of length would be stretching only 0.001 in. Yet, the steel may be so highly stressed that it is ready to fail. Therefore, the usual instruments for measuring length such as rulers or even micrometers are not of great use when dealing with strain. A special instrument called an "extensometer," pictured in figure 2.21, can be used to measure extensions in a given gage length. However, extensometers are quite expensive and delicate, and are not practical outside the laboratory setting.

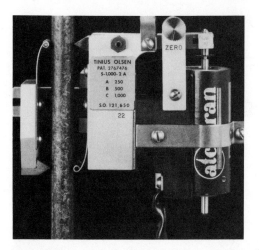

Figure 2.21   An extensometer is a mechanical instrument for measuring small elongations over a fixed gage length. The gage length is the distance between the two knife edges which bear against the specimen. The extensometer on the left is equipped for electronic readout. The extensometer on the right has a dial readout.

A more modern means of measuring strain is the electrical resistance strain gage shown in figure 2.22. The strain gage measures strain directly, because its electrical resistance changes in proportion to the strain. A strain gage is cemented to a structure and wired to an electronic strain indicator. The strain gage may not be removed from the surface to which it was cemented without destroying the gage. Therefore, strain gages must be regarded as expendable items, even though they are not inexpensive. Considerable expertise is required in their installation and use. Technicians and technologists familiar with strain gages and their applications make a very important contribution to the engineering team.

Figure 2.22   The electrical resistance strain gage. A metal foil of a precise grid pattern deposited on a plastic backing registers a change in resistance proportional to the strain when cemented to the structure or specimen. (Source: Courtesy Measurements Group, Inc., Raleigh, North Carolina, USA)

PROBLEM SET 2.5

2.18   A strain gage is mounted on a straight tension member in a railroad bridge. During the test of the bridge the strain gage indicator registers a strain of 813 micros. If the tension member is 30 ft long, what is the total elongation of the member?

2.19   A round tension specimen is machined with gripping lugs. It has a gage length of 20 cm. Therefore, most of the specimen elongation occurs over the gage length. The specimen is put in an electronic materials testing machine. The grip travel indicator shows a travel of 1.035 mm. What is the strain?

2.8 THE TENSILE TEST

A *tensile test* is a test of a material to destruction in tension during which force and elongation are measured. From a tensile test much can be learned about how a certain material will behave in tension. But, more important, information gathered from a tensile test can be used also for other forms of loading. The fundamental structural properties of many materials can be determined from a tensile test. Usually, metals and plastics are subjected to tensile tests. Brittle materials such as concrete are usually tested in compression, and fibrous materials such as wood in bending.

Specimens, or "test coupons," of materials are tested after the materials are manufactured. Frequently, they are also tested when received by a fabricator or on a job site. These tests are performed to ensure that the materials conform to specifications. Testing to check performance against specifications is known as "quality control." Specifications are prepared by the American Society for Testing Materials (ASTM).

Materials testing machines for conducting tensile tests have evolved through several generations of technology. Table 2.1 is a summary of this evolution. The earliest testing machines applied weights to the specimen. This is still the most straightforward means of testing, and for certain kinds of tests one may want to simply hang weights on a small specimen. Galileo was the first to use such a test to attempt the rational design of structures; he is considered the father of structural science.[2] But for most structural materials and reasonably sized specimens, thousands of pounds of weights would have to be moved and lifted. Therefore, practical "dead-weight testing machines" employ a series of levers to magnify the force of the weight. Force is measured by the amount and position of the weight on the lever. Dead-weight testing machines, still used, are easy to operate but take much space and have limited capacity. Also, there is insufficient control of the time rate of elongation. Many modern materials, like plastics, are sensitive to this rate.

In the 1930s, hydraulic testing machines became available. Such a machine appears in figure 2.23. Hydraulic cylinders permitted dramatic increases in the available testing force. Machine capacities of 300 000 lb are common, and 1 000 000 lb are frequently found. Good machines of this kind have a pneumatic force-measuring system independent of the hydraulic system. There is a subtle but highly important distinction between the control of a hydraulic and a dead-weight machine. In a hydraulic machine, the rate of fluid entry into the hydraulic cylinder is easily controllable. This results in a controlled rate of extension or stretch of the specimen. The force upon the specimen is a dependent variable. In other words, with a hydraulic machine, you apply deformation and take what you get in stress or load; whereas, with a dead-weight machine, you apply weight and have no control over the extension that results.

Table 2.1 Evolution of materials testing machines

Introduced	Force Application	Load Measurement	Control
17th century	dead weight	weights	force
1930s	hydraulic (figure 2.23)	mechanical/pneumatic	elongation
1950s	electrically driven screw (figure 2.24)	electronic load cell (strain gage)	elongation
1970s	hydraulic	electronic load cell (strain gage)	closed-loop feedback

Figure 2.23 A hydraulically driven universal materials testing machine.

Hydraulic testing machines are easy to operate and can perform most tests on most materials. Although technology has surpassed them, it is incorrect to call them obsolete. They are still the appropriate machine for most quality control tests. Their persistence is partially due to the unfamiliarity of testing personnel with techniques of strain gage electronic measurement upon which the new technologies are based.

The more sensitive control requirements of plastic materials introduced in the 1950s spurred the development of electrically ''screw-driven,'' electronically controlled machines such as shown in figure 2.24. In these machines force is measured by load cells. *Load cells* have metal elements that deform under load, producing an electrical signal proportional to force by means of strain gages. These machines provide a wide range of precisely controlled elongation rates and all sorts of automated data reduction and control features such as reversing or cycling. The force is obtained in the form of an electronic signal that can be plotted, recorded on magnetic tape, divided by area to give stress, fed to a computer, or simply read out in digital form. These machines require some knowledge of electronic measurements to be fully utilized. They are also limited to load capacities of about 50 000 lb.

The development of microelectronics in the 1970s permitted electronic control and measurement techniques to be applied to hydraulic systems. This combined the sensitivity and intelligence of electronics with the muscle of hydraulics. The operator can choose from force, elongation, stress, or strain signals to feed back into the controller. This is called a *closed-loop machine*. The electronic controller can be used to effect an increase of any one of the variables at a precisely controlled rate. It can also cycle the machine and permit all sorts of programmed time variations. Closed-loop feedback can simulate, on a piece of a structure or a specimen, the variations of force or elongation anticipated in the actual operation of the structure. However, using these machines requires substantial

Figure 2.24 An electrically driven screw type universal testing machine.

training. A limited knowledge of electronics and computers is necessary to provide technologists with the confidence and ability to use this powerful new testing equipment.

2.9 CONDUCTING A TENSILE TEST

Tensile test specimens may be either flat or round. Specimens of uniform cross section occasionally break in the grips of the testing machine due to the stress concentrations caused by the gripping action. It is desirable to shape specimens so that they are likely to break in their gage length. This is done by shaping the specimens so that they have gripping lugs (figure 2.25). Testing machines are normally equipped with jaws for gripping either flat or round specimens. Round specimens threaded on upset ends (figure 2.26) can also be used. Flat specimens are sometimes drilled through the lugs and attached to the testing machine with clevis pins, also shown in figure 2.26.

The purpose of a tensile test is to obtain a *stress-strain diagram* for the material. Therefore, it is necessary to measure values of load and of extension. While the tension load upon the specimen is increased, an extensometer can be used to measure extension at discrete points until strains of about 0.020 are reached. The extensometer is usually removed at that point to avoid potential damage from the sudden rupturing of the specimen. Electrical strain gages could also be used for the low strain measurements, but a

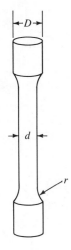

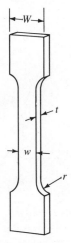

Figure 2.25
Typical tensile
specimens, or
test coupons,
with lugs for
round and
flat stock
requiring no
special grips.

strain gage is always destroyed with the specimen. Strain gages are too expensive for routine quality control tests.

In order to obtain large strain readings above 0.020, the specimen is usually marked before the test at $\frac{1}{2}$-in. intervals along its length using a center punch or a scribe. Then, during the test, two of these marks 2 in. apart are chosen. When a measurement is desired during the test, machinists' dividers are used to measure the new distance between the marks which were originally separated by 2 in. For high strain readings above 0.020, dividers are sufficiently accurate. Some machines have devices for measuring the total travel of the moving grip of the testing machine, eliminating the need for dividers when using specimens with gripping lugs.

A stress-strain diagram obtained from the test of a specimen of structural steel is shown in figure 2.27. Up to a strain of 0.0012 (point *A*), the strain increases linearly with stress. In other words, if you double the stress, you double the strain. The points on the stress-strain diagram are on a straight line.

At a stress of 36 000 psi, or a strain of 0.0012 (point *A*), the stress-strain curve begins to deviate from a straight line. As the strain increases, the stress falls off a bit and becomes unsteady. This is called "yielding." The *yield point* is the stress at which yielding begins. The yield point of the material shown is 36 000 psi.

At stresses below the yield point, the specimen will go back to its original length if the load is removed. In other words, the strain will disappear. At stresses above the yield point, the specimen would not assume its original length if the load were removed. It would have a *permanent set,* or a permanent change of length.

Beyond the yield point at a strain of 0.010 (or 1 percent)—(point *B*), the stress again begins to stabilize and increase with increasing strain. This continues until 58 000 psi. At a stress level of 58 000 psi and a strain of 0.18 (or 18 percent)—(point *C*), the stress-strain curve levels off and stress begins to decrease. This decrease occurs because the cross-sectional area of the specimen begins to decrease, as shown in figure 2.28b. This is called *necking down*. There is less area in the necked down region to resist the load. The true stress based upon the reduced area probably continues to increase. But since stress is calculated based upon the original cross-sectional area, *apparent stress* decreases. Finally, the area reduces to the point that it suddenly ruptures (point *D* in figure 2.27).

The fracture will have a "cup and a cone" with a cone angle of 45°. A perfect cup and cone is depicted in figure 2.28c. In fact, a perfect cup and cone is rare. More often, parts of the rim of the cup will form on both pieces.

The highest stress on the stress-strain diagram is called the *ultimate strength* of the material. For the structural steel diagrammed in figure 2.27, the ultimate strength is 58 000 psi. The ultimate strength is occasionally referred to as the *tensile strength* of the material.

Ductility is the ability of a material to be deformed permanently, a very important property for a structural material. Final elongation expressed as a percent is an index of the amount of ductility a material possesses. *Percent elongation* is determined by fitting back together the ruptured specimen. Then one selects two of the premarked lines or punch marks that were originally two inches apart. Test specifications prescribe the choice of two marks that straddle the break and that now are further apart than any other 2-in. pair of marks. The marks selected in figure 2.28d are labelled *A* and *B*. Since the specimen's rupture location is unknown, the markings cannot be preselected. That is the reason for marking the specimen on $\frac{1}{2}$-in. intervals over the entire gage length. The deformed distance between *A* and *B* is measured to be 2.60 in. The original length between *A* and *B* was 2 in. Therefore, the percent elongation is given by the following relationship:

$$\% \text{ elongation} = \delta/L \times 100 \qquad (2.4)$$

In figure 2.27, we have

$$\% \text{ elongation} = \frac{0.60}{2.00} \times 100 = 30$$

The material depicted has 30 percent elongation.

Percent elongation as an index of ductility depends upon the specimen diameter or cross-sectional shape. The same material in different diameters will show different per-

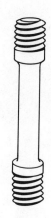

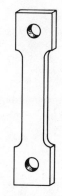

Figure 2.26 Tensile specimens or coupons with threaded or clevis pin attachments require special grips.

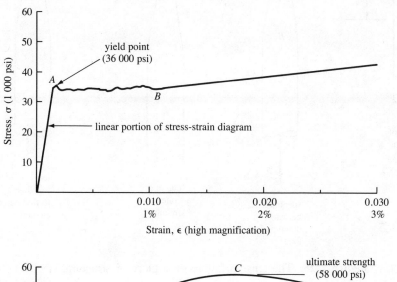

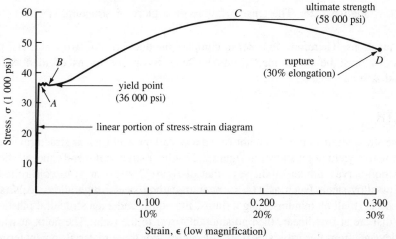

Figure 2.27 The stress-strain diagram for a structural steel.

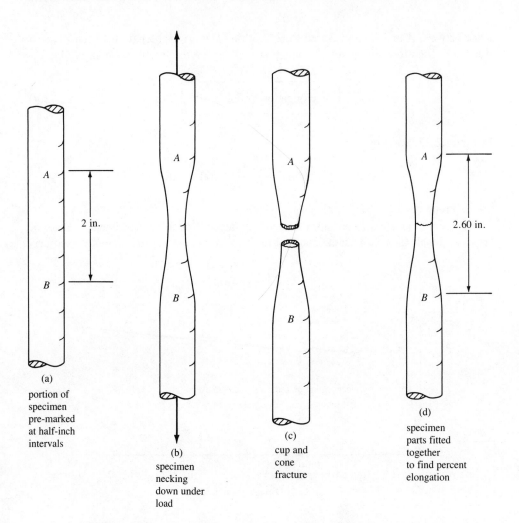

Figure 2.28 The tension failure on a piece of structural steel.

cent elongations. Therefore, in order to compare the ductilities of two materials, percent elongations must be determined from tensile tests on specimens of identical cross-sectional size and shape.

2.10 YIELD STRENGTH

Only the stress-strain curves from hot-rolled low-carbon steel such as structural steel have the distinctive yield point shown in figure 2.27. Other structural metals and plastics have stress-strain curves similar in shape to that of figure 2.29 for a 4140 cold-rolled steel.

Low-alloy steels (such as 4140), medium-carbon steels, high-alloy steels (such as stainless steel), all aluminums, magnesiums, titaniums, and even structural plastics have curves that are at first linear, then go smoothly over to the right. The point at which the curve deviates from linearity is difficult to specify. For these materials a new term, yield strength, is defined.

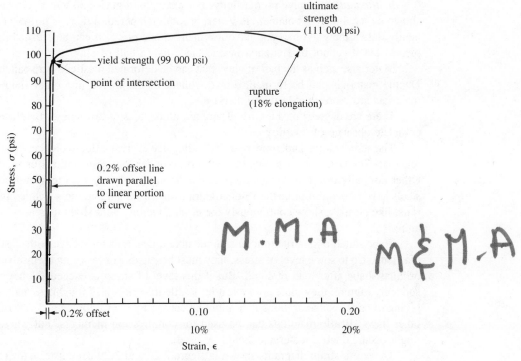

Figure 2.29 The stress-strain diagram for a specimen of 4140 steel.

Yield strength is the stress level at which the stress-strain curve is intersected by a line starting on the strain axis at a strain of 0.002 (0.2 percent) drawn parallel to the linear portion of the curve.

This method, called the 0.2-percent offset method, is illustrated in figure 2.29. A line parallel to the linear part of the curve is constructed through the abscissa at a strain of 0.002. The stress at which the line intersects the stress-strain curve (99 000 psi) is the yield strength.

2.11 ELASTICITY AND PLASTICITY

The ability of a material to return to its original shape after having been deformed by a load is called *elasticity*. *Plasticity* is the ability of a material to be permanently deformed by the application of a load.

All materials deform under load. Some deform elastically and some plastically. A rubber ball, when compressed and released, goes back to its original shape; it is said to be elastic. A ball of clay, when pressed under the heel of your hand, retains its compressed shape when you remove your hand. The clay is said to be a plastic material. It is the elasticity of the rubber ball that makes it bounce. A clay ball has no bounce whatsoever—it has no elasticity.

Brittle materials have no plasticity. For example, a glass ball will have elasticity as shown by its ability to bounce. However, at room temperature there is no way to permanently flatten the glass ball and have it remain in one piece. It can be crushed into many pieces, but it cannot be flattened in one piece like a ball of clay.

Materials, such as the ball of clay, that can be deformed plastically are called *ductile*. Ductile materials can be formed, drawn, rolled, or hammered into other shapes. Some materials are more ductile than others.

There are degrees of ductility. There are no degrees of brittleness. Brittleness is the complete absence of ductility.

The temperature and time rate of loading have great effects on the ductility of a material. Some ductile materials become brittle under low temperatures or shock loads. Other normally brittle materials become ductile at slightly elevated temperatures. Some steels have rather abrupt brittle-to-ductile transformation temperatures, while other materials, like plastics, show continuously increasing ductility with slight increases in temperature.

To be suitable for structural tension members, materials must have both elasticity and plasticity. Up to some level of stress, they must be capable of restoring themselves to their original shape after load removal. But if this level of stress is exceeded, they must not suddenly rupture: they must not be brittle. Brittle materials lack toughness and are unforgiving of stress concentrations that might be caused by microscopic cracks, scratches, or other flaws. Brittle materials can be used for compression members, but where tension might occur, ductile materials should be used.

The stress-strain diagrams shown in figure 2.27 and 2.29 demonstrate the characteristics of good tension materials. They are both elastic and plastic. The *elastic range* is that portion of the curve which is linear. The *yield point* or yield strength marks the end of the elastic range. The portion of the diagram between yield and rupture is the *plastic range*.

PROBLEM SET 2.6

2.20 A flat tension specimen as shown in figure 2.25 is cut from a sheet of 3.00-mm-thick rigid plastic ($w = 2$ cm). Stress concentration due to the radius is negligible, because r is large. An extensometer is used to measure strain up to 2 percent. Above that level, elongations over a 5-cm gage length are measured with a scale until rupture. What is the ultimate strength from the test data tabulated below?

Force (N)	Strain (micros)	Force (N)	Elongation (m)
0	0		
215	167	2 270	1.05E−3
413	333	2 480	1.35E−3
620	514	2 680	1.75E−3
827	673	2 890	3.15E−3
1 033	1 340	2 770	4.85E−3
1 240	3 010	2 604	6.85E−3
1 447	5 030		
1 653	7 520		
1 860	11 080		
2 060	16 100		

2.21 For the data of problem 2.20, plot a stress-strain diagram on a piece of graph paper. What is the yield strength for 0.2-percent offset?

2.22 For the data of problem 2.20, plotted in problem 2.21, what is the elongation?

2.23 A ½-in.-diameter steel bar is subjected to a tensile test. It ruptures at a load level of 17 600 lb. What is the ultimate strength?

2.24 A ½-in.-diameter aluminum bar was marked at ½-in. intervals along its length before a tension stress. After the test, the two pieces are fitted back together. Two marks straddling the break, formerly 2 in. apart, are measured to be 2.83 in. apart. What is the percent elongation?

2.25 A strain gage is mounted on a flat tension specimen of cold-rolled steel ¼-in. thick ($w = 0.500$ in.). On a piece of graph paper, plot the stress-strain diagram and determine the yield strength.

Force (lb)	Strain (micros)
0	0
1 000	260
2 000	530
3 000	790
4 000	1 060
5 000	1 560
6 000	3 530

2.12 ELASTIC LIMIT

The stress level that if exceeded causes a material to have a permanent strain is called the *elastic limit*. For example, if a steel tension specimen has an elastic limit of 30 000 psi, it will behave elastically when stresses remain below that level. But if stressed to 40 000 psi, the specimen will not go back to its original length if the load is removed. It will permanently have a new longer length.

The elastic limit of a material is a difficult property to test. The only way to tell if the elastic limit has been exceeded is to remove the load and measure the length. If the specimen has no permanent set upon load removal, the elastic limit has not been exceeded. If a change in length is detected, the elastic limit was exceeded. Using this logic, it is possible to bracket the elastic limit stress value with successive load applications and specimens until the elastic limit is determined to the precision required. Several specimens may be required to determine the elastic limit.

Fortunately, for most materials, the elastic limit is near enough to the yield strength for most structural purposes. The 0.2-percent offset used in defining yield strength does not result in an appreciable permanent set. Thus, though defined differently, yield strength is thought of as the elastic limit of the material and is easier to determine by test.

2.13 THE ELASTIC MODULUS

The *elastic modulus* is defined as the ratio of stress to strain in the elastic range. In mathematical form, the elastic modulus, E, is expressed as

$$E = \frac{\sigma}{\epsilon}$$

(2.5)

Relationship 2.5 applies only in the elastic range. The elastic modulus has the same units as stress. In SI units, E may be expressed in units of N/m^2 or Pa.

The elastic modulus, also called *Young's Modulus* or the *modulus of elasticity,* is the measure of the stiffness of a material. It is a very important structural property. The greater the elastic modulus, the stiffer is the material. The lower the elastic modulus, the more flexible is the material. Structural metals have elastic moduli in the neighborhood of 1E7 psi. Concretes are about ten times more flexible with elastic moduli around 1E6 psi. Structural plastics are in the neighborhood of 1E5, again ten times more flexible. And elastomers, like rubbers, might be ten times again more flexible. Many elastomers in the form of foams are even more flexible. Elastic moduli vary widely depending on the exact material within these broad classifications of structural metals, concretes, structural plastics, and elastomers.

Elastic moduli of plastics and elastomers vary greatly with temperature, composition, and previous working of the material. Structural metals, however, have rather invariant elastic moduli. Structural steel, for example, has an elastic modulus of 29.0E6 psi at 70° F. Increasing the temperature to 400° F reduces the elastic modulus by only 4.8 percent to 27.6E6 psi. Wide variations in carbon content and other alloying agents have little effect on the elastic modulus of carbon steel. Cold-working, annealing, and heat treating produce indetectable changes in elastic modulus. The elastic modulus of a structural metal is one of the most unchangeable of material properties.

One can find the elastic modulus for most metals or alloys in reference works. Appendix I contains a sample. Knowing the elastic modulus, it is possible to make accurate predictions of the stretch of tension members. From the definitions of stress, strain, and elastic modulus ($\sigma = F/A$, $\epsilon = \delta/L$, and $E = \sigma/\epsilon$),

$$E = \frac{F/A}{\delta/L} = \frac{FL}{A\delta}$$

Solving for δ, we can obtain a very useful relationship:

$$\delta = \frac{FL}{AE} \qquad (2.6)$$

Relationship 2.6 applies only in the elastic range.

ILLUSTRATIVE PROBLEMS

2.7 When tested in tension to a stress of 5 000 psi, a plastic registers a strain of 0.0010 on a strain gage. When the load is removed, the strain gage reading returns to 0. What is the elastic modulus?

Solution:

Because the specimen returned to a strain of 0 upon load removal, the elastic limit has not been exceeded. Therefore, the elastic modulus can be calculated from relationship 2.5:

$$E = \frac{\sigma}{\epsilon} = \frac{5\ 000 \text{ psi}}{0.0010} = \frac{5E3 \text{ psi}}{1E-3} = 5E6 \text{ psi}$$

The elastic modulus is 5 million psi.

2.8 How much will a 1-in.-diameter steel bar, 10 ft long, stretch under a tension load of 5.00 T? The elastic modulus is 29.0E6 psi.

Solution:

$$\delta = \frac{FL}{AE} = \frac{5.00\ \text{T} \times 10\ \text{ft}}{0.785(1.000)^2\ \text{in.}^2 \times 29.0\text{E}6\ \text{lb/in.}^2}$$

$$\delta = 2.196\text{E}{-}6 \frac{\text{T} \cdot \text{ft}}{\text{lb}} \times \frac{2\ 000\ \text{lb}}{1\ \text{T}} \times \frac{12\ \text{in.}}{1\ \text{ft}}$$

$$\delta = 0.0527\ \text{in.}$$

However, this presumes that the bar stays in the elastic range when 5 T of load is applied. Checking this with relationship 2.1, we get

$$\sigma = F/A$$

$$\sigma = 5\ \text{T}/0.785\ \text{in.}^2 \times \frac{2\ 000\ \text{lb}}{1\ \text{T}}$$

$$\sigma = 12\ 739\ \text{psi}$$

This result is well below the yield for any steel. Therefore, the elongation calculation is valid.

2.9 A steel bolt for retaining the cylinder head on an air-cooled engine is 30 cm long. It has upset threads. The diameter of the bolt shank is 0.500 cm. It has M8 × 1.25 threads. After the bolt is "snugged" by being turned by hand, how many additional turns must be applied to the bolt with a wrench in order to develop a tension stress in the shank of the bolt of 600 MPa?

Solution:
From relationship 2.6,

$$\delta = \frac{FL}{AE}$$

From the definition of stress (relationship 2.1),

$$\sigma = F/A$$

If we substitute for F/A in relationship 2.6, we get

$$\delta = \frac{\sigma L}{E}$$

For steel, $E = 200\text{E}3$ MPa. L and σ are given as 30 cm and 600 MPa, respectively. Therefore,

$$\delta = \frac{600\ \text{MPa} \times 30\ \text{cm}}{200\text{E}3\ \text{MPa}}$$

$$\delta = 90\text{E}{-}3\ \text{cm} = 0.090\ \text{cm} = 0.90\ \text{mm}$$

The elastic stretch of the bolt will be 0.90 mm.

By the definition of thread pitch, one turn advances the thread one pitch. Therefore, for the M8 × 1.25 thread one turn advances the thread 1.25 mm. If 1.25 mm = 1 turn,

$$\delta = 0.90 \text{ mm} \times \frac{1 \text{ turn}}{1.25 \text{ mm}}$$

$$\delta = 0.72 \text{ turn}$$

Therefore, after snugging, 0.72 turn of the wrench will be required to properly torque the bolt.

2.14 MEASUREMENT OF THE ELASTIC MODULUS

The elastic modulus is the slope of the elastic line of the stress-strain diagram as shown in figure 2.30. Two points, *a* and *b,* are chosen on the line. The vertical distance between the two points is then divided by the horizontal distance to give the elastic modulus.

$$E = \text{slope of the elastic line of the stress-strain diagram} \qquad \textbf{(2.7)}$$

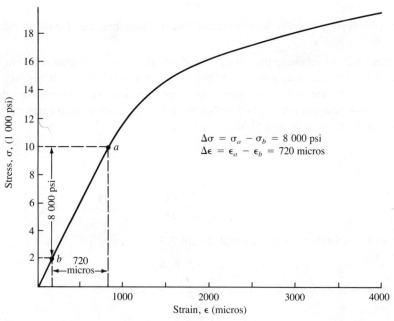

$$\Delta\sigma = \sigma_a - \sigma_b = 8\,000 \text{ psi}$$
$$\Delta\epsilon = \epsilon_a - \epsilon_b = 720 \text{ micros}$$

Two points, *a* and *b*, are selected on the elastic line.

$$E = \text{slope of line} = \frac{\Delta\sigma}{\Delta\epsilon}$$

$$E = \frac{8\,000 \text{ psi}}{720\text{E}-6} = 11.11\text{E6 psi}$$

Figure 2.30 The elastic modulus is the slope of the stress-strain diagram in the elastic range.

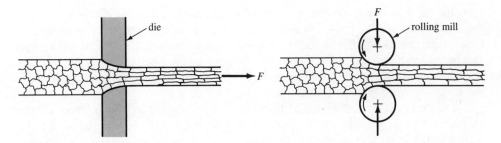

Figure 2.31 Cold-working can be done either by drawing (stretching) or by rolling. Both cause a similar reduction in cross section and elongation of the microstructure of the material.

2.15 STRAIN-HARDENING

The yield strength of a metal can be increased by "cold-working." Metallurgically speaking, *cold* means a temperature below the recrystallization temperature of the metal. That can be hundreds of degrees above room temperature. The term *cold* comes from blacksmiths, to whom anything not glowing red was cold.

Cold-working can be done by rolling or drawing. As shown in figure 2.31, either rolling or drawing produces reduction of the cross section and lengthening of the piece. On the microscopic level, both have the same effect of elongating the metallic crystals. Such elongation on the microscopic level causes things to happen at the sub-microscopic level within the crystal, which explains to metallurgists the phenomenon of strain-hardening. In structures we take a "macroscopic," as opposed to a microscopic, view of things. But it must be remembered that what happens on the microscopic level has a profound effect upon overall structural properties of materials.

From the structural point of view, strain-hardening can be explained in terms of the stress-strain diagram shown in figure 2.32. The stress-strain diagram for a hot-rolled

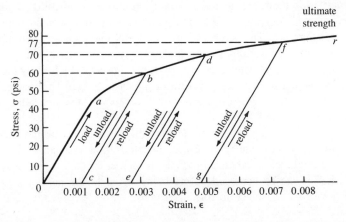

Figure 2.32 Strain hardening explained in terms of the stress-strain diagram.

low-alloy steel is shown. The diagram for the virgin material is the solid curve, *or,* from the origin to the rupture point. By hot-rolling, we mean that the material was rolled at a temperature that allows the formation of a new crystal structure. Crudely, it can be said that the material has no "memory" of events that happened before the last time it was heated above the recrystallization temperature. It doesn't "remember" rolling, because the crystals formed when it cooled after rolling.

The stress-strain curve begins to deviate from a straight line at point *a*. Point *a* is the elastic limit of the material and is approximately 45 000 psi. If the material were to be loaded to any stress level below the 45 000-psi elastic limit, it would assume its original shape after the load is removed. In other words, it would have zero residual strain or no permanent set. Loading or unloading the material below the elastic limit can be described by moving up and down the elastic line from the origin to point *a*.

Now suppose that the material is loaded beyond the elastic limit to a stress level of 60 000 psi, point *b*. What would happen when the load is removed?

It would unload along a line, *bc*. The line has the same slope as the original elastic line *oa*. When the load is completely removed, the specimen will have a new length. It will have a permanent set of 0.0012, the strain at point *c*.

If the specimen were loaded again, stress-strain data would fall on line *bc*. It would not go back to the original starter line *oa*. The line *bc* would become the new elastic line. It would have the same slope, or modulus, of elasticity as the original elastic line.

If the load increase were continued, the specimen would no longer begin to yield at the 45 000-psi elastic limit. Stress and strain data points would continue to be on the new elastic line *bc* until the stress level reached point *b*. Then the data would deviate from the line *bc* and once again follow the curve of the material, *or*. In effect, the elastic limit of the material would have been increased 15 000 psi because of strain-hardening. The material "remembers" the previous maximum stress to which it has been subjected and that becomes its new elastic limit.

If the load were then increased further to point *d,* the line *de* would become the new elastic line. This line would have a 70 000-psi elastic limit.

One could further cold-work the material to any new elastic limit up to the ultimate stress of 80 000 psi. For example, the line *fg* would have an elastic limit of 77 000 psi. In order to achieve this limit, the material would have to be cold-rolled or drawn to an effective permanent set of almost 0.005. The gain in elastic limit would be obtained at the expense of some ductility. The percent elongation would be reduced from 0.8 percent to 0.3 percent. In some applications such a trade-off might be appropriate.

The strain-hardening effect can be removed, or "erased," by heating the material to a temperature at which the crystal structure becomes unstable. The material will then recrystallize when cooled. This process is called *annealing*. The annealing temperature for a carbon steel is around 1 000° F.

Sometimes strain-hardening occurs inadvertently in manufacturing. This is particularly true in forming sheet metal. The material becomes more resistant and less ductile as it is formed, and might tear if stretched too far. In such cases the forming may be done in steps. The material can be annealed before being taken to the next step to restore its ductility and lower its resistance to forming forces.

2.16 FOUR ESSENTIAL STRUCTURAL PROPERTIES

In summary, the four properties that can be determined from tensile test data are (1) yield strength, (2) ultimate strength, (3) elastic modulus, and (4) percent elongation. Knowledge of these properties is essential in using a material intelligently for a specific purpose.

Yield strength is the stress level at which some offset line (0.2 percent for common structural metals) crosses the stress-strain curve. It is, in effect, the elastic limit of the material. Above this stress level the material will have a significant permanent set.

Ultimate strength is the maximum stress level on the stress-strain diagram. The material will be well into the plastic range and have an uncontrolled amount of permanent deformation at this stress. If there is no restraint to the deformation, rupture will occur.

Elastic modulus is the slope of the stress-strain diagram in the elastic range. It is a measure of the stiffness of the material and is used to estimate the elastic deformation of the structure. The elastic modulus, unlike the other three properties, is not measurably effected by heat-treating and cold-working.

Percent elongation is the overall elongation of the ruptured specimen, expressed as a percentage of the gage length. It is an index of ductility when compared to values from other specimens having the same cross-sectional area and shape. The greater the percent elongation, the more ductile the specimen.

These four structural properties of materials are given in Appendix I for a wide variety of materials used in tension structures. Appendix I also includes the materials' specific gravity. *Specific gravity* is the density of the material divided by the density of water. The density of water is 0.0361 lb/in.^{3} in EGS units or $1E-3$ kg/cm^3 in SI units. One can obtain the density of the material by multiplying the density of water by the specific gravity of the material.

The data in Appendix I is for room temperature and reasonably slow strain rates. At elevated temperatures and suddenly applied or shock loads, properties change as shown in the graph of figure 2.33. Strengths and elastic moduli are reduced at increased temperature. Ductility increases with temperature. However, just the opposite occurs when the rate of load application increases. Strengths and elastic moduli increase; ductility decreases.

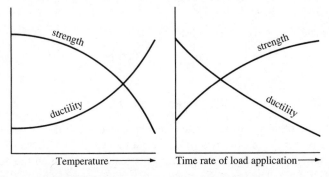

Figure 2.33   Properties of materials change with temperature and rate of loading.

2.17 FACTORS OF SAFETY

A *factor of safety* is a factor, greater than one, divided into either the yield or the ultimate strength of a material in order to give an allowable stress for design purposes.

A factor of safety is used in the design of structures to allow for (1) uncertainties of loading, (2) the statistical variation of material strengths, (3) inaccuracies in geometry and theory, and (4) the grave consequences of failure of some structures. Where permanent deformation is unacceptable, the factor of safety is based upon yield strength. In situations where permanent deformation is acceptable, the factor of safety is based upon the ultimate strength of the material. The factor of safety applies to the strength property most involved in the expected mode of failure. Factors of safety are obtained from design codes or experience specifically related to the kind of machine or structure in question. Codes are the accumulation of the experience gained with numerous previous designs. A *design stress* or a *code-allowable stress* is the yield or ultimate strength, as specified, divided by the factor of safety.

ILLUSTRATIVE PROBLEM

2.10 Section VIII, Division I, of the *Boiler and Pressure Vessel Code* of the American Society of Mechanical Engineers[3] specifies a factor of safety of one-fourth of the ultimate strength or two-thirds the yield strength, whichever is lowest. For a certain material the yield strength is 75 000 psi and the ultimate strength is 96 000 psi. What is the allowable stress for that material?

Solution:
First, we compute stress using ultimate strength:

$$\sigma_{\text{allow}} = \frac{\sigma_{\text{ultimate}}}{4} = \frac{96\ 000}{4}\ \text{psi} = 24\ 000\ \text{psi}$$

Then, we use yield strength:

$$\frac{2}{3}\ \sigma_{\text{yield}} = \frac{2}{3} \times 75\ 000\ \text{psi} = 50\ 000\ \text{psi}$$

The ultimate will govern. Therefore,

$$\sigma_{\text{allow}} = 24\ 000\ \text{psi}$$

PROBLEM SET 2.7

2.26 An electronic plotter is used in a tensile test to determine the elastic modulus of a specimen of a new plastic material. The plotter yields the curve of force versus deformation shown in figure 2.34. The gage length over which the deformation is measured is 5 cm. The specimen cross section is measured to be 4.07 mm thick and 2.13 cm wide. What is the elastic modulus of the material?

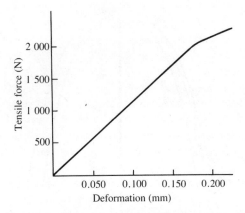

Figure 2.34 Force/deformation graph from plotter for problem 2.26.

2.27 Strain gages are used to measure the elastic modulus of a steel specimen. Two gages are cemented opposite each other on the 0.247-in.-thick specimen. The width of the specimen is 0.503 in. Readings are taken from the two gages at four load levels.

Tension Force (lb)	Left Strain Gage (micros)	Right Strain Gage (micros)
0	0	0
937	241	258
1 883	482	509
2 820	770	786
3 782	993	1 002
0	10	42

Determine the modulus of elasticity either by plotting the data points and measuring the slope of the best-fitting straight line *or* by using a calculator with a "least squares" (also called "linear regression") routine for obtaining the slope of the statistically best-fitting straight line to the data.

2.28 A rod 20 m long is made of steel having a modulus of elasticity of 206 840 MPa. How much does it stretch when carrying a tension stress of 205 MPa? (The yield strength is 300 MPa.)

2.29 A steel tape measure is used to measure a horizontal distance of 97 ft 4.04 in. The tape is 0.407 in. wide and 0.032 in. thick. While making the measurement, 25 lb of tension were applied to the tape. How much is the tape stretched during the measurement? Give the answer in inch units. E = 30E6 psi.

2.30 Suppose, in order to measure its height, an aluminum wire was suspended from the Jefferson National Expansion Memorial in St. Louis (figure 2.35). The arch is 620 ft high. The wire extends straight down from the apex of the arch and just touches the ground. Estimate the amount of stretch in the wire due to its own weight. Use data from Appendix I.

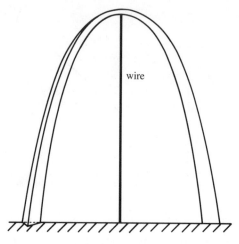

Figure 2.35 Problem 2.30.

2.31 A draw bar for a tractor has a length of 8 ft. It is 2.370 in. in diameter and pulls a load of 60 T. What is the stress and the elongation? The material is low-carbon steel.

2.32 How many millimeters must one stretch a steel guitar string 0.200 mm in diameter and 84.0 cm long in order to develop a stress of 4.82E8 Pa?

2.33 A tube 3.5 m long of 6061-T6 aluminum is fitted with welded end fittings and used as a member in a space truss. It has an outside diameter of 36 mm and a wall thickness of 4 mm. How much will the tube stretch elastically under a tension load of 5 000 kg?

2.34 A structural code sets an allowable stress of 22 000 psi for A36 structural steel, which has a yield of 36 000 psi. What is the factor of safety based upon yield?

2.35 What three materials in Appendix I have the greatest density?

2.36 Which material in Appendix I has the greatest stiffness?

2.37 Which material in Appendix I has the greatest ductility? Which has the least?

2.38 Which material in Appendix I has the greatest ultimate strength? Which has the greatest yield strength?

2.39 Which material in Appendix I has the greatest strength/weight ratio?

2.40 Which material in Appendix I has the lowest strength/weight ratio?

2.41 What is the maximum that a 20-m length of cold-drawn Type 316 stainless steel wire can be stretched without yielding?

2.42 How much force would be required to yield a 1-in.-diameter bar of 155A titanium alloy?

2.43 What percentage of load capacity increase would be gained by replacing a 4140 steel tie rod in the cold-rolled condition with one that is heat-treated and tempered?

2.44 An M8 × 1.25 thread on a 7075-T6 aluminum tie rod was found separated in a ductile tension fashion. At issue is the amount of force required to cause the rod to fail. Estimate the force.

2.45 Assuming the upset threads to be stronger than the pipe, how far could a string of hot-rolled 1040 carbon steel pipe be suspended into a hole in the earth?

2.46 How much would the drill pipe of problem 2.45 stretch elastically if suspended to the depth calculated?

2.47 A 6-in.-diameter steel bolt for a steam turbine casing has an 8-pitch thread. The shank is 36 in. long. For proper pre-tensioning, it is to have a stress of 15 000 psi. How many turns on the nut are required to take the bolt from the "snug" to the "tight" condition?

SUMMARY

☐ Straight-line structural elements are stable in tension. The forces on the free-body diagram of a straight-line tension member must satisfy equilibrium in the direction of the axis.

☐ *Stress* in tension members is defined by the relationship
  (2.1) $\sigma = F/A$

☐ Although stress and pressure have similar definitions and units, they are very different because stress does not follow *Pascal's Principle*.

☐ When abrupt changes in cross section occur, *stress concentration factors* from tables must be used to modify stress relationship.
  (2.2) $\sigma = kF/A$

☐ The effective cross section is reduced by the presence of threads. There are four common types of threads: coarse, fine, 8-thread, and metric.

☐ *Strain* is defined by the relationship
  (2.3) $\epsilon = \delta/L$

☐ *Elasticity* is the ability of a material to assume its original shape after load removal.

☐ *Plasticity* is the ability of a material to be permanently deformed after load removal.

☐ The stress-strain diagram for a ductile material will have a linear portion called the *elastic range* and a curved portion called the *plastic range*.

☐ Controlled destructive tests during which tension stress and strain are measured, called *tensile tests,* are used to determine the four fundamental structural properties of *ductile* materials. Such tests are performed on *universal testing machines*. The four properties are determined from the *stress-strain diagram* that results.

☐ *Yield strength* is the stress level at which the stress-strain diagram intersects with the line parallel to the elastic line, but offset 0.2 percent on the strain axis.

☐ The *ultimate strength* is the highest stress attained.

☐ The *percent elongation* is the maximum strain attained in a 2-in. gage length expressed as a percent.
  (2.4) % elongation $= (\delta/L) \times 100$

☐ The *elastic modulus* is the slope of the elastic line. The elastic modulus is the ratio of stress to strain in the elastic range.
  (2.5) $E = \sigma/\epsilon$

☐ The elastic elongation of tension members can be computed from the important relationship derived from the definitions of σ and ϵ.

  (2.6) $\delta = \dfrac{FL}{AE}$

☐ *Strain-hardening* can be used to advantage to increase the yield strength of a material.

☐ A *factor of safety* is a factor greater than one divided into either the yield or the ultimate strength of the material to give an *allowable stress* for design purposes.

REFERENCES

1. Peterson, R. E. *Stress Concentration Factors*. New York: John Wiley & Sons, 1953.

2. Galilei, Galileo. *Two New Sciences*. Translated by Stillman Drake. Madison, WI: University of Wisconsin Press, 1974.

3. *Boiler and Pressure Vessel Code*, Section VIII, Division I. American Society of Mechanical Engineers, 1980.

CHAPTER THREE

 # Compression and Bearing

Beginning with a comparison of compression members and tension members, this chapter defines compression stress in the case of short columns. Strain and modulus of elasticity are examined in relationship to compression members. A discussion of the testing of brittle materials, specifically the case of concrete, is included with a physical description of that important compression material. Since concrete columns are usually reinforced with steel, this discussion leads naturally to the consideration of members of two materials, which may be in tension or compression. This in turn complements the topics of thermal expansion and thermal stress. The chapter concludes with hardness and bearing stress, which are treated as localized compressions.

3.1 COMPRESSION AND BEARING

Compression is caused by a pushing force. It is the opposite of tension which is caused by a pull. But, there is more to compression than simply a tension force of opposite sign. Brittle materials, usually a poor choice for tension, may be the superior choice for compression. Buckling, which is not a consideration in tension, can be significant in compression. From the standpoint of the behavior of structures, compression brings many factors into play and warrants an entire chapter.

Compression is almost always present in a structural member because it is difficult to transmit a force into a member without having at least localized compression, called *bearing,* at the point of contact. This is true whether or not the member is predominantly a compression member. There would be compression under either the head or the shank of a bolt used to apply a pull to a member, for example. Hence compression and bearing are intimately related.

Other structural members may transmit forces to a compression member merely by resting against it. Fasteners or cements may not be necessary. In fact, compression forces, unlike those of tension, may be transmitted even across a crack.

3.2 STRAIGHT-LINE COMPRESSION MEMBERS

Compression members have many shapes, such as arches. But, in this chapter, only straight-line compression members will be considered. Figure 3.1 shows several examples of compression members.

   The posts in the ancient post-and-lintel system in figure 3.1 are straight-line compression members, as are the columns in the modern structural steel beam and column system. The legs in the water tower and the pillars of the highway bridge are further examples. Straight-line compression members needn't be static; they may also occur in mechanisms. The connecting rod of an engine or compressor becomes a compression member during the compression or the power strokes. The piston rod of the fluid power cylinder is a straight-line compression member when it is extending against resistance. Straight-line compression members, such as the boom of a crane, need not be vertical. In the crane, the boom is the compression member, and the cable system, which raises the boom, is the tension member. Although the boom is subjected to the same order of magnitude of force as the cable system, the boom is more massive and complicated, because it must resist compression.

   The only similarity between straight-line compression and tension members is that the lines of action of the force coincide with their geometric axes. Other than that, the two behave quite differently. Tension members can pull themselves into shape, while compression members have the unfortunate tendency to buckle off to the side. Compression members must often be braced against this tendency. Since brittle materials such as concrete, stone, and masonry perform well in compression, compression members are often quite different from tension members in material of construction as well as in appearance.

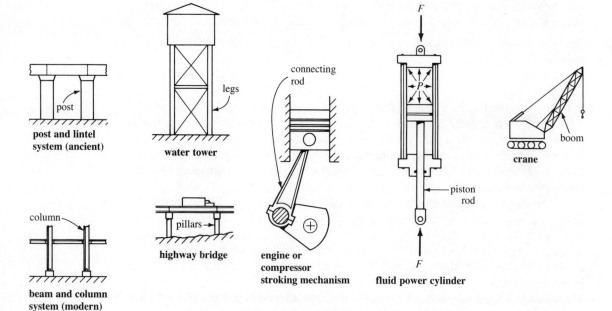

Figure 3.1 Examples of compression members.

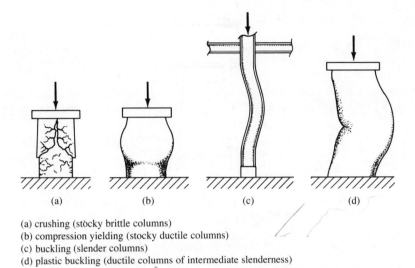

(a) crushing (stocky brittle columns)
(b) compression yielding (stocky ductile columns)
(c) buckling (slender columns)
(d) plastic buckling (ductile columns of intermediate slenderness)

Figure 3.2 Compression failures.

3.3 MATERIAL FAILURES VERSUS BUCKLING FAILURES

Short, stocky compression members will fail by crushing as shown in figure 3.2a if they are made of brittle materials. If constructed of ductile materials, short compression members are apt to yield in compression, as in figure 3.2b. Crushing and yielding are material failures.

A slender compression member, figure 3.2c, will probably fail by buckling. Buckling can occur without any rupture or yielding of the material. To illustrate this to yourself, take a hacksaw blade and apply a compression load with your hands as shown in figure 3.3. The blade will buckle under a very slight force. But if released after buckling, the blade will snap back to its original shape, showing that no failure of the material has taken place.

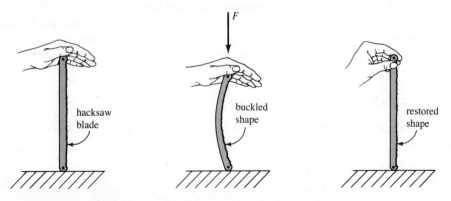

Figure 3.3 Buckling need not be accompanied by failure of the material. In the case of the hacksaw blade, which springs back to its original straightness after buckling, no yielding occurs.

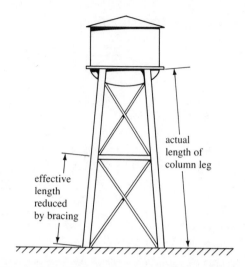

Figure 3.4 Diagonal bracing is used to reduce the effective length of compression members, thereby reducing their tendency to buckle.

When a compression member buckles, it fails at a load substantially below that which would cause a material failure. Therefore, compression members are usually stocky, or they are braced as shown in figure 3.4 to reduce their free length. When the length of a slender column is reduced by half, the buckling load will be increased by a factor of four. By shortening the effective length, the load at which a member will buckle can be increased to the point where a material failure would occur first.

Slender columns apt to fail by buckling with no accompanying material failure are classified as *long* columns. Those which are apt to fail by yielding or crushing are classified as *short* columns. Columns which will fail by a combination of buckling and material failure are called *intermediate* columns. Intermediate columns are usually the most economical in their use of material under load. Unfortunately, they are the most complicated phenomenon to understand and to mathematically describe.

Buckling is actually bending deformation produced by a compression load. Bending deformation is covered in chapter 12. After that, buckling is treated in chapter 14.

For now it is sufficient to restrict our attention and treatment of compression members to the category of short columns.

3.4 COMPRESSIVE STRESS IN SHORT COLUMNS

Compressive stress is defined exactly as it was in the case of tension. Stress is force per unit area, or

$$\sigma = F/A \tag{3.1}$$

where F is the compressive force and A is the cross-sectional area. As in the case of tension members, the cross-sectional area is the area of the member exposed if the member were cut perpendicular to its axis.

In some fields of work, the compression force, F, is given a negative sign. If this is the case, compressive stress, σ, will be negative. But there are other fields and applications where compression force and stress are given positive signs. Although it would be much simpler if everyone were consistent, this is not the case. The important thing to remember is that a plus or a minus sign is merely a method of bookkeeping. Structures don't know whether forces applied to them are positive or negative. But they do behave much differently in compression from the way they do in the case of tension.

A consistent sign convention requires that if the positive sign is used for tension, a negative sign must indicate compression. Or, if a negative sign is used to indicate tension, then the plus sign indicates compression. Before starting a problem, choose the sign convention and keep it throughout the problem. The most important thing is not whether the answer is positive or negative, but whether the stresses or forces are tension or compression. A problem is not complete until the positive or negative signs are converted to statements of compression or tension.

The units of compressive force or stress are identical to those for tension. There is one addition in compression, psf, which denotes pounds per square foot.

$$\boxed{1 \text{ psi} = 144 \text{ psf}}$$

The unit psf is frequently used for materials such as soils that have relatively low stress-carrying capability.

3.5 FOUNDATIONS

All buildings and other stationary structures must transfer their loads to the earth. They do this through foundations. Since the loads are mostly due to gravity, the primary loads are vertical and compressive.

There are four basic types of foundations: the pier, the footing, the slab, and the piling. These are shown in figure 3.5. In the pier, the compression load is transmitted through a column that extends directly to bedrock, which finally supports the structure. A hole might be drilled and the pier cast within the hole. Or, the soil might be excavated, and the concrete cast into a columnar or even a wall shape through the use of forms.

A footing is essentially a pad used to spread the foundation load over a larger area of soil. The area is made large enough to reduce the "bearing stress" on the soil to a safe value. Such safe values are determined by soil engineers who base their field and laboratory tests and calculations upon the principles of soil mechanics. Since soils are widely variable in their resistive properties, soil mechanics is an important engineering specialty.

Footings must be placed lower than the expected depth of freezing. Otherwise, the water in the soil might freeze and expand beneath the footing, moving it vertically. Since such displacement is apt to be uneven over the entire base of the structure, freezing under the footing may cause the foundation to crack or the structure to deform. In temperate climates, the depth of the "frost line" depends upon locality. In locations where it doesn't get cold enough to freeze the ground, slab foundations may be used.

A pile foundation is constructed by using a pile driver to hammer pilings into the earth until the desired load resistance is achieved. Pilings are normally specially shaped columns of reinforced concrete. For some applications steel or timber may be used. During pile driving, the blows required to advance the pile a distance of one foot are

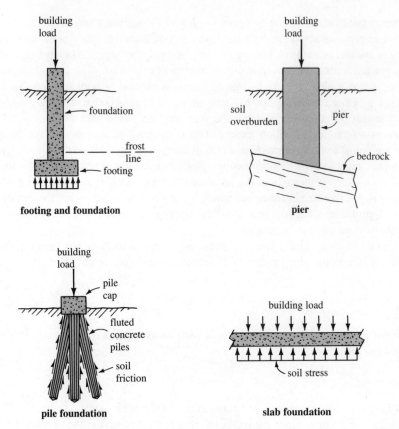

Figure 3.5 Types of foundations.

counted. Knowing the weight of the hammer and the height from which it is dropped in the pile driver, engineers are able to determine the pile resistance from the number of blows per foot. When the desired resistance is obtained, the pile is cut off at or near the surface. A pile cap is then formed over the top of the pile. Sometimes several piles are driven near one another and tied together with the pile cap and the structure is then built upon the pile cap.

ILLUSTRATIVE PROBLEMS

3.1 A short timber column is 30 cm square and supports a load of 100 000 kg. What is the compressive stress in units of kg/cm^2 and in MPa?

Solution:

The cross-sectional area is (30 cm)^2 or 900 cm^2. From relationship 3.1,

$$\sigma = F/A = \frac{100\ 000\ \text{kg}}{900\ \text{cm}^2} = 111.1\ \text{kg/cm}^2$$

$$\sigma = 111.1\ \text{kg/cm}^2 \times \frac{9.8\ \text{N}}{1\ \text{kg}} \times \frac{10^4\ \text{cm}^2}{1\ \text{m}^2} = 10.89\ \text{MPa}$$

The compressive stress is 111.1 kg/cm^2, or 10.89 MPa.

3.2 A short steel column (figure 3.6) on a concrete foundation and footing is to be designed to support a load of 75 kips. The column will be made of square steel tubing with a $\frac{1}{4}$-in. wall and an allowable stress of 20 ksi. The square concrete foundation will be made of concrete with an allowable stress of 1 500 psi. The soil is expected to be able to support a load of 2 000 psf. What must be the sizes of the column, the foundation, and the footing?

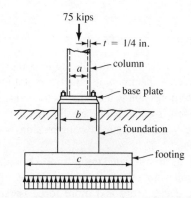

Figure 3.6 Illustrative problem 3.2.

Solution:

The column, foundation, and footing all transmit the same load of 75 kips. From relationship 3.1, we have

$$\sigma = F/A$$

Solving for A, we get

$$A = F/\sigma$$

For the column, σ equals 20 ksi. Therefore,

$$A = \frac{75 \text{ kips}}{20 \text{ ksi}} = 3.75 \text{ in.}^2$$

$$A = 4at \text{ (approximately)}$$

$$a = \frac{A}{4t} = \frac{3.75 \text{ in.}^2}{4 \times 0.250 \text{ in.}} = 3.75 \text{ in.}$$

For the foundation, σ equals 1 500 psi.

$$A = \frac{75\,000 \text{ lb}}{1\,500 \text{ lb/in.}^2} = 50 \text{ in.}^2$$

$$A = b^2$$

Therefore,

$$b = \sqrt{A} = \sqrt{50 \text{ in.}^2} = 7.07 \text{ in.}$$

For the footing, σ equals 2 000 psf.

$$A = \frac{75\ 000\ \text{lb}}{2\ 000\ \text{lb/ft}^2} = 37.5\ \text{ft}^2$$

$$A = c^2$$

$$c = \sqrt{A} = \sqrt{37.5\ \text{ft}^2} = 6.12\ \text{ft} = 73.4\ \text{in.}$$

The column should be 4 in. square, the foundation 8 in. square, and the footing 74 in. square.

3.6 THE COMPRESSION TEST OF DUCTILE MATERIALS

In order to avoid buckling when testing materials are in compression, short, stocky specimens are required. When such a specimen of a ductile metal is tested in compression, the slope of the elastic line in compression is the same as the slope in tension. In other words, the elastic modulus in tension is equal to the elastic modulus in compression. Also, for most ductile materials, the yield strength in compression is almost equal to the yield strength in tension.

There is no such thing as an ultimate strength for a ductile material in compression. As the short, stocky specimen goes into the plastic range, the cross-sectional area increases as shown in figure 3.2b. More and more force is required to further yield the specimen. The specimen appears to get ever stronger. But this is due to increased cross section, not increased material strength. Rupture does not occur. Plastic buckling, as shown in figure 3.2d, might occur. However, this gives no true information about the material, although it might cause the stress-strain diagram to peak out. Since rupture does not occur, there is no property similar to the percent elongation that can be determined from a compression test.

Since the elastic modulus and the yield stress of a ductile material in compression are for practical purposes identical to those properties in tension, compression tests are rarely used for ductile materials. Furthermore, no information about ultimate strength or percent elongation can be obtained from a compression test on a ductile material. Therefore, ductile materials are tested in tension.

3.7 COMPRESSION TEST ON BRITTLE MATERIALS

Just as compression tests tend to be impractical for ductile materials, tension tests are impractical for brittle materials. Brittle materials in tension are unforgiving of stress concentrations and of slight misalignments. Since in a tension test brittle materials tend to crack at the grips, it is very difficult to devise a suitable fixture for gripping brittle materials in a tension test. The compression test is the standard for brittle materials such as concrete or masonry.

The stress-strain diagram for concrete shown in figure 3.7 is an example of the information that can be obtained from a compression test on a brittle material. The standard specimen for compression tests on concrete is a cylinder 6 in. in diameter and 12 in. long. The test is conducted 28 days after the test specimen is poured. The strength based upon this test is called the *28-day cylinder strength*.

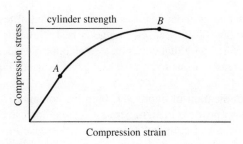

Figure 3.7 Stress-strain diagram for concrete.

Concrete is a building material made of portland cement, sand, gravel, and water. Portland cement is made by pulverizing limestone to an extremely small particle size. Then, in a kiln, any moisture and chemically bound water are driven off by heating the pulverized material. When the product comes from the kiln, the material is called *klinker*. The klinker is then pulverized again to create the finished product of portland cement.

Because portland cement is the most expensive component of concrete, sand and gravel are used as fillers. Sand and gravel are called the *aggregate*. The maximum particle size that can be used for aggregates depends upon the space in the forms or the distance between the reinforcing bars. In massive sections such as dams, the maximum-size particle may be 10 cm across. For typical building construction applications, the maximum particle size is usually in the neighborhood of 2 cm. The most efficient aggregate uses a gradation of particle sizes, from sand to coarse gravel. Since muddy or salty sands adversely affect the bond of the cement, it is important that the aggregate be clean.

The process of mixing and curing concrete is essentially one of returning the chemical water to the portland cement and converting it back into something much like the limestone from which it came (although, chemically and physically there are slight differences between the original limestone and the cured portland cement). Cured portland cement is the paste that holds concrete together. In good concrete it is not uncommon for the paste to be stronger than the gravel in the aggregate. Concrete tests frequently show cracks extending right through the gravel rocks.

It takes a certain amount of time for the water to be reabsorbed by the particles of portland cement during curing. Therefore, concrete does not become a solid immediately. In fact, even after becoming a solid, it takes many days for concrete to reach full strength. So long as the concrete is kept moist, it will continue to increase in strength, perhaps indefinitely. Most of the strength is achieved within 28 days. This is why the 28-day cylinder strength is used as the basis of concrete strength.

Quality control of concrete at the construction site is extremely important in the use of this building material. Although some tests can be applied to wet concrete at the time of pouring, no conclusive strength test can be conducted. The only reliable means of quality control is to cast samples of the concrete into cylinders and to test those cylinders 28 days later. The test cylinders are identified as to date and location on the structure and kept in a special moist-room. If construction procedure is carefully adhered to and the placed concrete is not allowed to dry out during curing, the strength of the concrete in the structure will be similar to that in the test cylinders.

The test of a concrete cylinder requires a fairly substantial testing machine. A hydraulic machine in the 300 000-lb-capacity range is best. Strain is determined from the decrease in length measured with dial gages located on opposite sides of the cylinder. The average of the two decreases measured on the dial gages is used. Figure 3.8 is a diagram of a cylinder under test.

In the stress-strain diagram in figure 3.7, both the compressive stress and the compressive strain are shown positive in sign. Stress is the compressive force divided by the cross-sectional area of the test cylinder. Strain is the decrease in length divided by the distance between points where the dial gages attach to the cylinder.

The stress-strain diagram will have a straight-line portion up to a certain point, A. The material can be presumed to be elastic over this linear portion. As the stress is increased beyond point A, the stress-strain curve begins to deviate off to the right of the straight line. But the deviation is not caused by plasticity as in ductile materials. Cured concrete is not ductile. The deviation is caused by cracking of the material, as shown in figure 3.9a.

Concrete has strength in compression even when it might contain cracks. Compression force can be transmitted across cracks. However, it is not good practice to use concrete in this form. Safety factors specified by code ensure that the concrete remains in the elastic range.[1]

The ultimate compressive strength of the concrete is reached at point B. Here the concrete fails in compression by crushing as shown in figure 3.9b. The load at which the cylinder crushes is the *cylinder strength* upon which safety factors are based. Cylinder strength depends upon the kind and the quality of the cement and the aggregate, the purity of the water, and, most important, the water/cement ratio. Concrete with high water/cement ratios is more porous and therefore has lower cylinder strengths. Twenty-eight-day cylinder strengths for typical concretes range from 2 500 psi to 5 000 psi. The allowable stress in direct compression for design purposes specified by code[1] is one-fourth of the 28-day cylinder strength.

The modulus of elasticity of concrete can be determined by finding the slope of the linear portion of the stress-strain diagram. The modulus of elasticity is related to the strength and porosity of the concrete. For design purposes, the modulus of elasticity of

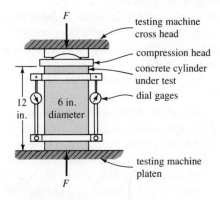

Figure 3.8 Concrete cylinder test.

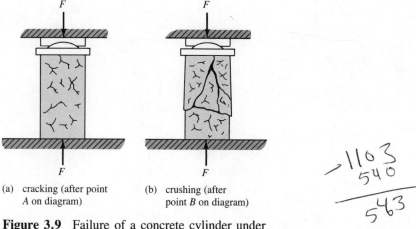

(a) cracking (after point
A on diagram)

(b) crushing (after
point B on diagram)

Figure 3.9 Failure of a concrete cylinder under test.

concrete, in psi, can be estimated by using the following formula from the code[1]:

$$E = 33w^{1.5}\sqrt{\sigma_{28}} \tag{3.2}$$

where w is the density of the concrete in pounds per cubic foot and σ_{28} is the 28-day cylinder strength in psi.

ILLUSTRATIVE PROBLEMS

3.3 A 6-in.-diameter concrete cylinder 12 in. long is cast on the construction site. A compression test 28 days later shows that the cylinder can withstand a maximum of 124 kips. The building was designed for a concrete having a cylinder strength of 3.00 ksi. Does the concrete meet the design requirements?

Solution:
From the definition of stress (relationship 3.1),

$$\sigma = F/A$$

The 28-day cylinder strength is

$$\sigma = \frac{124 \text{ kips}}{0.785 \times (6 \text{ in.})^2} = 4.39 \text{ ksi}$$

The concrete used exceeds the design requirement of 3.00 ksi.

3.4 A 6-in.-diameter test cylinder is cast during the pouring of a new building. The structural designers call for a cylinder strength of 4 000 psi. The dial gages are attached to fixtures giving a gage length of 10 in. The following data is obtained. Does the concrete meet specifications? What is the elastic modulus of the concrete as determined from the experimental data?

Force (lb)	Left Dial Gage Reading (in.)	Right Dial Gage Reading (in.)
0	0.153 2	0.208 1
20 000	0.151 3	0.205 4
40 000	0.149 0	0.203 1
60 000	0.146 7	0.200 8
80 000	0.142 4	0.196 5
100 000	0.138 1	0.192 2
120 000	0.131 8	0.185 9
123 000	crushed	crushed

Solution:

First the force data is reduced to stress by dividing by the cylinder area (28.27 in.²). Then the dial gage readings are reduced to strain readings by subtracting the zero reading from each, averaging the two, and then dividing by the gage length as follows:

Stress (psi)	Length Reduction (in.) (left)	(right)	(average)	Strain (in./in.)
0	0	0	0	0
707	0.001 9	0.002 7	0.002 3	0.000 23
1 415	0.004 2	0.005 0	0.004 6	0.000 46
2 122	0.006 5	0.007 3	0.006 9	0.000 69
2 830	0.010 8	0.011 6	0.011 2	0.001 12
3 537	0.015 1	0.015 9	0.015 5	0.001 55
4 245	0.021 4	0.022 2	0.021 8	0.002 18
4 351	—	—	—	—

The next step is to plot a graph of stress versus strain as shown in figure 3.10.

The maximum stress is 4 350 psi, which is the cylinder strength of the material. It exceeds the specification of 4 000 psi.

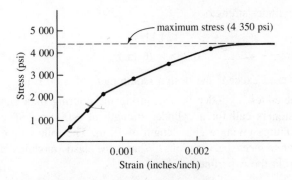

Figure 3.10 Illustrative problem 3.4.

The concrete appears to be elastic up to a stress level of 2 122 psi. This limit is determined from the fact that the stress-strain diagram appears to be linear up to that point. The elastic modulus can be computed from the slope of the elastic line as follows:

The elastic line appears to be through the origin and also the point (2 122, 0.000 68). Proceeding as in chapter 2,

$$E = \frac{\text{change in stress}}{\text{change in strain}}$$

$$E = \frac{(2\ 122 - 0)\ \text{psi}}{(0.000\ 68 - 0)\ \text{in./in.}}$$

$$E = 3.12\text{E}6\ \text{psi}$$

3.5 A concrete has a density of 125 lb/ft^3 and a 28-day cylinder strength of 5 500 psi. By CRSI code, what is the elastic modulus to be used for design purposes?

Solution:
Substituting the values into the code formula, we get

$$E = 33w^{1.5}\sqrt{\sigma_{28}}$$
$$E = 33 \times (125)^{1.5}\sqrt{5\ 500}$$
$$E = 33 \times 1\ 397 \times 74.16 = 3\ 42\text{E}6\ \text{psi}$$

PROBLEM SET 3.1

3.1 Timber columns made of 12 × 12's are used to shore up a mine ceiling. They are expected to carry a load of 500 000 lb each. Based on the actual dressed dimensions of the timbers, what is the anticipated stress in the columns? (See Appendix VI.)

3.2 Short columns are to be made of standard weight (Schedule 40) steel pipe having an allowable stress of 22 ksi. From Appendix IV, select the smallest standard-weight pipe that can support a load of 160 kips.

3.3 A square concrete column 500 mm × 500 mm carries a load of 5.27E6 kg. What is the stress in kg/cm^2 and in MPa?

3.4 A short brace made of a standard angle section 6 × 6 × $\frac{3}{8}$ carries a load of 40 T. What is the stress in psi? (See Appendix V.)

3.5 What is the load capacity of a W8 × 31 standard steel section used as a very short column if the allowable stress is 22 000 psi? (See Appendix V.)

3.6 An aluminum connecting rod for an air compressor is shown in figure 3.11. The material is 6061-T6. Estimate the compressive load required to yield the material. Where is the yielding likely to occur?

3.7 Concrete foundation walls for residential structures are normally 8 in. thick. If 2 000-psi concrete is used, what maximum load per foot of length can the foundation safely support by code, presuming that the soil and footings have adequate strength?

3.8 An oil-drilling rig is to be supported on a platform in a swamp in the Niger River Delta. The platform is intended to support a load of 6E6 lb. The platform will have a pile foundation. A 30-ft test pile is driven that provides a resistance of 140 T when

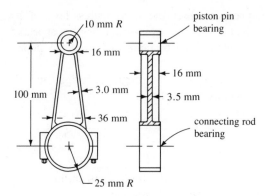

Figure 3.11 Problem 3.6.

driven to its full length. How many such piles will be required to support the platform?

3.9 A short wooden compression strut supporting the deck of a competition sailboat is to be made of a special German spruce having a crushing resistance of 4 800 N/cm². If the strut is to be square and be designed for 9 000 N with a safety factor of 3, what is the minimum size that will be satisfactory?

3.10 A short, round concrete column is to be used in a parking garage. The design load will be 2 000 kips and the design-allowable stress is 3 000 psi. What minimum diameter may the column have?

3.11 A short column, foundation, and footing similar to those shown in figure 3.6 is to carry a load of 500 000 lb. The allowable stresses for the standard pipe column, the concrete, and the supporting soil are 22 000 psi, 2 000 psi, and 15 000 psf, respectively. What must be the size of the pipe column? What must be the diameters of the foundation and the footing, which have circular cross sections?

3.12 A standard 6-in.-diameter concrete cylinder is taken during the pouring of the concrete for a tunnel. The test is conducted on the twenty-eighth day after the concrete was poured with the aid of a fixture having a gage length of 10 in. Two dial gages on opposite sides of the test cylinder are used to measure deflection. The following data is collected:

Force (lb)	Left Dial Gage Reading (in.)	Right Dial Gage Reading (in.)
0	0.206 0	0.183 0
20 000	0.203 7	0.181 5
40 000	0.201 5	0.178 5
60 000	0.198 0	0.176 3
80 000	0.194 7	0.172 2
100 000	0.190 2	0.167 5
120 000	0.175 0	0.162 5
130 000	0.160 0	0.155 0
137 000	crushed	

What is the 28-day cylinder strength for this concrete?

3.13 From the experimental data, what is the modulus of elasticity for the concrete tested in problem 3.12?

3.14 For design purposes, using the CRSI formula (relationship 3.2), what is the elastic modulus of the concrete tested in problem 3.12? The density of the concrete is 145 lb/ft^3.

3.15 A short non-reinforced column 14 in. square is cast from concrete having a design-allowable stress of 4 500 psi. What is the allowable load for the column?

3.16 A concrete column 24 in. in diameter and 10 ft long carries an axial compressive load of 1.5E6 pounds. Instruments show that under compression the column shortens 0.120 in. What is the elastic modulus of the concrete?

3.17 A 6061-T6 aluminum column 1.230 m long has a hollow rectangular cross section of outside dimensions 33 mm by 15 mm with a wall thickness of 2 mm. How much will the column shorten under a compressive load of 1 000 kg?

3.18 What will be the code-allowable stress for a column made of the concrete tested in problem 3.12?

3.19 A 6-in.-diameter test cylinder crushes under test at a load of 185 000 lb. What is the code-allowable load on a short, round concrete footing of the same diameter 18 in. in diameter?

3.20 Three strain gages are applied to a concrete column 12 ft long in a parking garage. The column diameter is 36 in. The three gages are located midway up the column and spaced equally around the column. Compressive strains of 0.0017, 0.0013, and 0.0019 are recorded from the strain indicator under a certain test load on the column. Based on the average strain reading, estimate the total shortening of the column under load.

3.8 THE EFFECT OF STRESS CONCENTRATIONS UPON BRITTLE MATERIALS

Stress concentrations in compression are identical to those for the same shape in tension. Stress concentration factors can be determined from graphs such as figure 2.12. For a more complete selection of stress concentrations, see reference 2.

Stress concentrations have very severe consequences in brittle materials. Brittle materials are unable to yield and flow plastically at points of high stress. Therefore, they are unable to relieve stress concentrations. Brittle materials in tension are particularly affected by stress concentrations due to microscopic surface cracks. A crack extending into the material, even a small distance can be regarded as a notch of extremely small radius of curvature. Because the notch radius at the tip of a crack is extremely small (in fact, not measurable), there is an enormous stress concentration factor at the root of a crack. Since a brittle material is unable to yield, the crack propagates. The propagation continues until the entire specimen separates catastrophically.

Surface cracks and flaws may cause no significant problems when the material is loaded in compression. Compression load can be transmitted across a crack. Even if the crack extended entirely through the cross section, the member could transmit compression load.

Brittle materials such as stone and concrete make excellent compression members. But they must be used very cautiously in tension. As an example, the code-allowable stress for a typical concrete in compression is 1 800 psi. In tension only 102 psi of stress is permitted in the same material.

3.9 MEMBERS OF TWO MATERIALS

If a short column is made of two materials, such as concrete and steel, it is not possible to determine at the outset which part of the force applied to the column is carried by the first material (concrete) and which part is carried by the second material (steel). But in order to satisfy static equilibrium,

$$F = F_1 + F_2 \tag{3.3}$$

where F is the total load on the column and F_1 and F_2 are the parts of the load carried by the first material and the second material, respectively, as shown in figure 3.12.

The steel reinforcing rods have the same length as the concrete column (L). The lengths will be equal even when the column is under load as shown in figure 3.12. In other words, because the two materials must act together, they must have the same axial deformation (δ).

Assuming that both materials remain in the elastic range during the loading of the member, the deformations δ_1 and δ_2, will be given by relationship 2.6 (from chapter 2):

$$\delta_1 = \frac{F_1 L}{A_1 E_1} \quad \text{and} \quad \delta_2 = \frac{F_2 L}{A_2 E_2} \tag{3.4}$$

where A_1 and A_2 are the respective areas of the two materials and E_1 and E_2 are the moduli of elasticity. But, since the materials must always have identical length loaded or unloaded,

$$\delta_1 = \delta_2$$

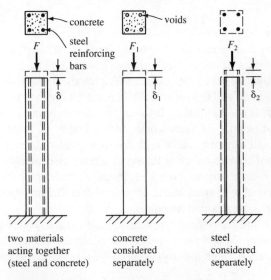

two materials
acting together
(steel and concrete)

concrete
considered
separately

steel
considered
separately

Figure 3.12 A compression member of two materials, steel and concrete, under load.

Or, substituting, we get

$$\frac{F_1 L}{A_1 E_1} = \frac{F_2 L}{A_2 E_2}$$

Therefore,

$$F_1 = F_2 \frac{A_1 E_1}{A_2 E_2} \tag{3.5}$$

This relationship can be used to express the force on the first material (concrete), F_1, in terms of the force on the second material (steel), F_2. Substituting for F_1 in relationship 3.3, we have

$$F = F_2 \frac{A_1 E_1}{A_2 E_2} + F_2$$

Solving for F_2, we get

$$F_2 = \frac{F}{\left(\dfrac{A_1 E_1}{A_2 E_2} + 1\right)} \tag{3.6}$$

One could solve similarly for F_1:

$$F_1 = \frac{F}{\left(\dfrac{A_2 E_2}{A_1 E_1} + 1\right)} \tag{3.7}$$

In order to get the stresses in the two materials, one must divide the forces by the proper area in each case.

$$\sigma_1 = F_1/A_1 \quad \text{and} \quad \sigma_2 = F_2/A_2 \tag{3.8}$$

The above relationships are valid for both compression and tension members. They are not valid if either material is stressed beyond the elastic range.

In determining stresses in members of two materials, one finds quickly that the greater stress tends to go to the least flexible material; that is, the one having the highest modulus of elasticity.

ILLUSTRATIVE PROBLEMS

3.6 A short 12-in.-square concrete column is reinforced with four longitudinal #6 bars. The column carries a load of 100 kips, which is well within the elastic range of both materials. The modulus of elasticity of the steel is 29 000 ksi, and that of the concrete is 3 000 ksi. What portion of the load is carried by the concrete and what portion by the steel? What are the stresses in the steel and the concrete?

Solution:
From Appendix VIII, the cross-sectional area of a #6 reinforcing bar is 0.44 in.^{2} Therefore, the total steel area for four bars is 1.76 in.^{2} The concrete area is 144 minus 1.76, which equals 142.24 in.^{2} Letting the concrete be material 1 and the steel be

material 2,

$$A_1 = 142.24 \text{ in.}^2 \text{ of concrete}$$

and

$$A_2 = 1.76 \text{ in.}^2 \text{ of steel}$$

(Note that the 142.24 is not rounded off at this point; it is best to carry the decimal places on your calculator through the problem and to round the final answer.)

$$E_1 = 3\ 000 \text{ ksi} \quad \text{and} \quad E_2 = 29\ 000 \text{ ksi}$$

From relationship 3.7,

$$F_1 = \frac{100 \text{ kips}}{\left(\dfrac{1.76 \text{ in.}^2 \times 29\ 000 \text{ ksi}}{142.24 \text{ in.}^2 \times 3\ 000 \text{ ksi}} + 1\right)}$$

$$F_1 = \frac{100 \text{ kips}}{(0.1196 + 1)} = 89.3 \text{ kips}$$

From relationship 3.6,

$$F_2 = \frac{100 \text{ kips}}{\left(\dfrac{142.24 \text{ in.}^2 \times 3\ 000 \text{ ksi}}{1.76 \text{ in.}^2 \times 29\ 000 \text{ ksi}} + 1\right)}$$

$$F_2 = \frac{100 \text{ kips}}{(8.3612 + 1)} = 10.68 \text{ kips}$$

Checking with relationship 3.3, does $F = F_1 + F_2$?

$$100 \text{ kips} = 89.3 \text{ kips} + 10.68 \text{ kips}$$
$$100 \text{ kips} = 99.98$$

The answers check. One hundred kips is equal to 99.98 rounded off to an appropriate number of significant places.

The concrete carries 89.3 kips of the load. The steel carries the remaining 10.68 kips of the 100-kip load. The steel carries 10.68 percent of the load, even though it comprises only 1.22 percent of the cross-sectional area. Load does not distribute itself uniformly over the area when more than one material is involved. The stiffer material will gather a greater share of the load.

The stresses are

$$\sigma_1 = F_1/A_1 = \frac{89.3 \text{ kips}}{142.24 \text{ in.}^2} = 0.628 \text{ ksi} = 628 \text{ psi}$$

$$\sigma_2 = F_2/A_2 = \frac{10.68 \text{ kips}}{1.76 \text{ in.}^2} = 6.070 \text{ ksi} = 6\ 070 \text{ psi}$$

The concrete has a stress of 628 psi and the steel is stressed to a level of 6 070 psi.

3.7 A titanium tube (155A alloy) is fit tightly over an aluminum rod (7075T6 alloy) and used as a tension rod for an aircraft control. The outside diameter of the tube is

12 mm. The wall thickness is 1 mm. When 2 000 kg of tension is applied to the assembly, what is the stress in the tube and what is the stress in the rod?

Solution:
The filled tube is a member of two materials. Relationships 3.3 through 3.8 apply so long as both materials are in the elastic range *whether the member is in tension or in compression.* We will use the relationships to calculate stresses and then check to see if the materials are in their elastic ranges, thereby checking the validity of the calculations.

Taking titanium as material 1 and aluminum as material 2, we have

	Titanium *(material 1)*	*Aluminum* *(material 2)*
Elastic Modulus	113E9 Pa	69E9 Pa
Yield Strength	930 MPa	492 MPa

$$A_1 = 0.785[(12 \text{ mm})^2 - (10 \text{ mm})^2] = 34.5 \text{ mm}^2$$
$$A_2 = 0.785(10 \text{ mm})^2 = 78.5 \text{ mm}^2$$

Using relationship 3.6,

$$F_2 = \frac{F}{\left(\dfrac{A_1 E_1}{A_2 E_2} + 1\right)}$$

$$F_2 = \frac{2\ 000 \text{ kg}}{\left(\dfrac{34.5 \text{ mm}^2 \times 113\text{E}9 \text{ Pa}}{78.5 \text{ mm}^2 \times 69\text{E}9 \text{ Pa}} + 1\right)}$$

$$F_2 = \frac{2\ 000 \text{ kg}}{(0.719 + 1)} = 1\ 163 \text{ kg}$$

The aluminum core carries 1 163 kg of the 2 000-kg load. From relationship 3.3,

$$F = F_1 + F_2$$

Therefore,

$$F_1 = F - F_2 = 2\ 000 \text{ kg} - 1\ 163 \text{ kg} = 837 \text{ kg}$$

Converting to newtons by multiplying by 9.81 N/1 kg,

$$F_1 = 8\ 210 \text{ N} \quad \text{and} \quad F_2 = 11\ 410 \text{ N}$$
$$\sigma_1 = F_1/A_1 \quad \text{and} \quad \sigma_2 = F_2/A_2$$
$$\sigma_1 = \frac{8\ 210 \text{ N}}{34.5 \text{ mm}^2} \quad \text{and} \quad \sigma_2 = \frac{11\ 410 \text{ N}}{78.5 \text{ mm}^2}$$
$$\sigma_1 = 238 \text{ MPa} \quad \text{and} \quad \sigma_2 = 145.4 \text{ MPa}$$

The stresses are 237 MPa in the titanium tube and 145.4 MPa in the aluminum. The yield strengths are 930 and 492 MPa, respectively. The stresses are well within the elastic range. Therefore, the calculations are valid.

3.10 THERMAL EXPANSION

When a material is heated, it expands uniformly in all directions. But, if the material is a straight tension or compression member, the most important direction is along its axis. The amount of expansion, δ, is given by the following relationship.

$$\boxed{\delta = \alpha L \Delta T} \tag{3.9}$$

where α equals a constant for the material called the *coefficient of thermal expansion, L* is the length of the member, and ΔT is the change in temperature.

Table. 3.1 lists the coefficients for a variety of materials.

Table 3.1 Coefficients of thermal expansion, α

Material	$1/F°$	$1/C°$
carbon steel	6.0E−6	10.8E−6
stainless steel	9.0E−6	16.2E−6
aluminum	13.0E−6	23.4E−6
copper	10.0E−6	18.0E−6
magnesium	14.0E−6	25.2E−6
titanium	5.0E−6	9.0E−6
concrete	6.0E−6	10.8E−6
delrin	4.5E−5	8.1E−5

ILLUSTRATIVE PROBLEMS

3.8 A straight run of 12-in. standard weight steel pipe carries high-pressure steam in an oil refinery. The run is 250 ft long. Because of an expansion joint, the pipe may expand freely. If the steam temperature is 350° F, what is the amount of thermal expansion of the steam pipe over its length at 70° F?

Solution:
From relationship 3.9,

$$\delta = \alpha L \Delta T$$

The temperature of the surroundings is usually assumed to be 70° F.

$$\Delta T = 350° \text{ F} - 70° \text{ F} = 280 \text{ F}°$$

The units of the temperature difference are Fahrenheit degrees.

From Table 3.1, the coefficient of thermal expansion for steel is 6.0E−6 1/F°. Note that the units are the reciprocal of degrees, 1/F°.

$$\delta = 6.0E−6 \text{ 1/F}° \times 250 \text{ ft} \times 280 \text{ F}°$$
$$\delta = 0.420 \text{ ft}$$

The pipe will expand 5.04 in.

3.9 An aluminum machined part with two holes is shown in figure 3.13. The dimensions are measured at 20° C. What are the width, length, hole spacing, and hole diameters after the part has been dropped into boiling water at 100° C?

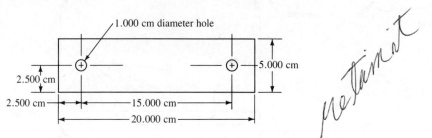

Figure 3.13 Illustrative problem 3.9.

Solution:

When uniformly heated, all the dimensions of the part will grow uniformly.

$$\Delta T = 80 \text{ C}°$$

$$\alpha = 23.4\text{E}{-}6 \ 1/\text{C}°$$

Starting with width, using relationship 3.9,

$$\delta = \alpha L \Delta T$$
$$\delta = 23.4\text{E}{-}6 \ 1/\text{C}° \times 5.000 \text{ cm} \times 80 \text{ C}°$$
$$\delta = 0.009 \ 360 \text{ cm}$$

The heated width is 5.009 cm. This is 0.18 percent larger than when the part was measured at 20° C. All dimensions will grow 0.18 percent. Therefore, the length, hole spacing, and hole dimensions will be 20.036, 15.027, and 1.0018 cm, respectively.

3.11 THERMAL STRESSES

Some materials are thermally compatible with other materials. By this we mean that their coefficients of thermal expansion are nearly equal. Examples of thermally compatible materials are copper with stainless steel, and concrete with plain carbon steel. When thermally compatible materials are assembled together in a structural member, they expand and contract in unison when temperature changes. There is little tendency for one to expand more than the other. Therefore, one material does not restrain the other and thermal stresses do not develop when the assembly is exposed to uniform temperature changes. Because the materials are thermally compatible, copper bottoms on stainless steel cookware can be expected to have a long life. For the same reason, there is little tendency for thermal stresses to build up in a steel-reinforced concrete member which is subjected to a temperature change.

Thermal stresses may develop when different members of a structure are heated to different temperatures. Bridges are good examples of this, because their temperature changes frequently with variations in weather or sunlight. Unfortunately, the foundation of a bridge is attached to the earth, whose surface changes temperature much more slowly than does the bridge. An example of this is shown in figure 3.14, which is a beamed highway bridge.

I Beams

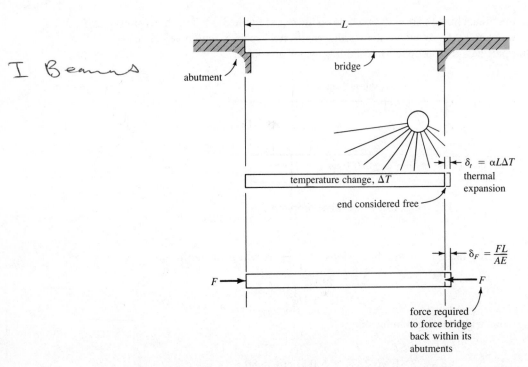

Figure 3.14 Thermal stresses are produced when a member is not free to expand and contract thermally.

If the bridge were free to expand, it would grow in length by the amount, δ_t, due to thermal expansion. If, however, the bridge were rigidly contained by its abutments, a compressive force, F, would develop that would be exactly equal to the force required to bring it back to its unheated length, L.

Envision a compressive force applied at the ends so that the heated bridge can be squeezed back between its abutments. The contraction due to the force, δ_F, must be equal to δ_t:

$$\delta_F = \delta_t$$

From relationships 2.6 and 3.9, respectively, we get

$$\delta_F = \frac{FL}{AE} \quad \text{and} \quad \delta_t = \alpha L \Delta T$$

Therefore,

$$\frac{FL}{AE} = \alpha L \Delta T$$

Dividing through by L and multiplying through by AE, the end force is

$$F = AE\alpha\Delta T \qquad \textbf{(3.10)}$$

The thermal stress is the force divided by the area:

Masood

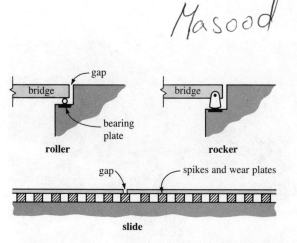

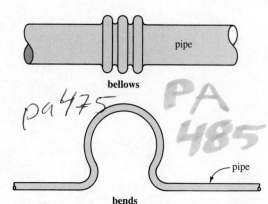

Figure 3.15 Types of expansion joints to reduce or eliminate thermal stresses.

$$\sigma_T = \alpha E \Delta T \qquad (3.11)$$

Note that the thermal stress is independent of the cross-sectional area or the length of the bridge. It depends only upon the change in temperature, ΔT, and the material properties, α and E. If the change in temperature is positive, the thermal stress will be compression. The compressive force, F, could easily be great enough to buckle the bridge or to damage the abutments. Therefore, expansion joints are normally built into bridges to allow them to expand freely.

Similar conditions occur with railroad tracks and with pipelines carrying hot or cold fluids. Thermal stresses of destructive magnitude can easily develop. Therefore, expansion joints are frequently provided when the conditions warrant. An expansion joint is a device which restrains the structure laterally but allows it to expand axially. Examples of expansion joints are shown in figure 3.15.

Figure 3.16 shows what can happen when the thermal expansion exceeds the travel of the expansion joint. In this case, the bridge was overheated by a fire caused by a traffic accident. The petroleum truck burned under the bridge, causing the bridge to expand further than the travel of the expansion joint. Large compression forces were produced as the bridge tried to expand against its abutments. These forces caused the bridge to buckle.

If the failure were due only to the fact that the strength of the material was reduced by the elevated temperature, the bridge would simply have sagged downward over its supports. But in this case the beam bent upward, indicating buckling due to an enormous compressive force.

ILLUSTRATIVE PROBLEMS

3.10 If the fire caused a temperature rise of 700 F° in the steel bridge in the accident shown in figure 3.16, what compressive thermal stresses would have developed in the beam, assuming that the material remained elastic?

Solution:
Using relationship 3.11,

$$\sigma_T = \alpha E \Delta T$$

Figure 3.16 Highway bridge buckled by heat from a tank truck fire. (Copyright *Houston Chronicle*. Reprinted by permission)

For steel, $\alpha = 6E-6$ 1/F° and E = 30E6 psi, so

$$\sigma_T = 6E-6 \text{ 1/F}° \times 30E6 \text{ psi} \times 700 \text{ F}°$$
$$\sigma_T = 126\ 000 \text{ psi}$$

3.11 Large-diameter bolts used on equipment such as steam turbine casings are too difficult to tighten, even with power wrenches. Therefore, electrical heating rods are inserted in lengthwise holes through the bolts. While the bolts are hot, they are snugged using an impact wrench. Then the bolts are allowed to cool. Cooling tightens the bolts.

If 6-in.-diameter low-alloy steel bolts 30 in. long are to be tightened to a stress level of 20 000 psi, how hot will the heating rods need to be?

Solution:

Low-alloy steel has an α of approximately $6E-6$ 1/F° and an E of 30E6 psi.

Relationships 3.10 and 3.11 apply equally in tension and in compression for elastic materials. In this case the bolts are in tension because they will be snugged hot, then allowed to cool. From relationship 3.11,

$$\sigma_T = \alpha E \Delta T$$
$$\Delta T = \frac{\sigma_T}{\alpha E} = \frac{20\ 000 \text{ psi}}{6E-6 \text{ 1/F}° \times 30E6 \text{ psi}}$$
$$\Delta T = 111 \text{ F}°$$

If the ambient or room temperature is 70° F, the bolts will need to be heated to 70° F + 111° F, or 181° F.

PROBLEM SET 3.2

3.21 A concrete column 30 in. in diameter has reinforcement consisting of ten longitudinal #10 steel bars. It carries 800 kips of compressive load. What is the stress in the concrete and the stress in the steel? (The elastic moduli are 30E6 and 2.7E6 psi for the steel and concrete, respectively.)

3.22 A concrete column 12 in. square contains a tied reinforcement bundle of four #8 steel bars. The elastic moduli are 30E6 and 3.3E6 psi. What are the stresses in the concrete and in the steel when the column carries a compressive load of 200 kips?

3.23 A 12-in. standard weight steel pipe is filled with concrete ($E = 3.0E6$ psi). What are the stresses in the steel and in the concrete when the column supports a compressive load of 200 kips?

3.24 A fiberglass rod 1 cm in diameter consists of longitudinal glass fibers in a polyester resin. The ratio of the elastic modulus of the glass to that of the resin is 1:10. The glass constitutes 70 percent of the cross-sectional area. What is the stress in the glass and that in the polyester when the rod carries a tension load of 5 000 N?

3.25 How much does an aluminum rod 10 cm long expand when heated 100° C?

3.26 A surveying party measures with a steel tape a distance of 120.18 ft on a summer day when it is 104° F. If the party came back on a 40° day in the winter and measured the same distance with the same tape, what is the measurement likely to be?

3.27 What travel should be specified for an expansion joint for a 90-m-long concrete highway bridge where temperatures of 40° C are expected in the summer sun? Assume that the bridge is constructed at an ambient temperature of 21° C.

3.28 A 10.000-in.-diameter stainless steel fry pan has a copper bottom. What is the diameter of the stainless steel pan, if allowed to expand freely, when heated 350° F? What would be the diameter of the copper bottom?

3.29 A railroad track has continuous steel rails with no expansion gaps. The rails were laid at a temperature of 50° F. In the summer sun they are heated to 105° F. What compressive stress develops in the rail?

3.30 A steel piano wire needs to be stressed to 23 000 psi to be in tune. The wire cools and heats rapidly with changes in air temperature. But the piano harp is massive and changes dimensions slowly. During a concert, the air conditioning control system permits a 4° F temperature variation. If the piano was tuned when the temperature was on the low side, what will be the stress when the air conditioning comes on during the concert? Will the pitch raise or lower as the piano goes out of tune?

3.12 THE HARDNESS TEST

Hardness tests are localized compressions tests. They consist of forcing a spherical or conical indentor into the surface of the object under test, as shown in figure 3.17. They are essentially nondestructive because they deform permanently only the small region of the object in the neighborhood of the indentor. The material must deform plastically. Therefore, hardness tests cannot be used on brittle materials. However, they can be used on very hard materials, including hardened tool steels that have only the slightest trace of ductility. Diamond indentors are used for such very hard materials.

Because the hardness test affects only a small locality of the specimen, it can be used on finished structural and mechanical parts. Hardness tests are extremely valuable in

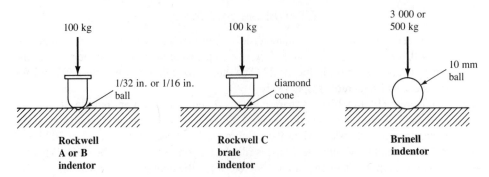

Figure 3.17 Penetrators for hardness tests.

ensuring that parts have been made of the correct material and have been properly heat-treated or otherwise conditioned.

Several kinds of hardness tests are used, but two are most common. They are the Brinell and the Rockwell tests. The advantage of the Brinell test over the Rockwell is that it can be performed on almost any compression testing machine. All one needs is a 10-mm ball and some type of collar to hold it in place. The Brinell test is conducted by forcing a hardened steel ball 10 mm in diameter into the surface of the object under test. Either a 3 000-kg or a 500-kg force is used. The 3 000-kg force is used for harder materials such as steels, and the 500-kg force for rather soft materials like brass. Special fixtures can be designed for production tests. Standard Brinell testing machines are also available.

The Rockwell test must be conducted on a special machine, as shown in figure 3.18. The most common Rockwell tests are the Rockwell B and the Rockwell C. These tests use as indentors a 1/16-in.-diameter ball or a diamond cone, respectively, with a 100-kg load. The Rockwell test offers several advantages over the Brinell test. The Rockwell machine is direct-reading, so no calculations (which can be a source of error) need be performed. The Rockwell C test, which uses the diamond indentor, can be done on very hard materials. The Rockwell has a much broader range of applications than the Brinell. The Rockwell has versions which can be used successfully on very thin materials or on case-hardened materials where it is desired to measure the hardness of only a thin surface layer. One of these is the "T" test, which uses only a 30-kg load.

The hardness test yields what is known as a "hardness number." Table 3.2 (p. 96) shows approximate conversions between Brinell hardness numbers and the Rockwell numbers on the B and 30-T scale. A hardness number is not a true structural property of a material. Hardness number depends upon other structural properties such as yield strength, the elastic modulus, and the ultimate strength. It seems to be most heavily dependent upon the ultimate strength. The ultimate strength in psi is equal approximately to 500 times the Brinell Hardness Number. You will never need to use a hardness number in a truly theoretically based mathematical relationship. A hardness number is simply an index of strength. Higher hardness numbers indicate greater strength, but the relationship is not truly proportional. The primary value of the hardness number is that it can be conveniently obtained without destroying the part or structural component.

The Brinell Hardness Number is the indentation force in kilograms, as shown in figure 3.19, divided by the spherical area of the indentation in square millimeters. The

test material

F

← D →

d

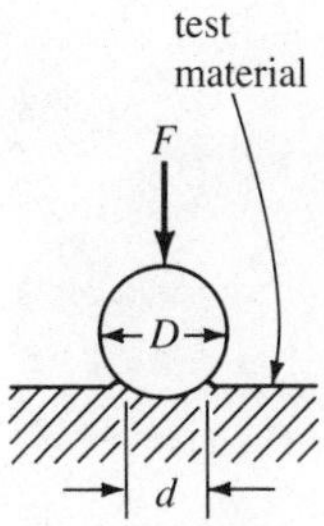

Figure 3.19 Brinell hardness test geometry.

Figure 3.18 A Rockwell hardness testing machine.

Brinell Hardness Number, abbreviated BHN, is essentially the bearing stress capacity of the material in units of kg/mm^2. After making the indentation using the 500-kg load or the 3 000-kg load as appropriate, the diameter, d, must be measured. On a rather hard material such as a cold-rolled carbon steel, the diameter of the impression is too small to measure with sufficient accuracy using a scale. In such cases a small portable microscope is needed to measure the diameter of the indentation. The spherical area of the indentation can be calculated from the relationship,

$$A = \frac{\pi}{2}(D^2 - D\sqrt{D^2 - d^2}) \tag{3.12}$$

where D is the spherical diameter and d is the impression diameter. When the indentations are very small, there is little difference between the spherical area and the area of a circle of diameter, d. In such cases it is hardly worth calculating the spherical area from relationship 3.12; the circular area, $(\pi/4)d^2$, will do.

For the standard test with a 10-mm ball, the Brinell Hardness Number is given by the formula

$$BHN = \frac{2F}{\pi(10^2 - 10\sqrt{10^2 - d^2})} \tag{3.13}$$

where F is the 500- or 3 000-kg load and d is the impression diameter in millimeters.

Table 3.2 Hardness test conversion chart (approximate)

Rockwell B 100-kg $\frac{1}{16}$-in. ball	Rockwell 30-T 30-kg $\frac{1}{16}$-in. ball	Brinell 500-kg 10-mm ball
100	82.0	201
97	80.5	184
94	78.5	171
91	77.0	160
88	75.0	151
85	73.5	142
82	71.5	135
79	69.5	128
76	67.5	122
73	65.5	116
70	63.5	110
67	61.5	106
64	59.5	101
61	57.0	96
58	55.0	92
55	53.0	89
52	51.0	85
49	49.0	82.0
46	47.0	79.5
43	45.0	77.0
40	43.0	74.5
37	40.5	72.0
34	38.5	70.0
31	36.5	68.0
28	34.5	66.0
25	32.5	64.0
22	30.5	62.5
19	28.5	61.0
16	26.0	59.5
13	24.0	58.0
10	22.0	57.0
7	20.0	56.0
4	18.0	55.0
1	16.0	53.5

For small indentations (less than 1 mm), the more simple formula 3.14 is sufficiently accurate:

$$\text{BHN} = \frac{4F}{\pi d^2}$$

(3.14)

ILLUSTRATIVE PROBLEMS

3.12 A Brinell test is conducted on a magnesium casting. A 500-kg load produces an indentation having a diameter measuring 3.2 mm across. What is the BHN?

Solution:
Using formula 3.13,

$$\text{BHN} = \frac{2F}{\pi(10^2 - 10\sqrt{10^2 - d^2})}$$

where $F = 500$ and $d = 3.2$.

$$\text{BHN} = \frac{2 \times 500}{\pi(10^2 - 10\sqrt{10^2 - 3.2^2})} = \frac{1\ 000}{\pi(10^2 - 94.741)} = 60.5$$

Note that 3.13 is a *formula* rather than a *relationship*. In a formula, units are not carried through the calculations. Instead, you must know what units are required by the formula and enter numbers in those units. In this formula F must be entered in kilograms and d must be entered in millimeters. The BHN has no units.

3.13 A hardened steel shaft is checked for proper heat-treating through the use of a hardness test. The 500-kg load produces an indentation measuring only 0.8 mm across. Compare the spherical area of the indentation to the projected circular area. Calculate the BHN using both. Is there a significant difference?

Solution:
Using relationship 3.12, the spherical area in mm^2 is

$$A = \frac{\pi}{2}(D^2 - D\sqrt{D^2 - d^2})$$

$$A = \frac{\pi}{2}(10^2 - 10\sqrt{10^2 - 0.8^2})$$

$$A = 0.503 \text{ mm}^2 \text{ (spherical area)}$$

The projected circular area is given by

$$A = \frac{\pi}{4}d^2 = \frac{\pi}{4}(0.8)^2$$

$$A = 0.503 \text{ mm}^2 \text{ (projected circular area)}$$

When rounded to three places, the spherical and the projected circular area are the same:

$$BHN = \frac{500 \text{ kg}}{0.503 \text{ mm}^2} = \underline{99\ 4.}$$

3.14 A hand-held Rockwell machine is used to check some structural steel ordered by a fabricating shop. The inspector measures a hardness of 41 Rockwell B. What is the approximate Brinell hardness of the material?

Solution:
In table 3.2 the closest thing to 41 in the Rockwell B column is 40. This number corresponds to 74.5 in the BHN column. We could interpolate, but that would imply more accuracy than is valid for this approximate conversion chart. The BHN is approximately equal to 75.

3.13 BEARING STRESS

In the structural sense, to *bear* means to push against. A bearing is a part on which something rests or on which a pin or shaft turns. It is usually at a point of load transfer between two larger members and is intended to transfer compression load between them without allowing permanent deformation of either. Usually, but not always, the matter is complicated by relative motion between the pieces.

There are sleeve bearings, roller bearings, ball bearings, bearing plates, and many other types. Here we consider in a very simple way some of the bearings relevant to structures, leaving such things as motion and lubrication to more advanced studies in mechanical design.

On a microscopic level, when an object such as the wheel in figure 3.20 bears upon a surface such as the rail, deformation occurs.

Deformation produces an area of contact. The size of the area of contact depends upon the elastic moduli of the two materials and the load applied. The force distributes itself over the contact area. Bearing stress is the force divided by the area. The bearing stress distribution varies over the contact area. This distribution can be described mathematically by complex relationships, found in the literature for elastic bodies of most shapes. (See reference 3.) It is important to understand that a contact area always develops when bearing force is applied and that the force is distributed over the area.

> Forces are always distributed through and upon structures in the form of pressure, internal stresses, and bearing stresses. The representation of a force as a vector, in the form of an arrow applied at a point on a structure, is an idealization.

When the load is not great enough to cause the maximum bearing stress to be greater than the yield strength or crushing strength of either material, the two surfaces will restore themselves to their original shapes when the load is removed. If the load is great enough to yield or crush the material of the wheel, the wheel will remain permanently flattened after the removal of the load. Or, if the rail material is the weaker of the two, the rail will remain indented, as in a Brinell test. In fact, permanent deformation due to bearing force is occasionally called ''Brinelling.'' It is possible to leave both objects permanently

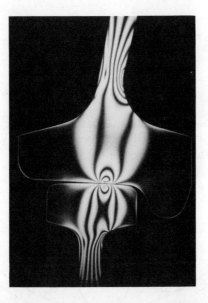

Figure 3.20 Contact area between railroad wheel and rail as shown by an experimental method called photoelasticity. (Source: Courtesy Measurements Group, Inc., Raleigh, North Carolina, USA)

deformed by overloading. Permanent deformation of either surface constitutes bearing failure.

ILLUSTRATIVE PROBLEMS

3.15 An expansion joint on a bridge consists of a cylindrical steel shoe bearing on a brass plate (figure 3.21). The cylinder is 10 in. long and has a 3-in. radius. The microscopic contact width is 0.012 in. under a bearing load of 15 000 lb. What is the average bearing stress over the contact area?

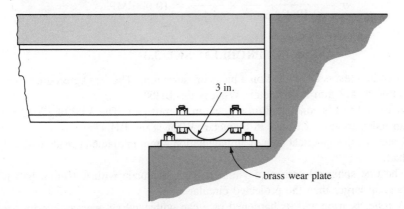

3 in.

brass wear plate

Figure 3.21 Illustrative problem 3.15.

Solution:
The contact area is 0.012 in. by 10 in., or 0.12 in.² The average bearing stress is given by

$$\sigma = \frac{15\ 000\ \text{lb}}{0.12\ \text{in.}^2} = 125\ 000\ \text{psi}$$

3.16 The ball bearing shown in figure 3.22 has a microscopic contact diameter of 0.50 mm between the bottom ball and the hardened steel race when the bearing carries a load of 200 kg. What is the average bearing stress?

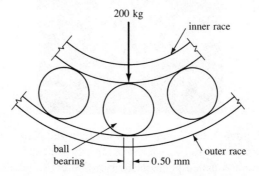

Figure 3.22 Illustrative problem 3.16.

Solution:
The contact area is

$$\frac{\pi}{4}d^2 = \frac{\pi}{4}(0.50\ \text{mm})^2 \times \frac{1\ \text{m}^2}{1\ 000\ \text{mm}^2}$$
$$= 0.196\text{E}{-}6\ \text{m}^2$$

The bearing stress is

$$\sigma = \frac{200\ \text{kg}}{0.196\text{E}{-}6\ \text{m}^2} \times \frac{9.8\ \text{N}}{1\ \text{kg}} = 10\ 000\ \text{MPa}$$

PROBLEM SET 3.3

3.31 A Brinell test is conducted on a piece of aluminum. The 500 kg produces an indentation of 4.2 mm in diameter. What is the BHN?

3.32 A Brinell test is conducted on a titanium aircraft part. The 3 000-kg load produces an indentation of 2.3 mm in diameter. What is the BHN?

3.33 Calculate the spherical area of an indentation 4 mm in diameter made with a 10-mm ball.

3.34 The true spherical area of a 1-mm indentation made with a 10-mm ball is what percent larger than the projected circular area?

3.35 A roller bearing and its hardened race can withstand an average bearing stress of 300 000 psi over a contact area. The rollers are 0.780 in. long. The contact area width is 0.020 in. What is the safe bearing load?

3.36 An expansion joint on a bridge has a steel roller 6 in. in diameter and 24 in. long. It bears on steel plates carrying a load of 10 kips. The microscopic contact width is 0.152 in. What is the average bearing stress?

SUMMARY

☐ Stress and elastic deformation in short straight-line compression members are the same as in tension members. Brittle materials may be used for compression members although they may not be suitable for tension. Length is a primary consideration for compression members. The member must be short for the relationships in this chapter to apply.

(3.1) $\sigma = F/A$

☐ Brittle materials are tested in compression in order to determine their structural properties.

☐ Concrete properties are usually determined from compression tests on standard-size cylinders cast 28 days before testing. The code-allowable stress is one-fourth of the *28-day cylinder strength*. The elastic modulus by code is:

(3.2) $E = 33w^{1.5}\sqrt{\sigma_{28}}$

☐ The stress concentration factors for tension also apply to compression. Brittle materials are much more susceptible to stress concentration, but in compression this may make no difference.

☐ A straight-line axially loaded member of two materials, 1 and 2, whether compression or tension, shares the load between members as follows:

(3.5) $F_1 = F_2\dfrac{A_1E_1}{A_2E_2}$

☐ The *coefficient of thermal expansion* is an indicator of the amount a material will expand when heated.

(3.9) $\delta = \alpha L\Delta T$

☐ *Thermal stress* is stress caused by restraint of thermal expansion.

(3.11) $\sigma_T = \alpha E\Delta T$

☐ The *hardness test* is a means of material verification. There are two basic types of hardness tests, the *Brinell* and the *Rockwell*.

☐ The ultimate tensile strength in psi is approximately equal to 500 times the BHN.

☐ *Bearing stress* is localized stress due to contact between two parts or members. A contact area forms that may be microscopic and may be elastic. The stress distributes itself over the contact area in a complex manner.

REFERENCES

1. *Design Handbook*. Kingsport, TN: Concrete Reinforced Steel Institute, 1968.

2. Peterson, R. E. *Stress Concentration Factors*. New York: John Wiley & Sons, 1966.

3. Griffel, W. J. *Handbook of Formulas for Stress and Strain*. New York: F. Ungar Publishing Co., 1966.

CHAPTER FOUR

 Shear Elements

This chapter defines shear force and shear stress, leading to the consideration of fasteners and welds. Shear strain is explained in terms of shear panels. The modulus of elasticity in shear is defined. Elastic panels are used to show the interrelationships among shear and tension or compression. Such directional interdependence leads naturally to the presentation of Poisson's ratio and the relationship between material properties in tension or compression and those in shear. The topic of friction, a shear force at the surface where two objects meet, closes the chapter. Even though an examination of friction along with shear elements is unique to this text's parallel approach, the subject of friction is addressed in a traditional manner.

4.1 SHEARING

The severing of a material by a sliding action is called *shearing*. In a pair of scissors, one blade slides against the other and *shears* the paper or cloth between the blades. In order to shear something, two parallel opposing forces (as shown in figure 4.1) are needed. A sharp edge concentrates the shearing action to a small volume of the material.

Shearing is the essence of many manufacturing processes such as drilling, milling, sawing, punching, turning, and shearing. Shear is intentional in shaping parts. But, in structures, shear can occur unintentionally and cause failure.

The term *structural element* is introduced to indicate a small part such as a bolt, pin, or other fastener. It may also denote a piece of a larger structural member, such as a lug, upset, or positioner. It may be only a selected piece of a structural member. An element is something less than a member in size and singularity of purpose but of equal importance.

Many structural elements are intended to resist shear action. Examples of shear elements are shown in figure 4.2. Most fasteners, such as the rivet and bolt, are essentially

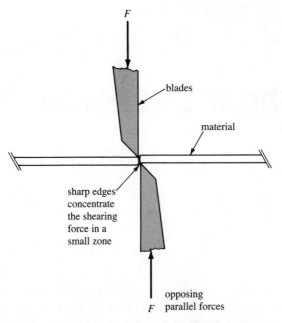

Figure 4.1 A close-up of a pair of shears.

Figure 4.2 The rivet, bolt, pin, key, and weld as loaded here are shear elements. Although the beam must resist shear, that shear is not concentrated in a small volume. The beam is a bending rather than a shear element.

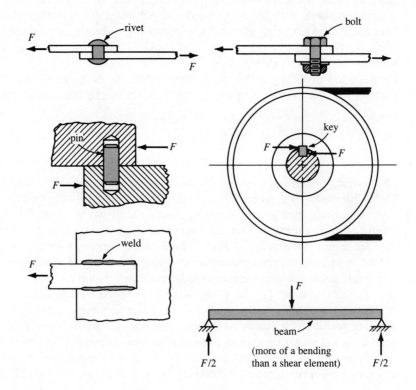

shear elements, although in another application bolts and rivets might be tension elements. The pin is a shear element, because it prevents sliding between the two parts. The shaft key is also a shear element; it prevents rotational sliding of the pulley over the shaft. A weld may be a shear element. Beams also must resist shear. But beams are more properly classified as bending members rather than shear elements, because the shear is not concentrated in a small volume of material. Shear is rarely the mode of failure of a beam.

4.2 SHEAR STRESS

The stress produced by a shear force is called *shear stress*. Shear stress is defined as force per unit area, similar to tension stress and compression stress.

$$\tau = F/A \qquad \textbf{(4.1)}$$

where τ, the Greek letter tau, represents the shear stress, F is the shear force, and A is the shear area shown in figure 4.3.

The shear area, A, is the area that would be exposed if the shear element were to fail. The shear area is parallel to the lines of action of the force. The area used to calculate shear stress is substantially different from that of tension or compression, where the area of consequence is perpendicular to the line of action of the force.

ILLUSTRATIVE PROBLEM

4.1 What is the shear stress in the rivet shown in figure 4.3?

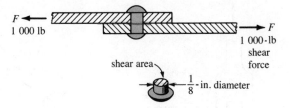

Figure 4.3 The shear area is the area that would be exposed if the element were to fail under the shear force applied (illustrative problem 4.1).

Solution:
The shear area is a circular area ⅛ in. in diameter.

$$A = \frac{\pi}{4}d^2$$

From relationship 4.1,

$$\tau = F/A = \frac{4F}{\pi d^2}$$

.0122669

$$\tau = \frac{4 \times 1\ 000\ \text{lb}}{\pi \times (0.125\ \text{in.})^2} = 81\ 500\ \text{psi}$$

The shear stress is 81 500 psi.

Ultimate Shear Strength

The maximum shear stress that a material can withstand is *ultimate shear strength*. It can be determined in several ways. Two direct ways are the punching shear test and the clevis test shown in figure 4.4.

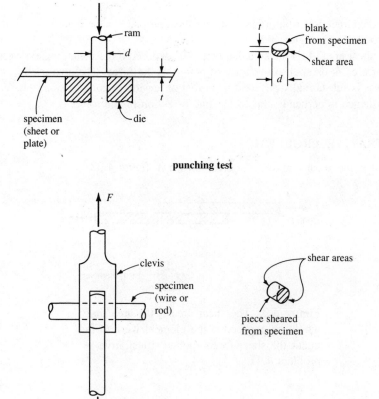

Figure 4.4 Tests for ultimate shear stress—punching and clevis.

In the punching shear test, load is applied by means of a circular ram and die to a sheet or plate specimen. Since stock thicknesses of material can be used, the preparation of specimens is inexpensive. Load is applied through the fixture in a testing machine rigged for compression. Some sort of indicator is needed to record the maximum stress, because the specimen fails without warning. Suddenly, when the ultimate shear strength is reached, a blank is punched out of the specimen. The shear area is *not* the circular cross section of the ram. It *is* the cylindrical surface area formed on the blank by the punch. Remember, shear area is *parallel* to the line of action of the force. The shear area is given by the relationship,

$$A = \pi dt \tag{4.2}$$

where d is the diameter of the blank, and t is its thickness.

In the clevis test, the fixture consists of a clevis. The specimen is a piece of rod or wire used to form the clevis pin. Again, stock material diameters may be used. But the clevis hole diameter must fit not too loosely over the specimen diameter. The fixture is pulled in a testing machine rigged for tension. The pin is sheared suddenly when the ultimate shear strength is reached.

In the clevis pin, area is circular. However, the specimen pin is in what is known as "double shear" rather than "single shear." The pin must be sheared in two places when failure occurs. In both places, a circular area is exposed.

ILLUSTRATIVE PROBLEMS

4.2 A testing machine in a student laboratory is used to test a piece of 6061-T6 aluminum stock $\frac{1}{16}$ in. thick. The testing fixture shown in figure 4.5 has a ram 0.250 in. in

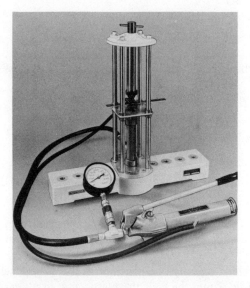

Figure 4.5 The punching shear test fixture in a materials testing machine. (Photo courtesy of Technovate, Inc.)

diameter. The specimen thickness measured with a micrometer is 0.063 in. The failure load is 1 510 lb. What is the ultimate shear stress?

Solution:
The shear area is a cylindrical surface 0.250 in. in diameter and 0.063 in. long. From relationship 4.2,

$$A = \pi dt$$

Substituting in relationship 4.1,

$$\tau = \frac{F}{A} = \frac{F}{\pi dt}$$

$$\tau = \frac{1\ 510\ \text{lb}}{\pi \times 0.250\ \text{in.} \times 0.063\ \text{in.}} = 30\ 500\ \text{psi}$$

The ultimate shear stress is 30 500 psi.

4.3 A clevis pin is made of a material having an ultimate shear stress of 250 MPa. The pin has a diameter of 4.00 mm. How much tensile load can the clevis assembly support before the pin shears?

Solution:
The shear area consists of two circular areas having a 4.00-mm diameter.

$$A = 2\left(\frac{\pi}{4}d^2\right) = \frac{\pi}{2}d^2$$

From relationship 4.1,

$$\tau = \frac{F}{A} = \frac{2F}{\pi d^2}$$

Solving for F, where in this case τ is the ultimate shear stress,

$$F = \frac{\tau \pi d^2}{2} = \frac{250\ \text{MPa} \times \pi \times (4.00\ \text{mm})^2}{2}$$

Since 1 MPa = 1 N/mm^2,

$$F = 6\ 280\ \text{N or } 6.28\ \text{kN}$$

PROBLEM SET 4.1

4.1 A steel rivet has a shank diameter of 0.375 in. It is made of a material having an ultimate shear stress of 37 500 psi. How much shear load can be carried in single shear by the rivet?

4.2 What must be the load capacity of a punch press that is to punch a 12-mm diameter hole in sheet plastic 3 mm thick, if the plastic has an ultimate shearing stress of 40.0 MPa?

4.3 What must be the diameter of a pin for a clevis made of stainless steel having an ultimate shear stress of 37 500 psi? The clevis must carry a load of 14.7 T and have a factor of safety of 2.

4.4 What is the shear stress acting on the keys preventing two timber beams from sliding upon each other, thereby causing them to act as a single beam, as shown in figure 4.6? The timbers are 4 × 4's. The steel keys are $\frac{1}{2}$-in. square rods. It was determined that each key must carry 40 kips of shear load.

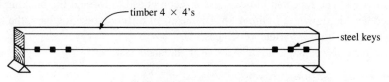

Figure 4.6 Problem 4.4.

4.5 Square metal stops, 2 cm on a side, are bonded to a steel mechanism housing with an epoxy adhesive rated for a shear strength of 200 kg/cm^2 as in figure 4.7. Using a factor of safety of 3, what amount of shear force could each stop be expected to resist?

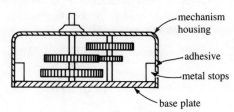

Figure 4.7 Problem 4.5.

4.6 What metric bolt diameter is required to resist a shear load of 1.89 kN, if the material has an allowable stress of 187 MPa?

4.7 A clevis pin has a diameter of 8 mm and carries a load of 40 kN. What is the shear stress?

4.8 A punching shear test is used on a piece of 1040 steel sheet 0.122 in. thick. The 0.250-in.-diameter ram punches a blank at a load of 2 900 lb. What is the ultimate shear strength?

4.3 FASTENERS

Fasteners are devices used to attach members together to form structures. There are numerous kinds of fasteners: nails, wood screws, machine screws and bolts, and rivets are the main categories. Nails and wood screws, because of the relative softness of wood, rarely fail in shear. Therefore, they are not considered here. Of more immediate interest are rivets and bolts.

Rivets

Rivets are deformable fasteners frequently used on metallic structures. One head of a rivet is formed during manufacturing. The other head is formed during assembly. The most common means of producing the formed head is by "bucking," as shown in figure 4.8. A rivet gun is a pneumatically driven hammer. The hammer is equipped with a driving head that fits the manufactured head of the rivet. The rivet gun applies repeated blows to the rivet, driving it against the bucking bar, which is hand-held to the opposite end of the rivet as in figure 4.8a. The repeated blows cause a second head to form at the bucking bar. A flat-formed head is shown in figure 4.8b. By means of different bucking bars, the formed head can be made into a variety of shapes.

Rivets have three major advantages over threaded fasteners:

1. Vibration does not loosen rivets.

2. Rivets can be visually inspected to see that they are properly installed.

3. The bucking action causes the shank of the rivet to expand to fill its hole completely. Rivets are structurally tight and sealed.

However, rivets have the following disadvantages:

1. Rivets must often be installed at other than the surrounding temperature. Steel rivets for structural steel are bucked while red hot. Some aluminum rivets for aircraft, called "ice box rivets," must be bucked while cold.

2. Rivets must be made of a soft material in order to be formable. High-strength steels cannot be used.

3. Riveted joints are difficult to disassemble. The rivets must be drilled and then punched out.

4. Most rivets require at least two people to install them—one to operate the rivet gun, and another to hold the bucking bar.

Due to their disadvantages, rivets have been replaced by bolts in most structural steel applications and due to recent advances in welding, rivets have been replaced by welding

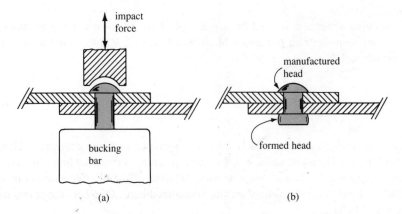

Figure 4.8 Installation of rivets.

in virtually all shipbuilding and pressure vessel applications. Inroads are being made by adhesives in aircraft structures. But due to their reliability in high vibration environments and their ease of quality control, rivets will probably always find application in airframe manufacturing.

Bolts

Bolts are threaded fasteners. They may have the same threads as shown in Appendix II and discussed in chapter 2: coarse threads (NC), fine threads (NF), 8-thread series (8 N), or metric threads (M). When used as shear elements, bolts should be chosen that do not have threads at the shear plane so that the full shank area is available to resist shear (see figure 4.9).

Bolts with nuts, as opposed to blind threads, are preferred for shear applications in structures. A washer should be inserted under the nut to help prevent damage to the plate when the nut is turning. If it is likely that the bolt head will be turned when tightened, then it too should have a washer.

Threaded fasteners need to be precisely torqued for good structural applications. A torque wrench or an impact wrench must be used. A properly torqued bolt is elastically stretched, maintaining an axial force on the nut and locking the threads. Properly torqued bolts in a well-designed joint will not come loose. But since it is impossible to visually determine whether a bolt has been torqued to specifications, all sorts of locking devices are used when vibration is expected. Among these devices are adhesives, inserts, special locking threads, lock washers, and cotter pins. By some codes, the friction in the joint caused by the bolt tension can be taken into account as additional joint strength. Here we will consider only the shear strength of the bolts.

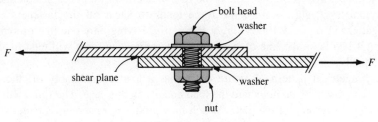

poor: threads in shear plane

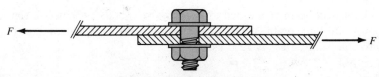

better: no threads in shear plane

Figure 4.9 Bolts used as shear elements.

Bolts have the following advantages:

1. The joint can easily be disassembled if necessary.

2. Bolts can be made of high-strength materials.

3. Bolts normally do not require heating or cooling at installation.

4. Installation tools can be relatively simple.

The disadvantages of bolts are:

1. They can vibrate loose if not properly torqued. Installers occasionally forget to torque a bolt.

2. It is difficult to determine by inspection whether all bolts have been properly torqued. The only reliable inspection technique is to completely retorque the bolts.

The objective in working with bolts and rivets is to provide a sufficient amount of shear area to carry the load.

It is assumed that all fasteners in the pattern, attaching the same member, carry the same shear stress. This assumption is valid when (a) the geometry of the pattern is symmetrical and when (b) the load is a symmetrical shear load in the plane of the pattern. Designers should avoid nonsymmetrical patterns except when absolutely necessary. When nonsymmetrical patterns are used or when the load is nonsymmetrical, a more elaborate analysis technique is required than that discussed in this chapter. Off-center or twisting loads are always nonsymmetrical.

A *symmetrical fastener pattern* is one in which a line can be drawn such that for every fastener on one side of the line there is a fastener of the same size equidistant on the other side.

The pattern in figure 4.10a shows one horizontal and one vertical line of symmetry; thus, it is a *doubly symmetric* pattern. If the line of action of the force lies on one of the axes of symmetry (figure 4.10c), the force tends to shear all the fasteners equally.

The pattern in figure 4.10b is *singly symmetric,* having only one line of symmetry. The load in 4.10d is on the line of symmetry. Therefore, the force will tend to shear all fasteners equally.

The symmetrical pattern of figure 4.10e has a load considerably off the axis of symmetry. The fasteners do not share the load equally in this case. The force will tend to cause a clockwise turning of the plate, which may produce greater stress than the downward shear force alone.

The pattern in 4.10f is completely unsymmetrical.

ILLUSTRATIVE PROBLEMS

4.4 A structural steel beam is attached to a column with bolts as shown in figure 4.11. The downward force on the connection is 150 kips. What bolt sizes must be used if the allowable shear stress is 25.0 ksi?

Solution:
Bolts *A* transfer the load from the beam to the connectors. Three bolts are in double shear. Therefore, there are six shear areas.

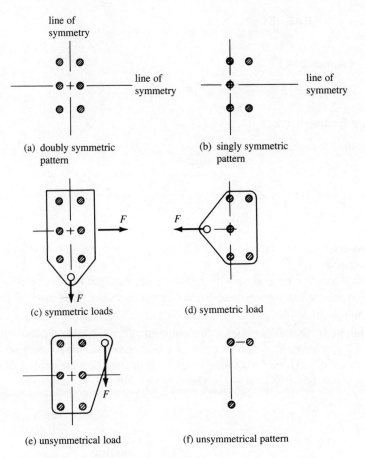

(a) doubly symmetric pattern

(b) singly symmetric pattern

(c) symmetric loads

(d) symmetric load

(e) unsymmetrical load

(f) unsymmetrical pattern

Figure 4.10 Symmetry in fastener patterns and loads.

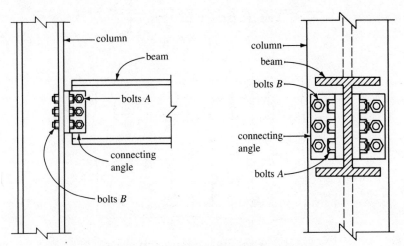

Figure 4.11 Illustrative problem 4.4.

From relationship 4.1,

$$\tau = \frac{F}{A} = \frac{F}{6 \times (\pi/4)d^2}$$

Solving for the diameter,

$$d = \sqrt{\frac{F}{1.5\pi\tau}}$$

$$d = \sqrt{\frac{150\ 000\ \text{lb}}{1.5\pi \times 25\ 000\ \text{lb/in.}^2}}$$

$$d = 1.128\ \text{in.}$$

The diameter must be at least 1.128 in. The next nominal-size bolt would be 1¼ for the web bolts *A*.

Bolts *B* transfer the load from the connectors to the columns. Six bolts are in single shear. Again, there are six areas to carry the same load. Therefore, the same size bolts should be used.

4.5 The skin of an aircraft is made of an aluminum material having a permissible tensile stress of 200 MPa. The skin is attached using high-shear rivets with an allowable shear stress of 500 MPa. What must be the rivet diameter for the joint shown in figure 4.12 to have the same strength in shear as the skin has in tension?

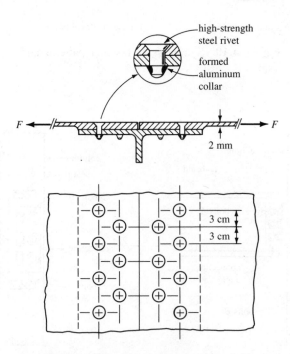

Figure 4.12 Illustrative problem 4.5.

Solution:
Considering a repeatable strip of the pattern, for example, one 3 cm wide, the strength of a 3-cm strip of the skin in tension is given by relationship 2.1,

$$\sigma = F/A$$
$$F = \sigma A = 200 \text{ MPa} \times 2 \text{ mm} \times 30 \text{ mm}$$
$$F = 12\,000 \text{ N}$$

A force of 12 000 N is the permissible tension load on a 3-cm strip of the skin material.

One full rivet shear area occurs in 3 cm of the pattern.

$$\tau = \frac{F}{A} = \frac{F}{(\pi/4)d^2}$$
$$d = \sqrt{\frac{F}{(\pi/4)\tau}}$$
$$d = \sqrt{\frac{12\,000 \text{ N}}{(\pi/4) \times 500 \text{ MPa}}} = 5.53 \text{ mm}$$

Rivets 6 mm in diameter will develop full skin strength.

PROBLEM SET 4.2

4.9 What is the shearing stress in the $\frac{1}{2}$-in. bolts of the pattern shown in figure 4.13, if a downward force of 230 000 lb is applied on its line of symmetry? The bolts are in single shear.

4.10 Which patterns in figure 4.14 are doubly symmetric, singly symmetric, or nonsymmetric?

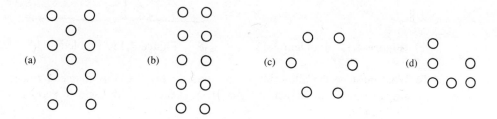

Figure 4.13
Problem 4.9.

Figure 4.14 Problem 4.10.

4.11 What size rivets are required to develop in shear the full tensile strength of the strap shown in figure 4.15? The ultimate tensile strength of the strap material is 400 MPa and the ultimate shear strength of the rivets is 250 MPa.

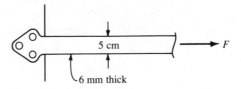

Figure 4.15   Problem 4.11.

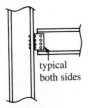

4.12 What is the load rating of the beam-to-column joint shown in figure 4.16 for a downward load, if the rivets are $\frac{7}{8}$ in. in diameter and have an allowable shear stress of 20 000 psi?

4.13 How many 10-mm bolts in a straight line are required to attach the angle to the connecting plate shown in figure 4.17, if the allowable bolt stress is 425 MPa?

Figure 4.16
Problem 4.12.

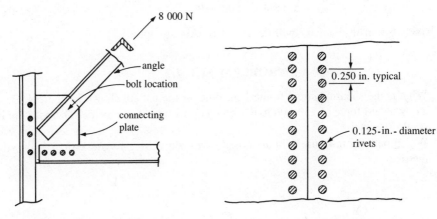

Figure 4.17   Problem 4.13.   **Figure 4.18**   Problem 4.14.

4.14 What must be the allowable shear stress for the 0.125-in.-diameter rivets in the aircraft skin seam shown in figure 4.18 in order for the skin to carry a load of 2 000 lb/in.?

4.4 WELDS

Another means of fastening structural members is welding. Welding can be done both in the fabrication shop and in the field.

Welding is the process of depositing molten filler metal of essentially the same chemical composition as the parent metal in the structural joint. The molten metal can be deposited by electric arc or gas flame processes. Structural welding has been advanced over some bolted and almost all riveted joints in stationary structures due to recent devel-

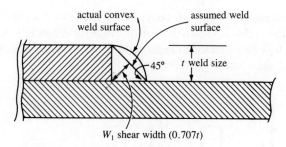

actual convex
weld surface

assumed weld
surface

45° t weld size

W_1 shear width (0.707t)

Figure 4.19 Fillet welds, the most common of structural shear welds.

opment in arc welding techniques. Most structural shear welds are fillet welds as shown in figure 4.19.

Fillet welds are essentially equal in size to plate thickness, t. To avoid cracks during cooling, the deposited weld surface must be convex as shown. For structural purposes the weld is assumed to be a 45° equilateral triangle, ignoring the material in the convexity. The assumed shear width of the weld is the distance from the 90° angle to the hypotenuse.

$$w = t \sin 45°$$
$$w = 0.707t$$

ILLUSTRATIVE PROBLEMS

4.6 The $\frac{1}{4}$-in.-thick steel strap shown in figure 4.20 is attached with fillet welds. The strap carries a tension load of 75 kips. What is the shear stress in the welds?

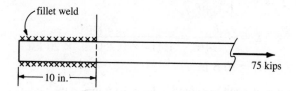

fillet weld

✗✗✗✗✗✗✗✗✗✗✗✗✗

✗✗✗✗✗✗✗✗✗✗✗✗✗

75 kips

├── 10 in. ──┤

Figure 4.20 Illustrative problem 4.6.

Solution:
The weld area,

$$A = 2 \times \tfrac{1}{4} \text{ in.} \times 0.707 \times 10 \text{ in.}$$
$$A = 3.535 \text{ in.}^2$$
$$\tau = \frac{F}{A} = \frac{75 \text{ kips}}{3.535 \text{ in.}^2}$$
$$\tau = 21.2 \text{ ksi}$$

There are 21.2 ksi of shear stress in the welds.

4.7 A water tank weighing 300 kips sits on 4 pipe legs. The load is transferred to the legs by means of shear welds as shown in figure 4.21. What distance, L, up the wall of the tank must the legs be extended in order to have an adequate amount of weld? The allowable weld shear stress is 20 000 psi.

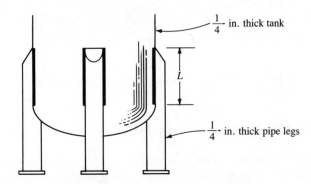

$\frac{1}{4}$- in. thick tank

L

$\frac{1}{4}$- in. thick pipe legs

Figure 4.21   Illustrative problem 4.7.

Solution:
The force on a single leg is

$$F = \frac{300 \text{ kips}}{4} = 75 \text{ kips}$$

The weld area on each leg is

$$A = 2 \times 0.707tL = 1.414tL$$

$$\tau = \frac{F}{1.414tL}$$

$$L = \frac{F}{1.414t\tau} = \frac{75 \text{ kips}}{1.414 \times 0.250 \text{ in.} \times 20 \text{ ksi}}$$

$$L = 10.6 \text{ in.}$$

The welds on the legs need to extend at least 10.6 in. up the tank wall to transmit the vertical shear load.

PROBLEM SET 4.3

4.15 A steel tension strap 10 mm thick is attached by fillet welds as shown in figure 4.22. What is the shear stress in the welds when the strap is loaded with 400 kN of tension?

4.16 Connectors are welded to a steel column during shop fabrication. The attachment to the beam is completed with bolts during field erection, as shown in figure 4.23. If the weld has an allowable shear stress of 18 000 psi and the bolts an allowable shear stress of 40 000 psi, how many bolts will be required to make the bolted joint as strong as the welded joint?

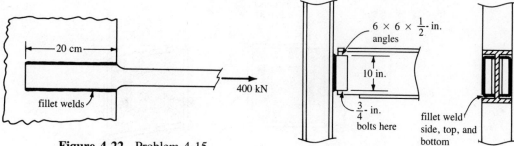

Figure 4.22 Problem 4.15.

Figure 4.23 Problem 4.16.

4.17 Four legs made of $8 \times 8 \times \frac{1}{2}$-in. angle support a petroleum tank weighing 600 000 lb. (See figure 4.24.) What length of shear weld is required if the weld-allowable stress is 10 000 psi?

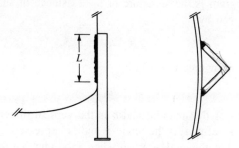

Figure 4.24 Problem 4.17.

4.5 SHEAR DEFORMATION AND STRAIN

So far in this chapter, no mention has been made of shear deformation. That is because in fasteners the shear stress occurs only in shear planes that are assumed to have no thickness. Without thickness, deformation cannot occur. Moving away from the subject of fasteners to that of general shear stress requires an understanding of shear deformation. Shear deformation is difficult to visualize because it rarely occurs by itself except in the special case of torsion, which is the subject of chapter 5.

In an attempt to visualize shear deformation, consider a frame pinned at its corners over which is suspended a sheet of an elastomer, such as rubber or foamed polyurethane (figure 4.25). This figure illustrates a simple experimental apparatus. In common structural materials, shear deformations are too small to be seen. In elastomers, however, shear deformations can be large enough to be seen. A square can be drawn on the elastomer material with a felt pen. The distortion of the square is easily visible as force is put on the frame.

A vertical shear force, V, can be applied to the frame causing shear stress, τ, through the material. The square under this stress will be distorted into a parallelogram of equal

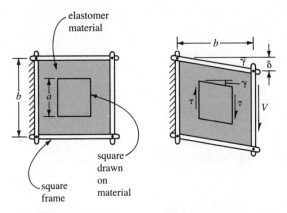

Figure 4.25   Shear deformation and strain.

sides. Such a parallelogram is the signature of the existence of shear stress. The shear stress in this case is given by

$$\tau = V/A \qquad \qquad (4.3)$$

For this geometry the shear area is given by

$$A = bt$$

where t is the thickness of the material and b is the vertical dimension.

The *shear strain* is defined as the angle γ and is expressed in radians.

A *radian* is defined as the angle formed when an arc having the same length as the radius is set off along the circumference of a circle, as shown in figure 4.26.

The angle in radians is the arc length divided by the radius of the circle.

$$\text{angle (in radians)} = \frac{\text{arc length}}{\text{radius}} \qquad (4.4)$$

Since a full circle has a circumference equal to $2\pi R$, the number of radians in a circle can be determined from relationship 4.4,

$$\text{radians in a full circle} = \frac{2\pi R}{R} = 2\pi$$

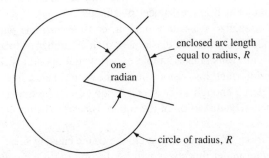

Figure 4.26   Definition of a radian.

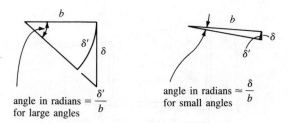

Figure 4.27 When the angle is small, there is little difference between the angle in radians and the quantity δ/b.

Therefore:

$$2\pi \text{ radians} = 360°$$
$$1 \text{ radian} = 57.3°$$

The radian is, in fact, a dimensionless unit, derived from a length unit divided by a length unit in relationship 4.4. In relationship 4.4, the length units of arc length are the same as those of the radius. The units may be meters, inches, kilometers, or any length unit. The length units will cancel, leaving no units. The word *radian* may be added or taken away from a calculation as needed.

Because it is dimensionless, a radian is an appropriate unit for shear strain. Linear strain in compression or tension, as recalled from relationship 2.3, is also dimensionless.

The shear deformation in figure 4.25 would be the quantity δ. For small angles the strain is assumed to be equal to the angle in radians, as shown in figure 4.27.

$$\gamma = \delta/b \tag{4.5}$$

This relation is an accurate assumption for small deformations. For example if the strain, γ, is 0.1 radian (approximately 6°), the error would be only 0.186 percent. As the angle gets smaller, the error decreases rapidly. In structural materials the strains in the elastic range are always very small. Metals have strains on the order of 0.002 radian.

The *elastic modulus in shear*, often called the modulus of rigidity, is the ratio of shear stress to shear strain in the elastic range.

$$\boxed{G = \frac{\tau}{\gamma}} \qquad G = \frac{E}{2(1 \times M)} \tag{4.6}$$

where G equals the elastic modulus in shear, τ is the shear stress, and γ is the shear strain.

Since γ is in radians and is dimensionless, the elastic modulus, G, has the same units as τ, which may be psi, ksi, Pa, MPa, or any stress unit.

ILLUSTRATIVE PROBLEM

4.8 A shear panel in a wing span of a cargo aircraft is 0.06 in. thick and carries a shear stress of 3 000 psi. It is made of aluminum with a modulus of rigidity of 5E6 psi. What is the downward deformation due to shear over the length of the panel shown in figure 4.28?

$\tau = 3000$
$t = .06 \text{ in}$

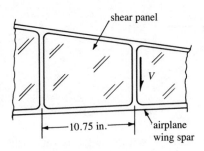

Figure 4.28 Illustrative problem 4.8.

Solution:
From relationships 4.5 and 4.6,

$$\gamma = \frac{\delta}{b}$$

and

$$G = \frac{\tau}{\gamma}$$

$$G = \frac{\tau b}{\delta}$$

$$\delta = \frac{\tau b}{G}$$

$$\delta = \frac{3\ 000\ \text{psi} \times 10.75\ \text{in.}}{5\text{E6 psi}}$$

$$\delta = 6.45\text{E}{-}3\ \text{in.}$$

The shear panel deforms vertically by 0.00645 in. due to shear. There will be additional deformation due to other forces.

PROBLEM SET 4.4

4.18 The frame shown in figure 4.25 suspends a 10-mm-thick sheet of foamed polyurethane. Under a shear load of 10 kg the elastic deformation, δ, is measured to be 1.25 cm. The frame is 45 cm square. What is the modulus of elasticity in shear for the material?

4.19 A railroad bridge has a uniformly loaded shear panel as shown in figure 4.29. The panel deforms when carrying a shear stress of 2 500 psi. What is the deformation due to shear over the length of the panel? $G = 11.5\text{E6}$ psi.

4.20 A steel material is loaded in shear almost to its yield stress of 400 MPa. G equals 11.5E6 psi for this material. What is the amount of shear strain?

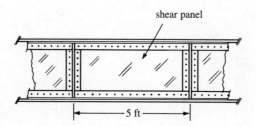

Figure 4.29 Problem 4.19.

4.6 DIRECTIONAL PROPERTIES OF STRAIN: POISSON'S RATIO

Stress in one direction will cause stress in other directions. Strain in one direction will cause strain in other directions. And, in the elastic range, when stress is proportional to strain, stress in one direction will cause strain in other directions. To examine this relationship, it is necessary to look first at tension and compression.

Figure 4.30 shows what happens in the lateral direction when an axial force is applied. Figures 4.30a and 4.30b show tension applied through a frame to an elastomer sheet. A square was drawn before the load application.

During load application the square elongates by the amount δ_A in the direction of load application (axial direction). But the square also contracts by the amount δ_L in the lateral direction. In 4.30c and 4.30d, a compression load is applied to a cube of foam rubber. The cube shortens by an amount, δ_A, in the direction of the load (axial) and gets fatter by an amount, δ_L, in the lateral direction. The elastic objects seem to try to keep the volume constant. But, it should be noted, there is a net gain of volume in the case of tension and loss of volume in compression.

The relationship between axial strain, ϵ_A, and lateral strain, ϵ_L, when the material is in elastic range is given by a constant for the material called Poisson's ratio. *Poisson's ratio* is the negative of the lateral strain divided by the strain in the direction of the load.

Poisson's ratio is indicated by the Greek letter μ. Poisson's has a French pronunciation, like *pwahssons*. It is capitalized because it is named after the French mathematician S. D. Poisson. It has no units, because it is a ratio of two dimensionless quantities.

$$\mu = -\frac{\epsilon_L}{\epsilon_A} \tag{4.7}$$

Poisson's ratio has a value slightly less than 0.30 for most structural metals.

ILLUSTRATIVE PROBLEM

4.9 Dimension *a* in figure 4.30 is 10 cm. $\delta_A = 1.0$ mm and $\delta_L = 0.33$ mm for a sheet-rubber tension material. $\delta_A = 1.2$ mm and $\delta_L = 0.21$ mm for a polyurethane foam compression material. What is Poisson's ratio in each case?

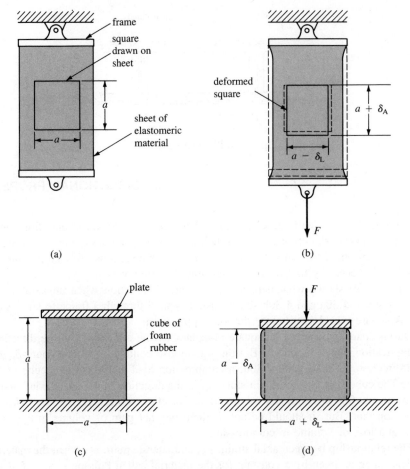

Figure 4.30 Axial and lateral strain. Axial stress, tension, or compression causes lateral as well as axial strain.

Solution:
Sheet rubber:

$$\epsilon_A = \frac{1.0 \text{ mm}}{10 \text{ cm}} = 0.010$$

$$\epsilon_L = \frac{-0.33 \text{ mm}}{10 \text{ cm}} = -0.0033$$

$$\mu = \frac{-(-0.0033)}{0.010} = 0.33$$

Polyurethane foam block:

$$\epsilon_A = \frac{-1.2 \text{ mm}}{10 \text{ cm}} = -0.012$$

$$\epsilon_L = \frac{0.21 \text{ mm}}{10 \text{ cm}} = 0.0021$$

$$\mu = \frac{-0.0021}{-0.012} = 0.175$$

The Poisson's ratios for the sheet rubber and for the polyurethane foam are 0.33 and 0.18, respectively.

PROBLEM SET 4.5

4.21 In a tensile test on a structural steel, strain gages are used to determine Poisson's ratio. An axially oriented strain gage reads 1 200 micros when a laterally oriented gage located near the same position reads −360 micros. What is Poisson's ratio?

4.22 An aluminum bar 1.2 mm square and 10 m long has an elastic modulus of 69E9 Pa and a Poisson's ratio of 0.29. Under a tension stress of 150 MPa, which is in the elastic range for this material, how much does the bar elongate and how much does its width contract?

4.23 A concrete column 14 in. in diameter and 10 ft long carries a force of 400 kips. It shortens by an amount 0.300 in. and expands diametrically 0.011 in. What is Poisson's ratio?

4.24 A rigid cube of rubber 10 cm on each side is compressed in a testing machine 1.00 mm in the axial direction. The block expands laterally 0.49 mm. How much change in volume was produced?

4.7 THE RELATIONSHIP BETWEEN AXIAL STRAIN AND SHEAR STRAIN

When an axial load such as tension is applied to an object, it not only produces a lateral compressive strain in a direction 90° from the axis, but it also produces shear strain in a direction 45° from the axis. This can be shown with an experiment such as shown in figure 4.31. In this experiment a square is drawn with a felt-tipped pen on an elastomeric foam sheet. Instead of aligning the square with the load axis, as was done in figure 4.30a, the square is rotated 45° from the tension axis. When the load is applied, the square deforms into a parallelogram rather than a rectangle. The parallelogram is similar in shape to that in figure 4.25, showing the presence of shear deformation at an angle 45° from the tension axis.

From advanced theory of elasticity,[1] the following relationship between the elastic modulus in shear, G, and that in tension, E, can be derived:

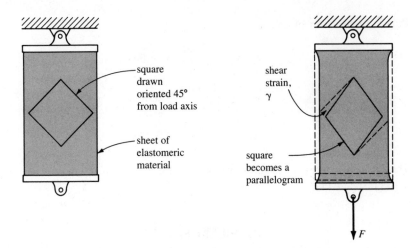

Figure 4.31 A square drawn at 45° from the tension load axis deforms into a parallelogram, showing the existence of shear deformation at that angle.

$$G = \frac{E}{2(1 + \mu)}$$ **(4.8)**

Therefore, a tensile test can be used to determine the elastic modulus in shear, G. Lateral strain as well as axial strain must be measured so that the Poisson's ratio can be determined.

ILLUSTRATIVE PROBLEM

4.10 A tensile test is conducted on a specimen of steel 0.250 in. thick and 1.000 in. wide. At a tensile force of 7 500 lb, two axial strain gages on opposite sides of the specimen give an average reading of 1 000 micros. Two lateral strain gages give an average reading of −300 micros. What is the elastic modulus in shear?

Solution:
From relationship 2.5,

$$E = \frac{\sigma}{\epsilon} = \frac{F}{A\epsilon}$$

$$E = \frac{7\ 500\ \text{lb}}{0.250\ \text{in.} \times 1\ \text{in.} \times 1\ 000\text{E}-6}$$

$$E = 30\text{E}6\ \text{psi}$$

From relationship 4.7,

$$\mu = -\frac{\epsilon_L}{\epsilon_A} = \frac{+300E-6}{1\ 000E-6} = 0.30$$

From relationship 4.8,

$$G = \frac{E}{2(1+\mu)} = \frac{30E6\ \text{psi}}{2(1.300)} = 11.5E6\ \text{psi}$$

The elastic modulus in shear is 11.5 million psi.

4.8 THE RELATIONSHIP BETWEEN AXIAL STRESS AND SHEAR STRESS

It will be shown in chapter 15 that an axial stress produces shear stress at any angle off the load axis. As shown in figure 4.32, *axial stress produces the maximum shear stress at an angle of 45° from the axis. The shear stress at this angle is one-half the value of the axial stress.*

Ductile materials actually fail by shear even when only tension load is applied. This can be shown experimentally by Lüder lines at yielding and by the failure surface at rupture.

Lüder lines are visible to the unaided eye on ductile materials with brittle surface coatings that flake off during yielding. Yielding does not occur uniformly throughout the specimen. Yielding is caused by a phenomenon which metallurgists call *slip*. Slip happens along the weakest planes through the material (figure 4.33). As yielding progresses, more and more slip planes form until the entire volume has yielded. But at the onset of yield, slipping is confined to a few planes through the specimen. And, near these planes, the local deformation is so large that brittle coatings crack. Brittle surface coatings that show

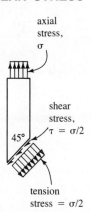

axial stress, σ

shear stress, $\tau = \sigma/2$

45°

tension stress = $\sigma/2$

Figure 4.32 Axial stress produces shear stress and tension stress at an angle of 45° equal to one-half the axial stress.

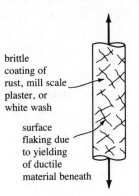

brittle coating of rust, mill scale plaster, or white wash

surface flaking due to yielding of ductile material beneath

Figure 4.33 Lüder lines show that yielding occurs by slipping along shear planes 45° to the load axis.

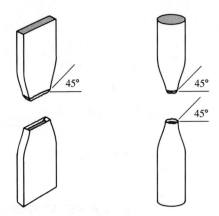

Figure 4.34 Ductile rupture surfaces showing shear parting lines at a 45° angle to the load axis.

Lüder lines include natural rust and mill scale and artificially applied coatings such as plaster or whitewash.

In tension, the rupture surface on a ductile material shows the ''cup and cone'' as discussed in chapter 2, if the specimen has a round cross section. If instead the specimen is flat stock, *shear lips* will form at the edge of the rupture surface as shown in figure 4.34. The rupture surfaces show clearly that ductile materials part by shearing at 45° from the load axis at the plane of maximum shear. Since the shear stress along the 45° plane is one-half the tension stress, the ultimate shear stress is one-half the ultimate tension stress. But, in making such an extrapolation, it is first necessary to correct the ultimate tension stress for the fact that the cross-sectional area was reduced by ''necking down'' in tension. When so corrected, ultimate shear stresses extrapolated from tension test data compare favorably to values determined from punching or clevis shear tests.

ILLUSTRATIVE PROBLEM

4.11 The following data for a structural steel was determined from a tensile test: $E = 29.0E6$ psi, $\mu = 0.290$, yield strength = 35 000 psi, and ultimate stress = 52 000 psi. The area was found to have been reduced to 60 percent of its initial value at the neck. Estimate the yield and ultimate strengths and the elastic modulus in shear.

Solution:
The yield stress in shear will be equal to one-half that in tension.

$$\tau = \frac{\sigma}{2} = \frac{35\ 000\ \text{psi}}{2} = 17\ 500\ \text{psi}$$

The ultimate stress in shear will be one-half the tension ultimate corrected for area reduction. The correction is done by dividing the tension ultimate stress by the cross-sectional area in the region where the specimen is necked down.

$$\sigma_{\text{corrected}} = \frac{\sigma}{0.60} = \frac{52\,000}{0.60} = 86\,700 \text{ psi}$$

$$\tau = \frac{\sigma_{\text{corrected}}}{2} = \frac{86\,700}{2} = 43\,400 \text{ psi}$$

The elastic modulus in shear, or modulus of rigidity, can be calculated from relationship 4.8.

$$G = \frac{E}{2(1 + \mu)} = \frac{29.0E6 \text{ psi}}{2(1 + 0.290)} = 11.24E6 \text{ psi}$$

The yield strength in shear is approximately 17 500 psi. The ultimate strength in shear is approximately 43 400 psi. The elastic modulus in shear is approximately 11.24E6 psi.

PROBLEM SET 4.6

4.25 A materials catalog gives values of elastic modulus in tension and in shear for 70 percent copper, 30 percent zinc cold-rolled cartridge brass of 17.0E6 psi and 6.4E6 psi, respectively. What is the Poisson's ratio?

4.26 Annealed titanium 115A alloy has a modulus of elasticity of 113.9E9 Pa and a Poisson's ratio equal to 0.29. What is the modulus of elasticity in shear?

4.27 A36 structural steel has a yield strength of 36 000 psi, an elastic modulus of 30E6 psi, and a Poisson's ratio of 0.3. Estimate the yield strength and the elastic modulus in shear.

4.28 A tensile test on a 10-mm-diameter rod of 6061-T6 aluminum alloy gives an ultimate strength of 310 MPa. The diameter after testing at the ''necked down'' rupture is 7.2 mm. Estimate the ultimate strength in shear as would be determined in a clevis test.

4.9 FRICTION

Friction is the resistance to sliding between two contacting surfaces. It results from a complex combination of surface roughness and adhesion. Friction acts to oppose motion between two surfaces.

Friction is a very complicated phenomenon about which much remains to be understood. We do know that it depends upon contact pressure. We also know that it depends upon the two materials involved, the surface condition, temperature, the relative velocity between the two surfaces, and any lubrication that may be present.

Friction can be a blessing or a curse. It is essential to the working of threaded fasteners, brakes, and clutches. In fasteners, no relative motion is desired; therefore, lubrication is not provided except as an aid during assembly.

On the other hand, friction is a source of energy loss and a hindrance to the proper operation of mechanisms where motion between parts is desired. In bearings for mechanisms, lubrication is used to separate the two surfaces with a film of liquid so that there is little or no contact between the solid surfaces. Lubrication is best left to the studies of mechanisms and machine design. *Dry friction* will be discussed in this book because it is relevant to the study of structures.

Dry friction force, F_F, is less than or equal to the contact force, N, times the coefficient of friction, f, as in relationship 4.9.

$$F_F \leq fN \tag{4.9}$$

The contact force, N, is called the *normal force*, meaning that it is perpendicular to the contact surface.

The coefficient of friction, f, is an experimentally determined constant that depends upon the materials involved and surface conditions. Table 4.1 gives several typical values for coefficients of friction. Variations of 25 to 100 percent from these values can be expected depending on surface condition, temperature, and velocity. The coefficient of friction is dimensionless.

Table 4.1   Coefficients of friction[2,3,4]

	Static	Dynamic
Steel on steel (dry)	0.6	0.4
Steel on steel (greasy)	0.1	0.05
Teflon on steel	0.04	0.04
Brass on steel (dry)	0.5	0.4
Brake lining on cast iron	0.4	0.3
Rubber tires on pavement* (dry)	0.9	0.8
Wire rope on iron pulley (dry)	0.2	0.15
Hemp rope on metal	0.3	0.2
Delrin on steel	0.3	0.1
Wood on wood	0.25 to 0.5	
Leather on wood	0.25 to 0.35	
Leather on metal	0.5	

*Tires on drag racing cars not included.

The friction force, F_F, shown in figure 4.35 is always in the direction opposing sliding between the surfaces. If there is no force tending to cause sliding, there is no friction force. The friction force will equilibrate the net force tending to cause sliding between surfaces, thereby preventing sliding. However, there is a maximum value, fN, which the friction force can reach, which is why the "less than or equal to" sign is used in relationship 4.9. If the maximum value is exceeded, breakaway occurs. At *breakaway*, sliding begins and the coefficient of friction changes abruptly from its static value to its

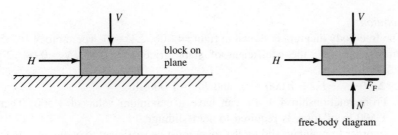

Figure 4.35 The normal force, N, is the force
tending to squeeze the surfaces together. In this
case it equals V. The friction force, F_F, always
acts to equilibriate the force, in this case H,
tending to cause sliding between the surfaces.

dynamic one. Dynamic friction is normally less than static friction. In other words, *it takes more force to start sliding than it does to maintain the sliding at a constant velocity.*

Note that the area of surface contact plays no role. Relationship 4.9 could be written in terms of contact stress and shear stress by dividing both sides by the contact area. But, since the areas can be cancelled, it is easier to express the relationship in terms of forces. The one hazard in omitting areas is that it may lead one to believe that the contact area has no effect in friction. In fact, the amount of contact area can be vital because area dissipates heat when sliding occurs. As area is reduced, temperature becomes higher, causing the coefficient of friction to be lower. Brake and tire designers are well aware of the need for sufficient contact area.

ILLUSTRATIVE PROBLEMS

4.12 A steel shipping container weighing 2 500 lb rests on the horizontal steel deck of a cargo ship. The loading crew attempts to move the container by attaching a rope and pulling horizontally with a capstan as shown in figure 4.36a.
   a. What is the shearing force between the deck and the container when 1 000 lb of force is applied by the capstan?
   b. What amount of force is required to start the container sliding on the deck?
   c. What amount of force is required to keep it sliding at a slow but constant speed?

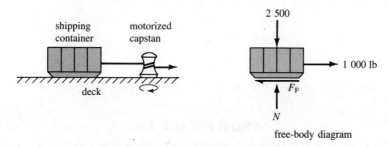

Figure 4.36 Illustrative problem 4.12.

Solution:

The free-body diagram is shown in figure 4.36b. $\Sigma V = 0$. Therefore $N = 2\,500$ lb. From table 4.1, the coefficient of static friction is 0.6. $fN = 0.6 \times 2\,500 = 1\,500$ lb.

a. $\Sigma H \rightarrow$: $-F_F + 1\,000 = 0$, and $F_F = 1\,000$ lb
b. From relationship 4.9, F_F can have a maximum value of 1 500. Therefore, 1 500 lb of force is required to start sliding.
c. In order to maintain sliding the force must be sufficient to overcome the force of dynamic friction. The coefficient of dynamic friction is 0.4.

$$F_F = 0.4 \times 2\,500 = 1\,000 \text{ lb}$$

One thousand pounds is insufficient to start sliding. But, once the force is increased to 1 500 lb to cause breakaway, 1 000 lb is sufficient to sustain constant slow motion.

4.13 A special apparatus with an extremely low-friction pulley is used to determine the coefficient of friction of a certain wood on steel (figure 4.37). The 1-kg block is made of the wood. The plane is made of steel. A weight of 240 gm was required in order to start the block moving. What is the coefficient of static friction?

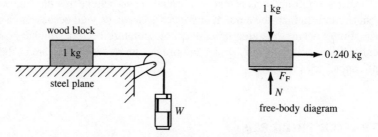

Figure 4.37 Illustrative problem 4.13.

Solution:

$$\Sigma V = 0: \quad N = 1 \text{ kg}$$
$$\Sigma H = 0: \quad F_F = 0.240 \text{ kg}$$

From relationship 4.9,

$$F_F = fN$$
$$f = \frac{F_F}{N} = \frac{0.240 \text{ kg}}{1 \text{ kg}}$$
$$f = 0.240$$

The coefficient of static friction for the wood on steel is 0.240.

PROBLEM SET 4.7

4.29 What horizontal force is required to start a 1.60-kg brass block moving on a horizontal steel plane? What force is required to sustain sliding?

4.30 A 2 000-lb automobile skids with locked brakes on dry horizontal pavement. Estimate the average shearing force between the tires and the pavement.

4.31 What additional vertical force must be added to a 10-kg steel block in order to prevent sliding on a greasy horizontal steel plane, if a horizontal force of 12 kg is applied?

4.32 An apparatus similar to that shown in figure 4.37 is used to measure the coefficient of friction of aluminum on steel. A force of 5.12 lb is required to start a 12.30-lb aluminum block moving on a steel plane. What is the coefficient of friction?

SUMMARY

☐ *Shearing* is the severing of a material by a sliding action. *Shear stress* is the stress due to the application of a shear force.
  (4.1) $\tau = F/A$

☐ The shear area is always parallel to the line of force.

☐ Rivets, bolts, welds, and other fasteners may be essentially shear elements. If the fastener pattern is symmetrical, relationship 4.1 applies.

☐ The *shearing strain* is the angle of deformation in radians.

☐ The *modulus of rigidity* is the modulus of elasticity in shear and is the ratio of stress to strain in the elastic range.
  (4.6) $G = \dfrac{\tau}{\gamma}$

☐ *Poisson's ratio* is the ratio of lateral strain to axial strain and has the value of about 0.3 for most structural metals.
  (4.7) $\mu = -\dfrac{\epsilon_L}{\epsilon_A}$
  (4.8) $G = \dfrac{E}{2(1 + \mu)}$

☐ Ductile materials in tension actually fail in shear. Shear stress occurs at an angle of 45° to the direction of tension.

☐ *Friction* is the sliding resistance between two surfaces. The *coefficient of friction* is a constant that depends upon the materials in contact and their conditions.
  (4.9) $F_F \le fN$

☐ The static coefficient of friction is greater than the dynamic.

REFERENCES

1. Timoshenko, S. P. and Goodier, J. N. *Theory of Elasticity*. 3rd ed. New York: McGraw-Hill, 1969.

2. Meriam, J. L. *Statics*. New York: John Wiley & Sons, 1966.

3. *Delrin Acetal Resin*. E. I. DuPont Co., 1963.

4. Hausman, E. and Slack, E. P. *Physics*, 4th ed. New York: D. Van Nostrand, 1959.

CHAPTER FIVE

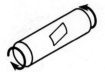 # Torsion

Forces under certain conditions can produce twisting. This chapter discusses the twisting effects of forces and extends the concepts of static equilibrium from chapters 2 and 3 to rotation. After the statics of torsion are set forth, it extends the concepts of shear stress and shear strain from chapter 4 to the study of torsional stresses and deformation in solid and hollow shafts. Torsional testing of ductile and brittle materials is discussed and compared to tests for tension and compression, so that the reaction of different kinds of materials to torsional stresses can be understood. The chapter closes with the practicalities involved in attaching torsional members to structures and in using the important box-shaped torsion member.

5.1 TORSION MEMBERS

A *torsion member* is a part of a machine or structure intended to resist or control twisting forces. Examples of torsion members are shafts, torsion bars, and torque boxes, as shown in figure 5.1.

Shafts are rotating torsion members designed to transmit power. Torsion bars are springs that control rotational motion or store energy. A coil spring is, in fact, a torsion bar wound about an axis. Torque boxes are efficient members for carrying torsional loads in stationary or vehicular structures. Due to their stiffness and ease of attachment, torque boxes are popular components in airplane wings and automobile frames.

5.2 MOMENT OF FORCE

The word *moment* comes from the Latin word *momen,* which may mean motion, weight, or time. Imprecise in its Latin origin, moment continues to have several meanings in modern usage. In the structural sense it means something entirely different from what the layperson would expect. Instead of having to do with time, moment is the twisting action

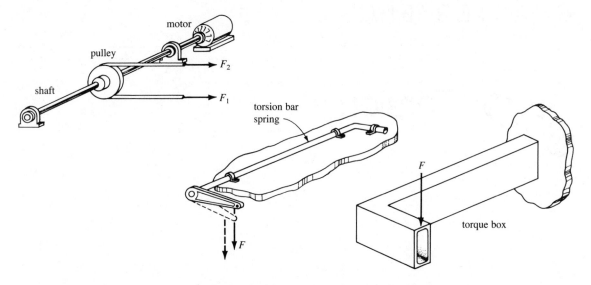

Figure 5.1 Examples of torsion members.

caused by a force. The words *twist* or *torque* work just as well and may be more descriptive to non-technical persons. This is something for technologists to bear in mind when they find themselves interpreting between engineers and field or shop personnel.

A force causes a moment (torque or twist) about any axis that does not lie on the line of action of the force itself. As shown in figure 5.2, a moment of a force may tend to turn the object clockwise or counterclockwise about an axis.

The magnitude of moment due to a force is the amount of force times the *moment arm*. As shown in figure 5.3, the moment arm is the *perpendicular* distance between the

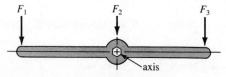

Figure 5.2 The force F_1 causes a counterclockwise moment about the axis. F_3 causes a clockwise moment. F_2 causes no moment at all, because its line of action passes through the axis.

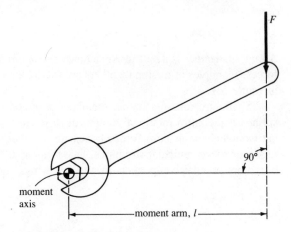

Figure 5.3 The moment arm is the perpendicular distance between the line of action of the force and the moment axis.

line of action of the force and the moment axis.

$$M = Fl \tag{5.1}$$

where M is the moment, F is the force, and l is the length of the moment arm.

The moment of a force will have units of force times units of distance, such as kg · m, N · m, lb · in., lb · ft, kip · ft.

ILLUSTRATIVE PROBLEM

5.1 What turning moment in N · m is caused when the crank assembly of a small engine is acted upon by a 315-N connecting-rod force as the crank is in the horizontal position shown schematically in figure 5.4a?

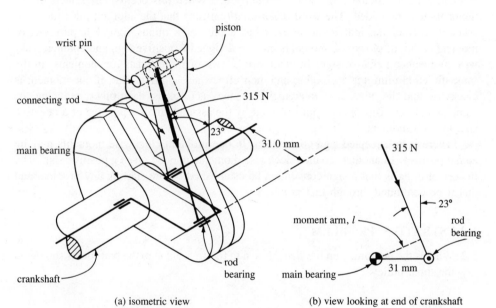

(a) isometric view   (b) view looking at end of crankshaft

Figure 5.4   Illustrative problem 5.1. A schematic diagram of a crank and connecting rod mechanism.

Solution:

As shown in figure 5.4b, the moment arm, l, forms a right triangle with the crank as the hypotenuse, h. There is a 23° angle between the crank arm and the moment arm. Using the definition of the cosine from trigonometry,

$$\text{cosine} = \frac{\text{adjacent side}}{\text{hypotenuse}}$$

$$\cos 23° = \frac{l}{31 \text{ mm}}$$

$$l = 31 \text{ mm} \cos 23° = 28.5 \text{ mm}$$

From relationship 5.1,

$$M = Fl = 315 \text{ N} \times 28.5 \text{ mm} \times \frac{1 \text{ m}}{1\ 000 \text{ mm}}$$

$$M = 8.99 \text{ N} \cdot \text{m}$$

The turning moment or torque on the crank in this configuration is $8.99 \text{ N} \cdot \text{m}$.

5.3 ADDITION OF MOMENTS

Moments sharing the same axis may be algebraically added. In order for moments of two or more forces to share an axis, the lines of action of those forces must be in planes perpendicular to the axis, as shown in figure 5.5. When this occurs, the effects of the moments may be added. The word *algebraically* means that the addition must take into account whether the individual moment has a plus or a minus sign. For purposes of moment addition, clockwise moments may be assigned a positive sign and counterclockwise moments a negative sign. In other words, two or more twists, or moments, in the same direction reinforce each other and their effects are additive. But, if one moment is clockwise and the other is counterclockwise, they act against each other, reducing the combined effect. This is taken into account by the assignment of a positive or a negative sign to each moment.

There is no accepted sign convention. In one problem clockwise moments may be *called* positive. In another, counterclockwise moments may be *given* a positive sign. The important thing is that a sign convention be established before starting any problem and that it be continued through that problem.

ILLUSTRATIVE PROBLEM

5.2 What is the net torque on the shaft shown in figure 5.5 due to the belt forces shown on the three pulleys?

Solution:
Assume that the clockwise direction, looking from the end of the shaft nearest pulley A, is positive.

$$
\begin{aligned}
M &= -31 \text{ kg} \times 15 \text{ cm} + 7 \times 15 - 75 \times 20 + 19 \times 20 - 50 \times 10 + 10 \times 10 \\
&= -1\ 880 \text{ kg} \cdot \text{cm}
\end{aligned}
$$

Although $\text{kg} \cdot \text{cm}$ units are frequently used for torque, it is better to express the answer in $\text{N} \cdot \text{m}$.

$$M = -1\ 880 \text{ kg} \cdot \text{cm} \times \frac{9.8 \text{ N}}{1 \text{ kg}} \times \frac{1 \text{ m}}{100 \text{ cm}} = -184 \text{ N} \cdot \text{m}$$

The net torque on the shaft due to the pulley forces is $1\ 880 \text{ kg} \cdot \text{cm}$ or $184 \text{ N} \cdot \text{m}$ in the *counterclockwise* direction.

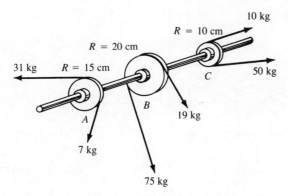

Figure 5.5 Illustrative problem 5.2. The
7- and 31-kg forces lie in the plane of pulley A,
the 75- and 19-kg forces in the plane of pulley
B, and the 10- and 50-kg forces in the plane of
pulley C. All three planes are perpendicular to
the axis. Therefore, the moments of these forces
about the axis may be algebraically added.

5.4 MOMENT EQUILIBRIUM

For static equilibrium the summation of moments on any object about any conceivable axis must be equal to zero.

$$\Sigma M = 0 \qquad (5.2)$$

where Σ indicates summation.

This is the principle of equilibrium introduced with forces in chapter 1. It applies also to moments of forces and is useful in understanding loads on torsional members.

A free-body diagram, as you may recall from chapter 1, is a diagram of a structure, or portion of a structure, drawn free of its supports. The supports are replaced with equivalent forces or, in this case, moments of forces. Application of the free-body diagram to a torsional structure is demonstrated in illustrative problem 5.3.

ILLUSTRATIVE PROBLEM

5.3 A torque box is used as the curb member on a tank truck loading platform as shown in figure 5.6. The platform has three ramps that can be hydraulically lowered to permit the operators to easily walk from the platform to the dome or domes of tank trucks. End E of the torque box is torsionally restrained by the platform structure. End A is presumed to be torsionally free. If a 300-lb person is assumed to be standing on the end of each platform and the platforms themselves weigh 750 lb, what is the maximum torsion on the member?

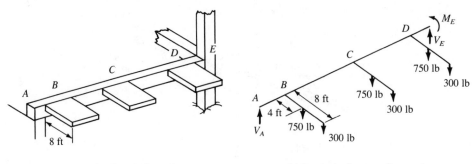

(a) tank truck loading platform shown
with guard rails omitted

(b) free-body diagram of torque box

Figure 5.6 Illustrative problem 5.3.

Solution:

The free-body diagram of the torque box with ramp loaded and extended is shown in figure 5.6b. The 300-lb loads are shown at the end of each ramp. The 750-lb ramp weights are presumed to be halfway out on the ramps.

End A can provide no torsional resistance, only a vertical reaction, V_A. The torsionally rigid end, E, can provide a reaction moment, M, and a vertical reaction force, V_E. Since the vertical reactions V_A and V_E can be presumed to act through the axis of the torque box, they contribute no moment.

From relationship 5.2, $\Sigma M = 0$. Taking clockwise moments as viewed from end A as positive,

$$\Sigma M = (750 \times 4 + 300 \times 8 + 750 \times 4 + 300 \times 8 + 750 \times 4 + 300 \times 8) \text{ lb} \cdot \text{ft} - M_E = 0$$
$$16\ 200 \text{ lb} \cdot \text{ft} - M_E = 0$$
$$M_E = 16\ 200 \text{ lb} \cdot \text{ft}$$

The reaction torque, M_E, is 16 200 lb · ft in the counterclockwise direction. The maximum torque extends over the length from D to E, because the length carries the effect of all three ramps.

If the free body were formed by a cutting plane between C and D, the moment equation would yield 10 800 lb · ft, and between B and C, 5 400 lb · ft. There is no torsion between A and B.

5.5 COUPLES

A *couple* is a pure torque. By *pure* we mean that it has no net linear force in any direction. A couple has only a twisting action. A couple can be produced by two equal parallel forces that are opposite in direction, as shown in figure 5.7. The two forces, because they are equal in magnitude and oppose each other, produce no net force in any direction. However, as shown here, they produce a counterclockwise moment or twisting action.

The moment of a couple can be computed by assuming an axis at point o. Since the downward force acts through that axis, it contributes nothing to the moment. Meanwhile, the upward force causes a moment of Fl. No matter where one assumes an axis in

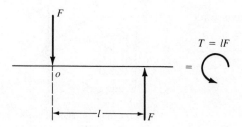

Figure 5.7 A couple is a pure torque and can be represented as a circular arrow.

relationship to the two forces, the same moment, *Fl*, will result. Try taking moments about the other force, for example.

The moment of a couple is equal to the magnitude of one of the forces times the distance between the forces.

5.6 RATIONALIZATION OF A FORCE AS A FORCE AND A COUPLE

The forces applied in figures 5.1 through 5.6 cause vertical and/or lateral forces as well as moments upon the members. The couple applied to the ratchet wrench in figure 5.8b causes no lateral force. A single force applied off the axis of a member can be thought of as a combination of a force on the axis and a couple about the axis. The force on the axis is unwanted because it tends to unseat the socket. The couple causes the desired torsion.

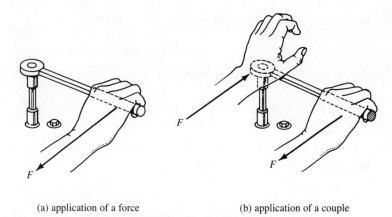

(a) application of a force   (b) application of a couple

Figure 5.8 The force (a) alone tends to unseat the socket. The couple (b) produces the same twisting moment without unseating the socket.

A force can be replaced by an equivalent force and a couple as shown in figure 5.9 on page 142. The basis of this assumption is that two equal and opposite forces applied at a point have no structural effect. Therefore, they can be added as necessary.

Figure 5.9 Rationalization of an off-axis force as an equivalent combination of a force and a couple.

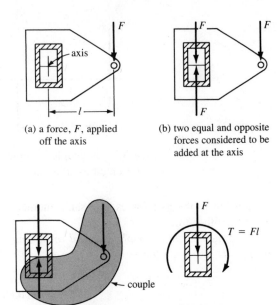

(a) a force, F, applied off the axis

(b) two equal and opposite forces considered to be added at the axis

(c) The lower of the added forces is taken with the original force as a couple. The upper force remains.

(d) The original force is, therefore, the same as a force, F, through the axis and a torque, Fl, about the axis.

PROBLEM SET 5.1

5.1–5.8
Find the moments about axis o due to the forces in figure 5.10.

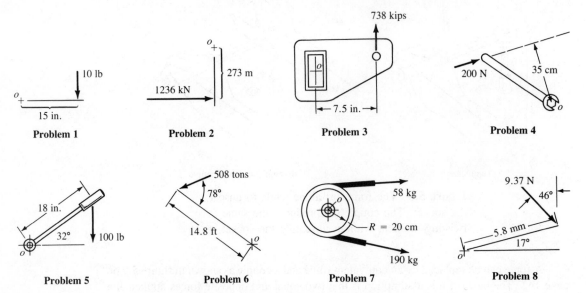

Figure 5.10 Problems 5.1 through 5.8.

5.9–5.14

Find the sum of the moments about axis o due to the applied forces in figure 5.11.

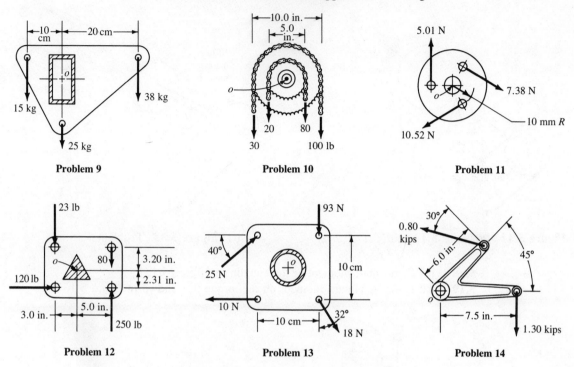

Problem 9

Problem 10

Problem 11

Problem 12

Problem 13

Problem 14

Figure 5.11 Problems 5.9 through 5.14.

5.15 What is the moment of force about the axis due to the applied forces in figure 5.12?

5.16 What is the moment of force on the crankshaft of the 4-cylinder engine due to the connecting-rod forces shown in figure 5.13? (The stroke of an engine is the total length of travel of the piston.)

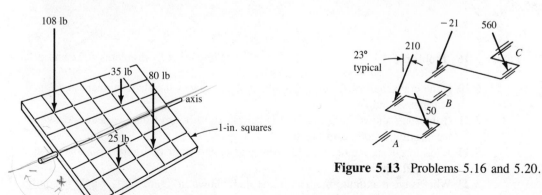

Figure 5.12 Problem 5.15.

Figure 5.13 Problems 5.16 and 5.20.

5.17 What is the moment of force produced on the shaft by the pulleys shown in figure 5.14?

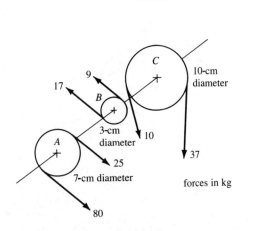

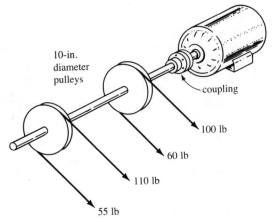

Figure 5.14   Problems 5.17 and 5.21.

Figure 5.15   Problem 5.18.

5.18 What is the shaft moment at the coupling in figure 5.15?

5.19 What is the maximum moment on the torque box shown in figure 5.16?

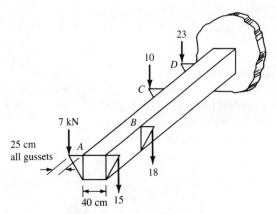

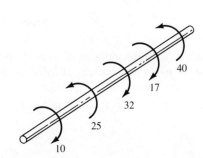

Figure 5.16   Problem 5.19.

Figure 5.17   Problems 5.22 and 5.23.

5.20 If the crankshaft of figure 5.13 is held at C, what is the shaft moment at the main bearings A and B?

5.21 If the shaft shown in figure 5.14 is held in equilibrium from the near end, what is the maximum torque in the shaft?

5.22 What torque is required on the near end of the shaft of figure 5.17 to hold it in rotational equilibrium?

5.23 What is the maximum moment in the shaft shown in figure 5.17, if held in equilibrium by a torque at the far end?

5.24 Resolve the engine thrust in figure 5.18 into a forward force and a couple about the wing elastic center shown.

5.25 An electric motor is producing $786 \, \text{kg} \cdot \text{cm}$ of torque, as shown in figure 5.19. What are the forces in the mounting bolts at A and B due to this torque? (Use couples.)

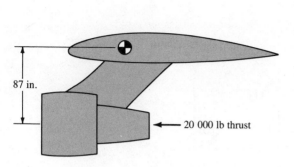

87 in.

20 000 lb thrust

Figure 5.18   Problem 5.24.

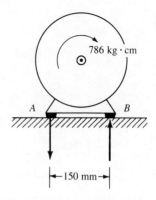

786 kg · cm

A

B

←150 mm→

Figure 5.19   Problem 5.25.

5.7 STRESS DUE TO TORSION IN A CIRCULAR CYLINDER

If a circular cylinder were made of foam rubber and marked with a felt-tipped pen in a grid pattern, it would deform in torsion as depicted in figure 5.20. A meridional line on the surface of the shaft parallel to the axis would deform into a helix. The circumferential circles remain circles. They do not deform. A square bounded by parallel lines on the surface and by circumferential circles deforms into a parallelogram. From figure 4.25 in chapter 4 we know this parallelogram deformation is the telltale sign of shear stress, τ. Shear stress is related to the amount of applied torque or moment, T. In pure torsion it is common practice to use the letter T to indicate the moment of the applied couples.

In the case of torsion, stress is not uniform over the cross section as it is in tension or compression or in the cross shear of chapter 4. The assumption is made that the shear stress increases linearly with radius from zero at the centerline of the cylinder to the

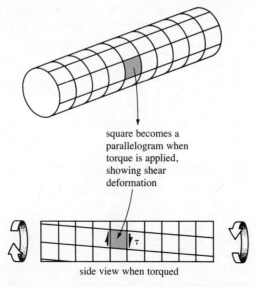

square becomes a parallelogram when torque is applied, showing shear deformation

τ

side view when torqued

Figure 5.20   Deformation of a cylinder under torque.

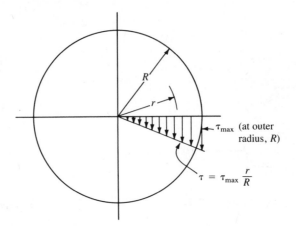

Figure 5.21 Shear stress distribution in the elastic range.

maximum shear stress at the surface. This case is shown in figure 5.21. The shear stress, τ, is related to the radius as follows:

$$\tau = \tau_{\max}\frac{r}{R} \tag{5.3}$$

where τ is the shear stress at any radius r, $\tau_{\max}$ is the maximum shear stress, r is the radius at which τ is being considered, and R is the outside or surface radius.

Experiments show that this stress distribution is an accurate assumption for most structural materials.

To relate the stress distribution to torque, it is necessary to consider the cross-sectional area to be divided into incremental concentric bands of radius r and very small thickness, Δr, as shown in figure 5.22. The delta sign (Δ) is used to indicate infinitesimal thickness. The area, ΔA, of one incremental band is also infinitesimally small.

$$\Delta A = 2\pi r \Delta r \tag{5.4}$$

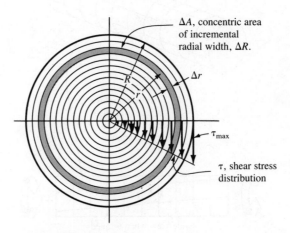

Figure 5.22 Division of cross section into incremental concentric areas.

The ΔA is arrived at by multiplying the circumference of the thin ring, $2\pi r$, by its thickness, Δr. This equation is accurate for thin bands of circular area.

Each incremental band of area carries an incremental amount of torque, ΔT, which can be expressed by multiplying the stress at that radius by the incremental area and by the moment arm, r.

$$\Delta T = \tau r \Delta A \tag{5.5}$$

From relationships 5.3 and 5.5,

$$\Delta T = \frac{\tau_{max}}{R} r^2 \Delta A \tag{5.6}$$

To calculate total torque, it is necessary to sum the incremental torques of all the incremental areas.

$$T = \sum \Delta T \tag{5.7}$$

From relationship 5.6,

$$T = \sum \frac{\tau_{max}}{R} r^2 \Delta A$$

Taking those things that do not vary with r forward of the summation sign,

$$T = \frac{\tau_{max}}{R} \sum r^2 \Delta A \tag{5.8}$$

Substituting the incremental area from relationship 5.4,

$$T = \frac{2\pi\tau_{max}}{R} \sum r^3 \Delta r \tag{5.9}$$

This equation is best evaluated by replacing the summation sign with an integral and the incremental sign, Δ, with a differential and using the methods of calculus.

$$T = \frac{2\pi\tau_{max}}{R} \int_0^R r^3 \, dr \tag{5.10}$$

Integrating from $r = 0$ to $r = R$,

$$T = \frac{2\pi\tau_{max}}{R} \frac{r^4}{4}\bigg]_0^R = \frac{2\pi\tau_{max}}{R} \frac{R^4}{4}$$

$$T = \frac{\pi\tau_{max}R^3}{2} \tag{5.11}$$

Solving for the maximum shearing stress yields the long-awaited relationship for the maximum shearing stress, τ_{max}, in terms of the torque, T.

$$\tau_{max} = \frac{2T}{\pi R^3} \tag{5.12}$$

The relationship is usually preferred in terms of the diameter of the shaft, D. Substituting $D/2$ for R,

$$\boxed{\tau_{max} = \frac{16T}{\pi D^3}} \tag{5.13}$$

This is the relationship for stress in an elastic circular cylinder of diameter D, when acted upon by a torque, T. The maximum shear stress, τ_{max}, occurs at the surface. Stress at an internal radius can be found from relationship 5.3, once the maximum stress is known.

Note from relationship 5.13 that the stress in a shaft due to torsion is inversely proportional to the cube of the diameter. If the diameter is doubled, the torque-carrying capacity is increased by a factor of eight. The material near the surface is most highly stressed and, therefore, most effective. This is why hollow shafts, discussed in the next section, are important.

ILLUSTRATIVE PROBLEM

5.4 A cylindrical shaft is to be constructed of a material with a yield stress in shear of 200 MPa. If it must carry a torque of 500 N · m, what is the minimum acceptable diameter?

Solution:
From relationship 5.13,

$$\tau_{max} = \frac{16T}{\pi D^3}$$

Solving for D,

$$D = \sqrt[3]{\frac{16T}{\pi \tau_{max}}} \tag{5.14}$$

Entering the values of T and τ_{max},

$$D = \sqrt[3]{\frac{16 \times 500 \text{ N} \cdot \text{m}}{\pi \times 200 \text{ MPa}}}$$

Since 1 MPa = 1 N/mm², we can obtain the answer in mm.

$$D = \sqrt[3]{\frac{16 \times 500 \text{ N} \cdot \text{m}}{\pi \times 200 \text{ MPa}} \times \frac{1 \text{ MPa}}{1 \text{ N/mm}^2} \times \frac{1\,000 \text{ mm}}{1 \text{ m}}}$$

$$D = \sqrt[3]{12.73\text{E}3 \text{ mm}^3}$$

$$D = 23.4 \text{ mm}$$

The shaft must be at least 23.4 mm in diameter to prevent yielding of the material.

5.8 STRESS DUE TO TORSION IN A HOLLOW CYLINDER

Since in torsion the material near the center of a solid cylinder does not get stressed as much as does the material near the surface, the core is not used efficiently. Sometimes it is economical to eliminate the core by using a hollow cross section.

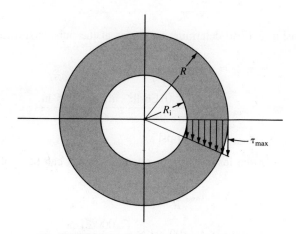

Figure 5.23 Shear-stress distribution in a hollow cylindrical shaft.

The stress distribution in the wall of a hollow section is shown in figure 5.23. It is linear, just as in a solid section. Therefore, relationship 5.3 applies also to hollow sections. Since a hollow section may also be described with incremental concentric areas, relationships 5.4 through 5.10 also apply.

However, relationship 5.10 must instead be integrated from $r = R_i$ to $r = R$, yielding the following:

$$T = \frac{2\pi\tau_{max}}{R}\left(\frac{R^4}{4} - \frac{R_i^4}{4}\right) \tag{5.15}$$

Solving for τ_{max},

$$\tau_{max} = \left(\frac{2TR}{\pi(R^4 - R_i^4)}\right) \tag{5.16}$$

In terms of diameters,

$$\boxed{\tau_{max} = \frac{16TD}{\pi(D^4 - d^4)}} \tag{5.17}$$

where d is used to indicate the inside diameter.

ILLUSTRATIVE PROBLEM

5.5 A hollow steel shaft carries a torque of 2.01E6 lb · in. The outside diameter is 8.065 in. and the inside diameter is 7.000 in. What is the stress at the outside and at the inside surfaces of the shaft?

Solution:

Using relationship 5.17 to determine the stress at the outside surface,

$$\tau_{max} = \frac{16TD}{\pi(D^4 - d^4)}$$

$$\tau_{max} = \frac{16 \times 2.01E6 \text{ lb} \cdot \text{in.} \times 8.065 \text{ in.}}{\pi(8.065^4 - 7.000^4) \text{ in.}^4}$$

$$\tau_{max} = 45\ 100 \text{ psi}$$

To find the stress at the inside surface, relationship 5.3 can be applied.

$$\tau = \tau_{max}\frac{r}{R}$$

$$\tau = 45\ 100 \text{ psi} \times \frac{(7.000/2) \text{ in.}}{(8.065/2) \text{ in.}}$$

$$\tau = 39\ 200 \text{ psi}$$

The stress at the outside surface is 45 100 psi and that at the inside surface is 39 200 psi.

Polar Moment of Inertia

The *polar moment of inertia, J,* is defined as

$$J = \sum r^2 \Delta A \tag{5.18}$$

It is also called the polar second moment of area. Second moment of area is a more correct nomenclature, as discussed in chapter 10, but not as widely used.

For a solid circular area,

$$J = \frac{\pi D^4}{32} \tag{5.19}$$

For a hollow circular area,

$$J = \frac{\pi}{32}(D^4 - d^4) \tag{5.20}$$

A relationship equivalent to 5.14 and 5.17 can be written in terms of J,

$$\tau_{max} = \frac{TR}{J} \tag{5.21}$$

This relationship proves to be identical to relationships 5.13 and 5.17 for the solid and the hollow circular shafts, respectively. Although J can be calculated for an area of any shape, relationship 5.21 is only valid for solid and hollow circular members. Be careful to use relationship 5.21 only on solid or hollow circular members.

5.9 TORSIONAL STRESS CONCENTRATION

Stress concentration occurs at discontinuities such as threads, holes, shoulders, and notches in torsional members just as it does in tension and compression members. In all instances it is handled identically by a stress concentration factor, k. The stress concentration factors for most shapes encountered have been determined by means of experiment or by theory of elasticity. The results of these findings are summarized in references such as Peterson.[1]

The shear stress at the discontinuity can be calculated for a round shaft under torsion using the stress concentration factor as follows:

$$\tau_{max} = \frac{kTR}{J} \qquad (5.22)$$

The stress concentration factor, k, for torsion is numerically different from that for tension and compression, even though the shapes may be identical. Care must be taken to draw the right factors from the references. Generally, stress concentrations for torsion are not as great as those for tension. The stress concentration factors caused by a hole or a shoulder in a solid shaft are given in figure 5.24 on page 152.

PROBLEM SET 5.2

5.26 What is the maximum shear stress in a 2-in.-diameter solid steel shaft when acted upon by a torque of 2 000 lb · ft?

5.27 What is the maximum shear stress in a solid aluminum shaft having a diameter of 10 mm when acted upon by a torque of 25 N · m?

5.28 A $\frac{3}{4}$-in.-diameter solid circular shaft is made of a material having an allowable shear stress of 47 500 psi. What is the torque capacity of the shaft?

5.29 A solid circular shaft is made of a stainless steel with a yield strength in shear of 150 MPa. It has a diameter of 5 cm. How much torque is required to yield the shaft?

5.30 A solid shaft intended to transmit a torque of 500 N · m must have what diameter if the allowable shear stress is 137 MPa?

5.31 What diameter solid 6061 aluminum shaft is required to carry a torque of 1 600 lb · ft without yielding? The yield stress for the material is 21 000 psi.

5.32 A torsion member in a structure is made of 10-in. schedule 40 steel pipe with an allowable shear stress of 18 000 psi. What is the allowable torque?

5.33 A hollow aluminum tube with a 4-cm outside diameter and a 3-cm inside diameter carries a torque of 200 N · m. What is the stress at the outside radius? What is the stress at the inside radius?

5.34 A solid shaft is to be replaced with one of the same material having a hole of half the outside diameter running down the center line. By what percent does the hole reduce the torque-carrying capacity? By what percent does it reduce the weight of the shaft?

5.35 Show that relationships 5.13 and 5.21 are identical for a solid shaft.

5.36 What is the polar moment of inertia of a hollow circular section 30 mm in outside diameter with an 8-mm-thick wall?

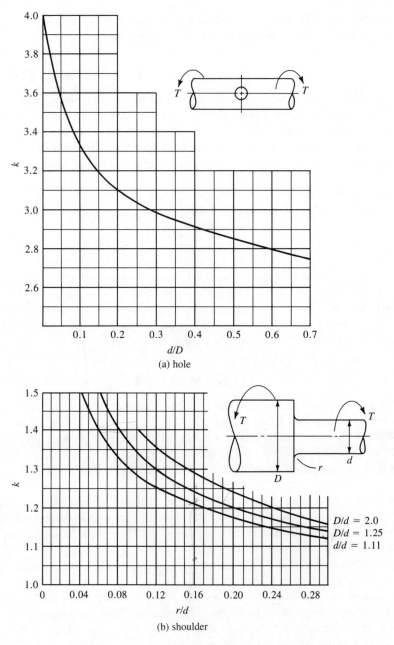

Figure 5.24 Stress concentration factors for torsion in a round shaft with hole or shoulder.

5.37 A shaft has a step reducing its diameter from 10 mm to 6 mm with a 1-mm corner radius. It carries a torque of 2.5 N · m. What is the stress at the concentration?

5.10 TORSIONAL DEFORMATION

When a torsional moment is applied over a circular shaft, it deforms as shown in figure 5.25. One end of the shaft turns an angle, θ, with respect to the other end. This angle of twist is the torsional deformation. It is the displacement in a rotational sense.

The angle γ is the shear strain shown in more detail previously in figure 5.20. The deformation angle, θ, is related to the shear strain, γ, by geometry. For structural materials such deformations are small and small angle approximations, as previously discussed (see figure 4.27), are accurate.

Therefore, two triangles, *obc* and *abc*, can be visualized. Side *bc* is common to both. It is actually an arc, but because angles θ and γ are so small, arc *bc* is practically a straight line.

For small angles expressed in radians,

$$\theta = \frac{bc}{R}$$

and also

$$\gamma = \frac{bc}{L}$$

Solving both for *bc* and equating,

$$\theta = \frac{\gamma L}{R} \tag{5.23}$$

From relationship 4.6 (introduced in the previous chapter, defining the elastic modulus in shear, G),

$$G = \frac{\tau}{\gamma}$$

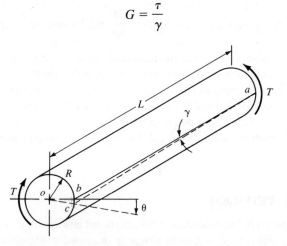

Figure 5.25 Rotational displacement of a shaft with an applied torsional moment.

Therefore,

$$\gamma = \frac{\tau}{G}$$

At the surface of the shaft,

$$\gamma = \frac{\tau_{max}}{G}$$

Recalling forward relationship 5.21,

$$\tau_{max} = \frac{TR}{J}$$

And substituting into the previous relationship,

$$\gamma = \frac{TR}{JG} \qquad (5.24)$$

This is a relationship for the surface strain, γ, in a circular shaft in the elastic range due to an applied torque.

Substituting relationship 5.24 into 5.23,

$$\boxed{\theta = \frac{TL}{JG}} \qquad (5.25)$$

where θ is the angular displacement in radians, T is the torque, L is the length of the shaft, J is the polar moment of inertia, and G is the elastic modulus in shear.

Relationship 5.25 is very important; it can be compared to relationship 2.6 for tension and compression.

$$\theta = \frac{TL}{JG} \text{ for torsion compares to } \delta = \frac{FL}{AE} \text{ for tension or compression}$$

Both θ and δ are displacements, θ in the rotational sense and δ in the longitudinal. T is the forcing quantity in torsion similar to F in tension. L is the length for both relationships. J is a property of area resisting torsion, similar to A in tension. Both G and E are elastic moduli. Both relationships 2.6 and 5.25 are valid *only* in the elastic range. But relationship 5.25 for torsion is different in one important way. It is valid only for solid or hollow circular members, whereas relationship 2.6 for tension or compression applies to cross sections of any shape.

ILLUSTRATIVE PROBLEM

5.6 Estimate the rotational displacement between the bit and the rotary table of an oil well drill string 15 000 ft long, when the torque is increased to the elastic strength of the drill pipe in an effort to free a stuck bit. The pipe is $4\frac{1}{2}$ in. in outside diameter with a $\frac{1}{2}$-in. wall. $G = 11.5E6$ psi. The yield stress in shear is 16 000 psi. Neglect the pipe joints by assuming the pipe to be uniform all the way down. The tension load is taken off the string during this unsticking operation.

Solution:

First it is necessary to solve for the elastic torque capacity of the drill pipe from relationship 5.21.

$$\tau_{max} = \frac{TR}{J}$$

Solving for T,

$$T = \frac{\tau_{max}J}{R}$$

Now, substituting this into relationship 5.25 and cancelling J,

$$\theta = \frac{\tau_{max}L}{RG}$$

$$\theta = \frac{16\,000 \text{ psi} \times 15\,000 \text{ ft}}{2.25 \text{ in.} \times 11.5E6 \text{ psi}} \frac{12 \text{ in.}}{1 \text{ ft}}$$

$$\theta = 111.3$$

All dimensions cancel. The dimensionless unit of radians can be inserted.

$$\theta = 111.3 \text{ radians}$$

The answer is more meaningful in revolutions.

$$\theta = 111.3 \text{ radians} \times \frac{1 \text{ revolution}}{2\pi \text{ radians}} = 17.7 \text{ revolutions}$$

The drill string deforms torsionally more than 17 revolutions over its length while remaining elastic.

5.11 TORSION BARS

A torsion bar is a very precise and efficient rotational spring. The spring constant is defined as the amount of torque required per radian of rotation.

$$\boxed{K_t = \frac{T}{\theta}}$$

$$\frac{T}{\frac{TL}{JG}} = \frac{JG}{TL} = \frac{JG}{L}$$

$$(5.26)$$

The ratio of T/θ for a circular shaft can be solved from relationship 5.25.

$$\boxed{K_t = \frac{JG}{L}}$$

$$(5.27)$$

ILLUSTRATIVE PROBLEM

5.7 A torsion bar is to be a counterbalance for an overhead door. The bar must have a torsional spring constant of 380 lb · ft and must be able to turn 90° without yielding. It will be made of high-strength steel. The shear stress may not exceed 25 000 psi. The modulus of rigidity is 11.5E6 psi. What must be its length and diameter?

Solution:
From 5.26, the torque at 90° is given by

$$T = K_t\theta$$

where K_t is given as 380 lb · ft and θ is 90° or $\pi/2$ radians. Torque capacity governs the diameter. Substituting $K_t\theta$ for T in relationship 5.14,

$$D = \sqrt[3]{\frac{16T}{\pi\tau_{max}}} = \sqrt[3]{\frac{16K_t\theta}{\pi\tau_{max}}}$$

$$D = \sqrt[3]{\frac{16 \times 380 \text{ lb} \cdot \text{ft} \times \pi/2}{\pi \times 25\ 000 \text{ lb/in.}^2} \times \frac{12 \text{ in.}}{1 \text{ ft.}}}$$

$$D = 1.134 \text{ in. at least.}$$

The spring constant governs the length. From relationship 5.19,

$$J = \frac{\pi D^4}{32}$$

And from relationship 5.27,

$$K_t = \frac{JG}{L} = \frac{\pi D^4 G}{32L}$$

Solving for L,

$$L = \frac{\pi D^4 G}{32K_t}$$

$$L = \frac{\pi \times (1.134 \text{ in.})^4 \times 11.5E6 \text{ lb/in.}^2}{32 \times 380 \text{ lb} \cdot \text{ft}} \times \frac{1 \text{ ft}}{12 \text{ in.}}$$

$$L = 409 \text{ in. or } 34.1 \text{ ft}$$

The torsion bar will need to be 34.1 ft long and 1.134 in. in diameter.

One may wish to go up to the next diameter bar stock available. But in so doing, the length must be increased with the fourth power of the diameter used. For example, if a $1\frac{1}{4}$-in. rod were used, the length would need to be increased from 34.1 to 50.3 ft. A special order may be economical.

PROBLEM SET 5.3

5.38 Calculate the torsional spring constant of a 2-in. standard weight (schedule 40) pipe 10 ft long. ($G = 11.5E6$ psi.)

5.39 What length of 0.500-mm-diameter tungsten wire is required for a torsion bar in an instrument in order that it have a torsional spring constant of 530 gm · cm/radian? (G = 140E9 Pa.)

5.12 TORSION TESTS OF DUCTILE MATERIALS

Torsion testing machines are devices for accurately twisting a cylindrical specimen of a material. Specimens may be round bars of stock diameters or they may be machined from hexagonal stock. Gripping devices for torsion are either chucks, not unlike those for a lathe, or hexagonal seats.

Torque is determined by measuring the force on a reaction lever. Angular deformation in the elastic range can be measured with sufficient accuracy using a dial gage on a special fixture. Such devices are called *torsiometers*. Torsional deformations are easier to measure than deformations in tension or compression because more movement is involved.

Torsion machines come in several types. They can be hydraulically driven or turned directly using a gear head motor. Others are hand-cranked devices. Still others are fixtures that are attached to a universal testing machine of the kind described in chapter 2. The torsion test requires a means of applying and accurately controlling the torsion on a specimen, a means of measuring the torque, directly or indirectly, and a means of measuring the angle of deformation.

A torsion test on a ductile material can yield a graph such as the one shown in figure 5.26. The line from A to B is elastic. On this line the relationships 5.13 and 5.23 can be used to convert the torque and the angle of twist readings from a testing machine to stress, τ_{max}, and strain, γ, data as follows:

$$\tau_{max} = \frac{16T}{\pi D^3} \qquad\qquad (5.28)$$

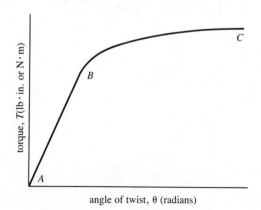

Figure 5.26   Torsion test data.

and

$$\gamma = \frac{\theta D}{2L}$$

where T and θ are measured directly during the test. These relationships produce values of stress and strain at the surface of the specimen that can be graphed as stress and strain as in figure 5.27. Internal stresses and strains are less than the surface values, reducing linearly with radius to zero at the axis.

The elastic modulus in shear, or modulus of rigidity, is the slope of the elastic line, as shown in figure 5.27.

The yield strength in shear can be determined by the 0.2-percent offset method as is done in tension tests.

Beyond point B relationship 5.13 is no longer exact, because some material begins to go into the plastic range. The further it is from the elastic range, the more inaccurate the relationship becomes.

The inaccuracy occurs because torsional members yield first at the surface, then gradually become plastic in a band that progresses inward as the angle of twist continues. This effect is shown in figure 5.28.

The linear stress distribution upon which relationship 5.13 is based is disrupted by the plastification in the outer zone, causing the inaccuracy. No relationship exactly governs the partially plastic range from B to C. It is also complicated by strain-hardening of the plastic material.

The elastic core continues to get smaller and smaller as rotations are increased past B. It could be argued that there will always be an elastic core, but it becomes infinitesimal at large angular deformations. The stress distribution will become much like that shown in

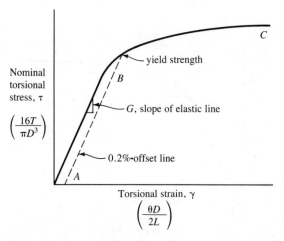

Figure 5.27 Torsion test data reduced to stress and strain.

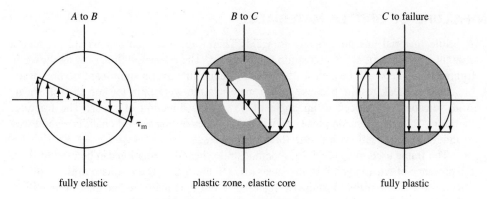

Figure 5.28 Progressive plastification of a
solid ductile material in torsion.

the fully plastic condition (figure 5.28). For such a stress distribution, the relationship
between torque and stress is

$$\tau_u = \frac{12T}{\pi D^3}$$ **(5.29)**

where τ_u is the ultimate shearing stress.

This condition occurs at final rupture of the specimen. The subscript u is used to
indicate ultimate shearing stress. Relationship 5.25 can be used to calculate the ultimate
shearing stress from the torque at rupture. A value of ultimate shearing stress so determined compares favorably with clevis shear or punching-shear test results.

For most structural materials the angle of twist at rupture becomes so large (many
revolutions) that it is not practical to plot the graph beyond C.

The appearance of the fracture sketched in figure 5.29 is demonstrated by a ductile
material failing in shear. It will be sheared clean at 90° to the torsion axis. No necking
down occurs as in a tension failure.

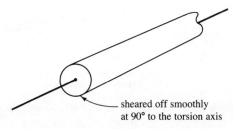

sheared off smoothly
at 90° to the torsion axis

Figure 5.29 Appearance of a ductile fracture
in torsion.

5.13 TORSION FAILURE OF BRITTLE MATERIALS

A brittle material has no capacity to yield. Therefore, a torsion test would produce a straight-line graph of stress and strain up to failure. The fracture will appear as shown in figure 5.30. There will be a clear 45° helical fracture line over a significant portion of the break. This break results because brittle materials are weaker in tension than in shear. And, tension is produced at an angle of 45° to the direction of shear. The fracture, therefore, follows a tension plane. Twisting off pieces of blackboard chalk is a good way to familiarize yourself with brittle torsion fractures.

The frame used in figure 4.25 to demonstrate shear deformation can be used to show the presence of tension at 45° to the shear axis (figure 5.31). If the square drawn with the felt-tipped pen upon the elastomeric material is put diagonally rather than vertically, it will deform as shown. This experiment demonstrates that shear is equivalent to tension at 45° one way from the shear direction and to compression at 45° the other way. It is the tension at 45° that fractures the brittle material in torsion.

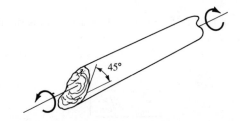

Figure 5.30 Appearance of a brittle fracture in torsion.

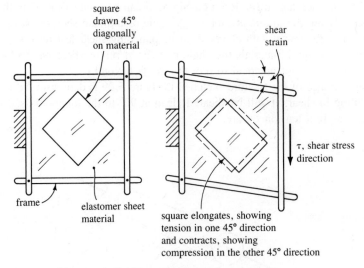

Figure 5.31 Square drawn at 45° to the shear stress elongates in one direction, showing tension, and contracts in the other direction, showing compression.

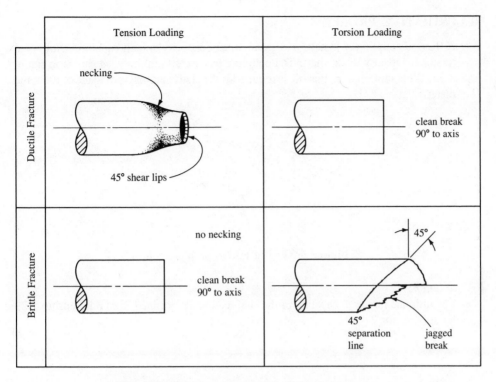

Figure 5.32 The appearance of the fracture can indicate type of loading.

Figure 5.32 gives a summary of the appearances of tension and shear fractures in ductile and brittle materials. These appearances result because brittle materials tend to fail in tension and ductile materials tend to fail in shear. Furthermore, tension produces shear stress at 45° from the tension direction. Correspondingly, shear produces tension at 45° from the shear direction.

5.14 ATTACHMENT OF TORSION MEMBERS

Loads are transferred to torsion members through shear connections. Such connections often have a narrow concentric band of shear area that may be perpendicular to the torsional axis (such as a flanged coupling), parallel to the axis (such as a spline or key), or at an angle with the axis (as in a fillet weld). When the shear area is essentially concentrated at a single radius from the moment axis, the shear stress, τ, can be expressed by the relationship

$$\tau = \frac{T}{AR} \tag{5.30}$$

where A is the shear area confined to a narrow concentric band of radius R.

ILLUSTRATIVE PROBLEM

5.8 A flanged coupling is to have a torsion moment capacity at least equal to the ultimate torsional capacity of its shaft. The coupling has six 10-mm bolts as shown in figure 5.33. What minimum radius, R, is required if the shaft and bolts are made of the same material?

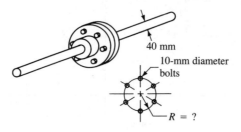

40 mm

10-mm diameter
bolts

$R = ?$

Figure 5.33 Illustrative problem 5.8.

Solution:

The ultimate torsional capacity of the shaft can be determined from relationship 5.29 by solving for T.

$$T = \frac{\pi D^3 \tau_u}{12}$$

Relationship 5.30 applies to the bolt pattern. The shear area, A, is equal to six times $\pi d^2/4$. Or,

$$A = 3\pi d^2/2$$

where d is the bolt diameter.

Therefore,

$$\tau = \frac{2T}{3\pi d^2 R}$$

Substituting for T and letting τ equal the ultimate shear strength,

$$\tau_u = \frac{2}{3\pi d^2 R} \frac{\pi D^3 \tau_u}{12}$$

Solving for R,

$$R = \frac{D^3}{18 d^2} = \frac{(40 \text{ mm})^3}{18(10 \text{ mm})^2}$$

$$R = 35.6 \text{ mm}$$

The minimum radius of the coupling bolt pattern in order for the bolts to have the same ultimate torque capacity as the shaft is 35.6 mm.

PROBLEM SET 5.4

5.40 A round rod of a new plastic material was tested in torsion. The torque-rotation curve had the appearance of figure 5.27. Point *B* occurred at a torque of 160 lb · in. and a deflection of 30°. The gage length of the $\frac{1}{4}$-in.-diameter specimen was 10 in. What is the elastic modulus in shear?

5.41 A steel shaft was tested in torsion. It failed in shear at a torque of 1 027 N · m. If the shaft diameter was 20 mm, what is the ultimate shear stress for the material?

5.42 A special structural bolt has a secondary head as shown in figure 5.34, which shears off clean when the proper fastening torque is reached. What must be the diameter of the throat if the ultimate shearing stress for the material is 30 000 psi and the bolt is intended to be tightened to 100 lb · ft of torque?

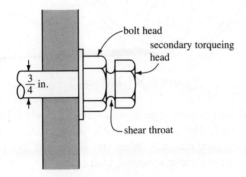

Figure 5.34 Problem 5.42.

5.43 Adhesives are now frequently used to attach machine elements to shafts. What must be the shear stress capability of the adhesive used to attach the gear to the shaft shown in figure 5.35 for the joint to be able to transmit a torque of 2.5 N · m?

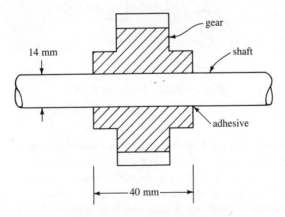

Figure 5.35 Problem 5.43.

5.44 A pulley is attached to a shaft by means of a $\frac{3}{16}$-in.-square key. What is the torque rating of the assembly shown in figure 5.36, if the key material can resist a shear stress of 20 000 psi? The hub is 1.20 in. long and the shaft is 1.000 in. in diameter.

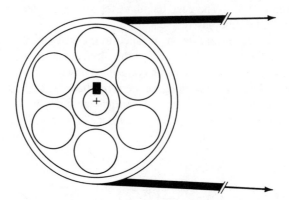

Figure 5.36   Problem 5.44.

5.45 A 3-in. standard-weight steel pipe torsion member is welded to a square column as shown in figure 5.37. The design-allowable stress in shear of the weld material is one-half that of the pipe material. What size fillet weld will be required so that the pipe and weld have the same torsional load rating?

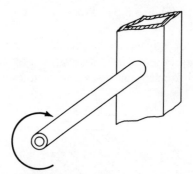

Figure 5.37   Problem 5.45.

5.15 TORQUE BOXES

Hollow square and rectangular shapes are efficient torsional members that have the added advantage of ease of attachment. They have great torsional rigidity.

A pipe or a box shape is called a *closed* section. A channel, zee, angle, I, or wide-flanged structural shape as shown in figure 5.38 is called an *open* section. Closed sections in torsion make optimal use of material. Furthermore, their load-carrying capacity and rigidity can be accurately and easily predicted by the methods explained in this chapter. Open sections require the services of an analytically trained engineer to properly evaluate each application in torsion. But since open sections are far too flexible for most torsion applications, they are usually avoided.

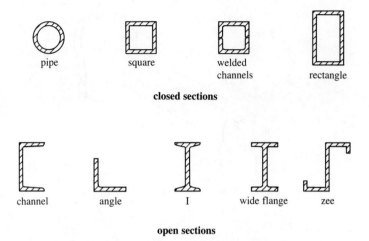

Figure 5.38 Hollow, closed sections make rigid and efficient torsion members. Open sections lack adequate torsional rigidity for most applications.

A rigorous understanding of torsion in any noncircular shape requires advanced mathematics and theory of elasticity.[2] However, the relationships that follow approximate the results of this theory. They are reasonable approximations for closed sections only.

Stress can be estimated from the relationship

$$\tau = \frac{T}{2At} \tag{5.31}$$

where T is the torsional moment, A is the cross-sectional area *enclosed* by the wall, and t is the thickness of the wall. Elastic angular deformation can be *estimated* from the relationship

$$\theta = \frac{TLs}{4GA^2t} \tag{5.32}$$

where θ is the angle of twist, G is the elastic modulus in shear, s is the peripheral length of the wall, and L is the length of the torque box. Relationships 5.31 and 5.32 can be used both for torque boxes and tubes or any closed, hollow thin-walled section. Calculations show little difference between these relationships and those for hollow circular sections when the wall thickness is less than one-tenth the diameter.

ILLUSTRATIVE PROBLEM

5.9 A structural steel torque box has a hollow rectangular section as shown in figure 5.39. The wall thickness is $\frac{1}{2}$ in. If the maximum allowable shear stress is 12 000 psi, what is the torque capacity of the box, and how many degrees of rotation would occur at the free end if the box were loaded to capacity?

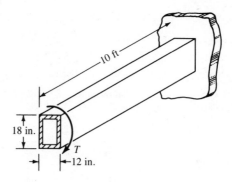

Figure 5.39 Illustrative problem 5.9.

Solution:
Solving for T in relationship 5.31,

$$T = 2At\tau$$

A is the inside area of the cross section and is equal to

$$17 \times 11 = 187 \text{ in.}^2$$

$$T = 2 \times 187 \text{ in.}^2 \times \frac{1}{2} \text{ in.} \times 12\ 000 \text{ psi}$$

$$T = 2.24\text{E6 lb} \cdot \text{in.}$$

From relationship 5.32,

$$\theta = \frac{TLs}{4GA^2t}$$

For most low-alloy steels, 11.5E6 psi is a good value for G.
  The peripheral length of the wall equals $18 + 12 + 18 + 12$ in.

$$\theta = \frac{2.24\text{E6 lb} \cdot \text{in.} \times 10 \text{ ft} \times 60 \text{ in.} \times 12 \text{ in./1 ft}}{4 \times 11.5\text{E6 lb/in.}^2 \times (187 \text{ in.}^2)^2 \times \frac{1}{2} \text{ in.}}$$

$$\theta = 0.02 \text{ radian}$$

The addition of the conversion factor of 12 in./1 ft in the previous calculation caused all units to cancel, giving the angle in radians. To change the units to degrees,

$$\theta = 0.02 \text{ radian} \times \frac{57.3 \text{ degrees}}{1 \text{ radian}}$$

$$\theta = 1.15 \text{ degrees}$$

The torque box may safely carry 2.24 million lb · in. of torque. The expected angular deflection under this amount of torque is 1.149 degrees.

PROBLEM SET 5.5

5.46 A structural torsion member is fabricated from 6061-T6 aluminum sheet 2 mm in thickness. The sheet is formed and welded into a closed square section 2 cm on the outside. If the yield strength in shear is 135 MPa, what torque could be carried before yield could occur?

5.47 What is the shearing stress in a 4-in.-square structural steel section with a wall thickness of 0.250 in. under a torque of 80 000 lb · in.?

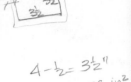

5.48 How much will a hollow rectangular section of cast magnesium 3.0 mm thick and 90 cm long deform under a torque of 100 N · m? The outside dimensions are 10 cm by 5 cm. (G = 43E9 Pa.)

5.49 For a round tube where the wall thickness is one-tenth the diameter, what percent difference in angular deformation under torque would be calculated by using relationship 5.32 rather than 5.25?

5.50 Olympic gold medalist Franz Klammer of Austria uses a slalom ski with a torsion box to give better edgehold on high-g turns. If the fiberglass torsion box, as shown in figure 5.40, has outside dimensions of 10 cm by 3 cm and a wall thickness of 5 mm, and it contributes 1.08 N · m/degree of torsion rigidity to the ski over its 205-cm length, what is the effective modulus of rigidity of the fiberglass material?

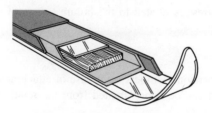

Figure 5.40 Problem 5.50.

SUMMARY

☐ The magnitude of the *moment* due to a force is the product of the force times its moment arm.

(5.1) $M = Fl$

☐ The *moment arm* is the perpendicular distance between the line of action of the force and the moment axis.

☐ For static equilibrium the summation of moments about any conceivable axis must be equal to zero.

(5.2) $\sum M = 0$

☐ A *couple* is a pure twist caused by two equal forces acting on parallel lines of action in opposing directions. The moment of a couple is equal to the magnitude of one of the forces times the perpendicular distance between the two lines of action. Sometimes it is useful to resolve a force into an equivalent force and a couple.

□ The *polar moment of inertia* is used as a geometric property of the cross sections of solid and hollow shafts.

(5.19) $J = \dfrac{\pi D^4}{32}$ for solid shafts

(5.20) $J = \dfrac{\pi}{32}(D^4 - d^4)$ for hollow shafts

□ Torsion produces shear stresses in circular shafts. The maximum stress occurs at the outer surface of the shaft.

(5.21) $\tau_{\max} = \dfrac{TR}{J}$ for both solid and hollow shafts

□ The deformation of a circular shaft loaded in torsion is the angle:

(5.25) $\theta = \dfrac{TL}{JG}$ for solid and hollow shafts

□ Relationships 5.19–5.32 apply only for shafts of circular cross sections and for materials in the elastic range.

□ Torsion-bar springs have a torsional spring constant of:

(5.26) $K_t = \dfrac{T}{\theta}$

□ The ultimate shear stress is related to the failure torque of a solid circular shaft of a ductile material by:

(5.29) $\tau_u = \dfrac{12T}{\pi D^3}$

□ Hollow box-shaped torsion members offer ease of attachment, lightness, and rigidity. Stress and deflection can be approximated from the relationships:

(5.31) $\tau = \dfrac{T}{2At}$

(5.32) $\theta = \dfrac{TLs}{4GA^2t}$

REFERENCES

1. Peterson, R. E. *Stress Concentration Design Factors*. New York: John Wiley & Sons, 1953.

2. Den Hartog, J. P. *Advanced Strength of Materials*, New York: McGraw-Hill, 1952.

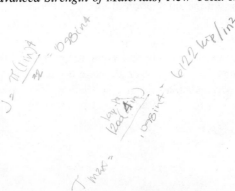

CHAPTER SIX

Angled Structures

The study of straight-line axially loaded members in the preceding chapters forms the basis for an understanding of more sophisticated structures. Directional properties of forces are now considered. The parallelogram law for resolving vectors is introduced and used to determine resultants and rectangular components. Graphical methods are discussed to aid visualization and to provide historical context. Trigonometry is introduced, and the method of summation of components is used for multiforce and non-orthogonal systems. The use of computers is demonstrated in an illustrative problem.

An understanding of the directional properties of forces leads naturally to the consideration of equilibrium of concurrent planar force systems. The free-body diagram concept is extended to include more kinds of forces due to supports and illustrated with applications from civil and mechanical engineering technology. After the special case of concurrent systems is discussed the general case of planar equilibrium is addressed, including two-force and bending members. Beginning with the inclined plane, friction is explored, both as a beneficial means of holding structural members together and as an unwelcome impediment to the motion of mechanisms. The chapter ends with the addition of friction forces to free-body diagrams.

6.1 STRUTS, GUYS, PROPS, AND BOOMS

The structures and members described in earlier chapters had forces with lines of action coinciding with the axis of the member. They illustrate the most efficient way to support a load: to get directly under it with a compression member or to suspend it from above with a tension member. But to do this, a stable foundation or attaching point must exist directly on the line of action of the load. Most often we are not so fortunate as to have the lines of action of the loads pass directly through the foundation area or the structural pick-up point. In some cases, the load can be carried by putting straight-line tension or compression members at angles to the load, in the form of struts, guys, props, or booms. These are angled structures.

6.2 DIRECTIONAL PROPERTIES OF FORCES AND MOMENTS AND VECTOR QUANTITIES

In order to understand angled structures, it is necessary to first understand the directional properties of forces.

Forces are *vector* quantities. This means that they have both magnitude and direction. Just as the magnitude of forces makes a difference upon a structure, so does the direction of the forces.

On the other hand, *scalar* quantities such as temperature and mass have no direction, only magnitude. Up to this chapter, forces were treated essentially as scalar quantities, because their direction was always obvious and had the same line of action as other forces and stresses under consideration. Now, as we consider angled structures, it is necessary to incorporate the angles of the lines of action of the forces.

Vector quantities can be graphically represented by an arrow. The length of the arrow indicates magnitude, and the angle shows the direction. In the case of forces, the length of the arrow indicates the number of pounds, newtons, kips, kilograms, or tons, and the angle shows the direction of action of the force.

Moments of forces are also vector quantities. The magnitude in lb · in., kip · ft, N · m, or kg · cm can be represented by the length of the arrow. The direction of the arrow is the direction of the moment axis according to the right-hand rule as shown in figure 6.1.

6.3 RESULTANT OF TWO FORCES OR MOMENTS

The combined effect of two vector quantities with intersecting lines of action is a third vector that forms the diagonal of the parallelogram constructed from the two vectors, as shown in figure 6.2. This combination is sometimes called the *parallelogram law*.

The combined effect of the two vectors is called the *resultant*. Applying the definition specifically to the case of forces, we can say that *the resultant is a single force that has the same effect as two or more forces*.

ILLUSTRATIVE PROBLEM

6.1 Two cables are attached to an eyebolt. The cables are at angles of 15° and 75° from the horizontal, as shown in figure 6.3, and carry tensions of 2 000 and 3 000 lb, respectively. What is the resulting force on the eyebolt?

Solution:
A graphical solution can be used to solve this problem. The two cable forces are drawn to a scale of your choice as vectors in the proper direction, as shown in figure 6.3b.

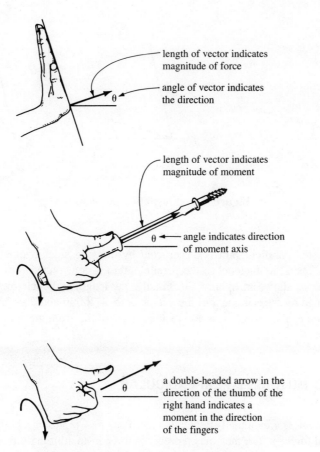

Figure 6.1 Representation of forces and moments by vectors.

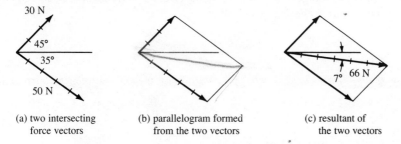

(a) two intersecting
force vectors

(b) parallelogram formed
from the two vectors

(c) resultant of
the two vectors

Figure 6.2 The parallelogram law. The resultant (c) has the same effect as the two forces forming intersecting sides of the parallelogram.

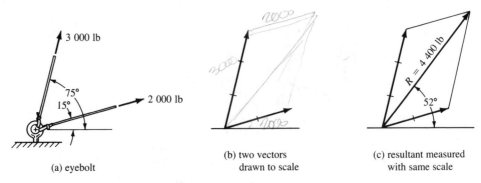

(a) eyebolt

(b) two vectors
drawn to scale

(c) resultant measured
with same scale

Figure 6.3 Illustrative problem 6.1.

Next, the parallelogram is constructed by drawing sides parallel to the original vectors. Then the diagonal of the parallelogram is constructed. This diagonal is the resultant, as shown in figure 6.3c. Finally, the length and angle of the resultant are determined by measurement of the vector. $R = 4\ 400$ lb and $\theta = 52°$. The resultant of the two eyebolt forces is 4 400 lb at an angle of 52° from the horizontal, up and to the right.

6.4 RESULTANT OF TWO MUTUALLY PERPENDICULAR VECTORS

Frequently, vectors are mutually perpendicular. They may represent a horizontal force, H, and a vertical force, V. Or, they may represent a force in an arbitrary direction, X, and one in a direction perpendicular, Y. When the forces are mutually perpendicular, the Pythagorean theorem can be used to determine the magnitude of the resultant, as follows:

$$R = \sqrt{V^2 + H^2} \quad \text{or} \quad R = \sqrt{X^2 + Y^2} \tag{6.1}$$

The definition of the tangent of an angle can be used to determine the acute angle, θ, from the horizontal or from the x axis as follows:

$$\tan \theta = \frac{V}{H} \quad \text{or} \quad \frac{Y}{X} \tag{6.2}$$

The angle, θ, can then be evaluated by determining the inverse tangent, also called the arctangent.

ILLUSTRATIVE PROBLEM

6.2 An upward vertical force of 13 500 N and a horizontal force of 21 300 N to the left act upon a foundation, as shown in figure 6.4. What is the resultant force upon the foundation?

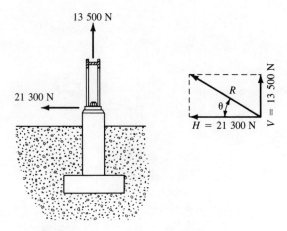

Figure 6.4 Illustrative problem 6.2.

Solution:
The parallelogram of forces in this case is a rectangle, because the forces are mutually perpendicular. From the Pythagorean theorem, relationship 6.1,

$$R = \sqrt{V^2 + H^2}$$
$$R = \sqrt{(13\ 500\ N)^2 + (21\ 300\ N)^2}$$
$$R = 25\ 200\ N$$

And from 6.2,

$$\tan \theta = \frac{V}{H} = \frac{13\ 500}{21\ 300} = 0.6338$$

$$\theta = 32.4°$$

The resultant force is 25 200 N up and to the left at an angle of 32.4° with the horizontal.

6.5 COMPONENTS OF FORCES

Components of a force are forces that have the same resultant as the force in question. In a plane, components of a force always come in pairs. The two directions of the pair of components of interest must always be defined.

Usually the directions of interest are perpendicular. Examples are the vertical and horizontal components or the parallel and normal (perpendicular) components shown in figure 6.5. When the components are perpendicular, they can be found from the definition of the sine and cosine of an angle as follows:

$$V = F \sin \theta \text{ and } H = F \cos \theta \quad \text{or} \quad Y = F \sin \theta' \text{ and } X = F \cos \theta' \quad \textbf{(6.3)}$$

where θ is the angle from the horizontal or the x axis. Alternatively, θ' is the angle from the y axis.

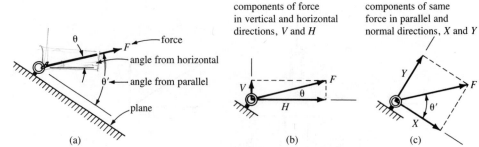

$V = F \sin \theta$
$X = F \cos \theta$

(a)   (b)   (c)

Figure 6.5   Perpendicular components of the force, F: (b) vertical and horizontal and (c) parallel and normal.

ILLUSTRATIVE PROBLEM

6.3 A force of 105 kN is applied to a large concrete block as shown in figure 6.6. What force tends to lift the block, and what is the tendency to slide the block along the ground?

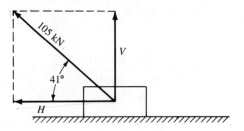

Figure 6.6   Illustrative problem 6.3.

Solution:
The lifting tendency and sliding tendency are the vertical and horizontal components, respectively. In this case, the vertical and horizontal components are identical to the normal and parallel components.

From relationship 6.3,

$$V = F \sin \theta \text{ and } H = F \cos \theta$$
$$V = 105 \text{ kN } \sin 41° \text{ and } H = 105 \text{ kN } \cos 41°$$
$$V = 68.9 \text{ kN and } H = 79.2 \text{ kN}$$

The uplifting force is 68.9 kN. The sliding force is 79.2 kN to the left.

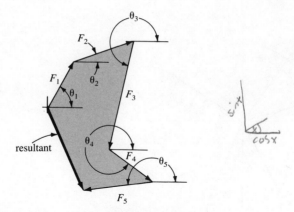

Figure 6.7 A string polygon for the resultant of five forces, F_1 through F_5. Note that the resultant vector is placed tail to tail and head to head.

6.6 THE STRING POLYGON

The string polygon is a graphical method of determining the resultant of several forces in a plane. As shown in figure 6.7, the force vectors are placed one by one, head to tail, in any order. As with the parallelogram method, the lengths of the vectors indicate the magnitude of the forces, and the angles of the vectors indicate the directions of the forces. When all the vectors have been placed, the diagram will be an incomplete, or open, polygon. The vector closing the polygon, placed tail to tail and head to head, is the resultant. The length and angle of the closing vector is the length and direction of the resultant force.

6.7 RESULTANTS BY THE SUMMATION OF COMPONENTS

Computers and hand-held calculators have rendered all graphical methods and some analytical methods obsolete. The method of summation of components is the most practical method of finding resultants. In the summation-of-components method, all forces are reduced to equivalent perpendicular components. The Pythagorean theorem is then used to find the resultant. This method is applicable when more than two forces are involved and when the forces are not at right angles to one another.

Consider a set of forces, F_1, F_2, F_3, F_4, and so on to F_i, as shown in figure 6.8. The net vertical effect of these forces is the summation of all the vertical components. The net horizontal effect is the summation of all the horizontal components. Using Σ to indicate the summation,

$$\Sigma V = \Sigma F_i \sin \theta_i \quad \text{and} \quad \Sigma H = \Sigma F_i \cos \theta_i \qquad (6.4)$$

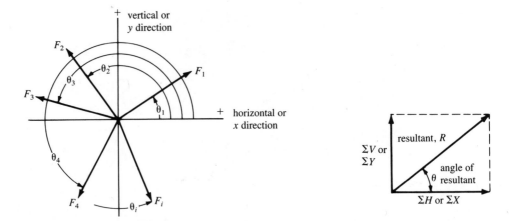

Figure 6.8 The summation-of-components method of computing resultants of any number of forces.

Or, this equation could be written in terms of any two mutually perpendicular directions, X and Y:

$$\Sigma\, Y = \Sigma\, F_i \sin\,\theta_i' \quad\text{and}\quad \Sigma\, X = \Sigma\, F_i \cos\,\theta_i' \qquad (6.5)$$

The summation reduces the forces to an equivalent rectangular system as shown in figure 6.8b. The Pythagorean theorem can then be applied to the equivalent system, where:

$$R = \sqrt{(\Sigma\, V)^2 + (\Sigma\, H)^2} \quad\text{or}\quad R = \sqrt{(\Sigma\, Y)^2 + (\Sigma\, X)^2} \qquad (6.6)$$

The angle of the resultant can be computed from

$$\tan\,\theta = \frac{\Sigma\, V}{\Sigma\, H} \quad\text{or}\quad \frac{\Sigma\, Y}{\Sigma\, X} \qquad (6.7)$$

ILLUSTRATIVE PROBLEMS

6.4 The shear forces shown in figure 6.9 act upon a bolt in a pattern. The 1 000-kg force is due to a twisting action on the bolt pattern. The vertical force is due to an overall downward force on the bolt pattern. What are the magnitude and direction of the resultant shearing force on the bolt?

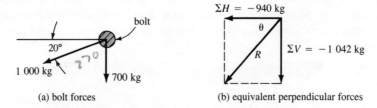

Figure 6.9 Illustrative problem 6.4.

Solution:

First, label the 1 000-kg force F_1 and the 700-kg force F_2. Then a table of components can be made using relationship 6.3.

Force	Magnitude (kg)	Direction	V (kg)	H (kg)
F_1	1 000	200°	−342	−940
F_2	700	270°	−700	0
			Σ −1 042	−940

A vertical force directly downward of 1 042 kg and a horizontal force to the left of 940 kg shown in figure 6.9b would have the same effect as the original forces.

From relationships 6.6 and 6.7,

$$R = \sqrt{(\Sigma\ V)^2 + (\Sigma\ H)^2} = \sqrt{(-1\ 042\ \text{kg})^2 + (-940\ \text{kg})^2}$$
$$R = 1\ 403\ \text{kg}$$
$$\tan\theta = \frac{\Sigma\ V}{\Sigma\ H} = \frac{-1\ 042}{-940} = 1.109$$
$$\theta = 48°$$

The angle θ is always the angle from the horizontal or from the x axis. In order to communicate with the widest variety of people, it is best to express the angle as a positive acute angle with a diagram such as in figure 6.9b showing the direction. Or, express the direction in words such as: "The resultant shear force on the bolt is 1 043 kg acting down and to the left at an angle of 48° from the horizontal."

6.5 The member forces shown in figure 6.10 act upon the structural joint. What is the resultant of the member forces?

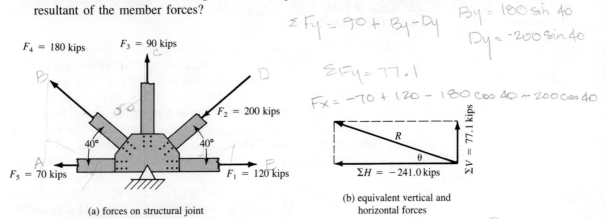

(a) forces on structural joint

(b) equivalent vertical and horizontal forces

Figure 6.10 Illustrative problem 6.5.

Solution:

The first step is to label the forces F_1 through F_5. Then, a table of vertical and horizontal components can be constructed.

	Force (kips)	Angle	V (kips)	H (kips)
F_1	120	0°	0	120.0
F_2	−200	40°	−128.6	−153.2
F_3	90	90°	90	0
F_4	180	140°	115.7	−137.9
F_5	70	180°	0	−70
			Σ 77.1	−241

The five forces have a net vertical effect of 77.1 kips upward and a net horizontal effect of 241 to the left as shown in figure 6.10b.

From relationships 6.6 and 6.7,

$$R = \sqrt{(\Sigma\ V)^2 + (\Sigma\ H)^2} = \sqrt{(77.1\ \text{kg})^2 + (-241\ \text{kg})^2}$$

$$R = 253\ \text{kips}$$

$$\tan\theta = \frac{\Sigma\ V}{\Sigma\ H} = \frac{77.1\ \text{kips}}{-241\ \text{kips}} = -0.320$$

Since both the angles of 162.3° and 342.3° have tangents of −0.320, it is best to use the equivalent rectangular system in figure 6.10b to determine the direction of the force. Such a diagram shows that the resultant must act up and to the left. Therefore, 162.3° is the correct answer. If the negative sign of the tangent of θ is ignored, θ will be the acute angle from the nearest horizontal. It is usually best to convert to an acute angle and clarify such as follows: "The resultant of the member forces upon the structural joint is 253 kips up and to the left at an angle of 17.7° from the horizontal."

6.8 COMPONENTS OF FORCES WHEN THE DIRECTION IS DEFINED IN TERMS OF THE SLOPE

Forces are often in the direction of structural members. Directions of structural members can normally be found from horizontal and vertical distances between various points on a structure. These are usually given on drawings and blueprints. Angles are rarely indicated because they are more difficult to measure accurately in the field than are vertical and horizontal distances. Of course, a vertical and horizontal distance can always be used to compute an angle, but this method is indirect. It is better to use the vertical and horizontal as a slope in the direct computation of vertical and horizontal components.

The slope of an inclined structural member is the vertical rise divided by the horizontal run.

For example, the slope of the member AB in the structural drawing of figure 6.11 can quickly be determined to be 7.2/10.8. It can be expressed in that form and need not be converted to an angle or the decimal equivalent. A *reference triangle* with a vertical side of 7.2 and a horizontal side of 10.8 can be drawn on the member.

When the direction of a force is defined in the form of a slope or a vertical and a horizontal distance, the vertical and horizontal components can be determined from the law of similar triangles, as shown in figure 6.12.

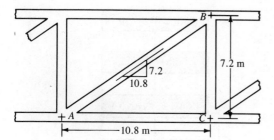

Figure 6.11   Slope of an inclined structural member.

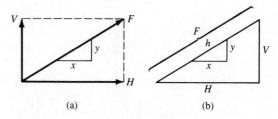

(a)

(b)

Figure 6.12   Using the law of similar triangles to compute vertical and horizontal components when the slope of a force is known.

First, h, the hypotenuse of the reference triangle, needs to be computed.

$$h = \sqrt{x^2 + y^2} \qquad (6.8)$$

where x is the horizontal side of the reference triangle, and y is the vertical side.

Then, using the law of similar triangles (that corresponding sides are proportional), the ratios in figure 6.12b can be written as:

$$\frac{V}{F} = \frac{y}{h} \quad \text{and} \quad \frac{H}{F} = \frac{x}{h}$$

where F is the force, H is the horizontal component, and V is the vertical component. Therefore:

$$V = y(F/h) \quad \text{and} \quad H = x(F/h) \qquad (6.9)$$

Note that the components are easily calculated by evaluating the ratio of F/h. For the vertical component, that ratio is multiplied by the vertical side of the reference triangle. To get the horizontal component, the same ratio is then multiplied by the horizontal side of the reference triangle.

ILLUSTRATIVE PROBLEM

6.6 Write a computer program in the BASIC language to solve for the resultant of several forces with the directions given in terms of slopes. Use the program to solve for the resultant of the four forces shown in figure 6.13a. (This problem is intended for those who have a background in BASIC or some experience with personal computers.)

Solution:
A computer program in BASIC can be coded and run on most personal computers. However, slight variations will be found among machines and versions of BASIC.

Since BASIC does not recognize Greek letters, SX is used in the program for $\Sigma\, X$, SY for $\Sigma\, Y$, and THETA for θ. Because lower-case letters are also not recognized by BASIC, H, X, and Y are used for h, x, and y, respectively, as the variables in relationship 6.9.

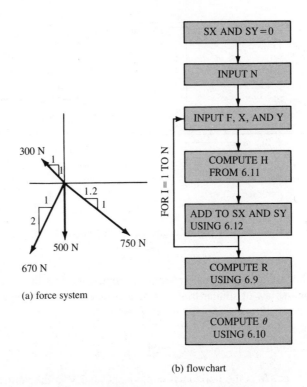

(a) force system

(b) flowchart

Figure 6.13 Illustrative problem 6.6.

The program begins by setting SX and SY equal to zero, as shown on the flowchart in figure 6.13b. Then the number of forces, indicated by N, must be entered.

For each force vector in turn, the program must be given the values of F, x, and y, from which it computes h from relationship 6.8 and uses relationship 6.9 to add that vector's components to SX and SY. When it has gone through this process N times, SX and SY represent ΣX and ΣY for the entire force system. The program then computes R and θ using relationships 6.6 and 6.7.

Program:
```
100 SX = 0: SY = 0
110 INPUT "HOW MANY FORCE VECTORS IN THE SYSTEM?"; N
120 FOR I = 1 TO N
130 INPUT "F=", F
140 INPUT "X=", X
150 INPUT "Y=", Y
160 H = SQR(X*X + Y*Y)
170 SX = SX + X*(F/H): SY = SY + Y(F/H)
180 NEXT I
190 R = SQR(SX*SX + SY*SY)
200 THETA = ATN(SY/SX)*57.3
210 IF THETA > 0 AND SY < 0 THEN THETA = THETA + 180
220 IF THETA < 0 AND SY > 0 THEN THETA = THETA + 180
```

230 PRINT "THE MAGNITUDE OF THE RESULTANT IS ",R
240 PRINT "THE DIRECTION IS ",THETA
250 GO TO 100
260 END

Use the program on the system of figure 6.13a.

The number 4 must be entered for N in this problem. Then each of the four forces, in any order, must be entered followed by their x and y, on cue. The X and Y entries for the 300-N force, for example, are -1 and 1, respectively.

For the 500-N force, which acts vertically downward, 0 must be entered for x and any negative number, such as -1, for y.

On an 8-bit personal computer, the program will yield the following:

$$R = 1\ 368.79$$
$$THETA = -87.3097$$

The resultant of the force system of figure 6.13a is 1 369 N at an angle of 87.3° from horizontal, down and to the right.

PROBLEM SET 6.1

6.1 Using a scale and a protractor, make parallelograms on separate sheets of paper to find the resultants of the force pairs shown in figure 6.14.

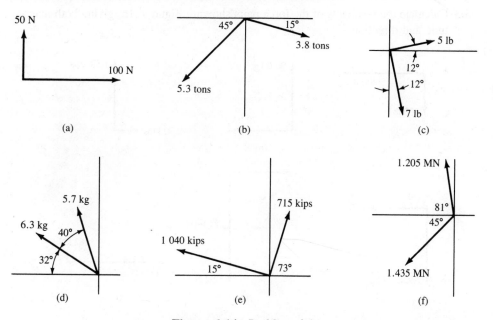

Figure 6.14 Problem 6.1.

6.2 Using a scale and a protractor, make parallelograms on separate sheets of paper to find the vertical and horizontal components (or the components in the x and y directions as indicated) shown in figure 6.15.

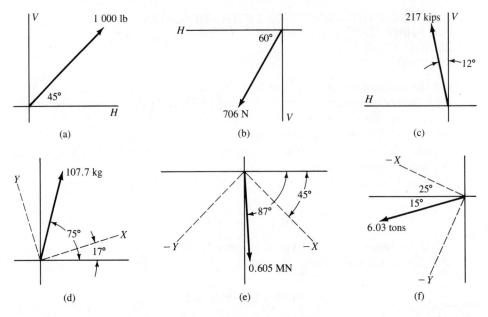

Figure 6.15 Problem 6.2.

6.3 Calculate the resultants of the force pairs shown in figure 6.16, giving both magnitude and direction.

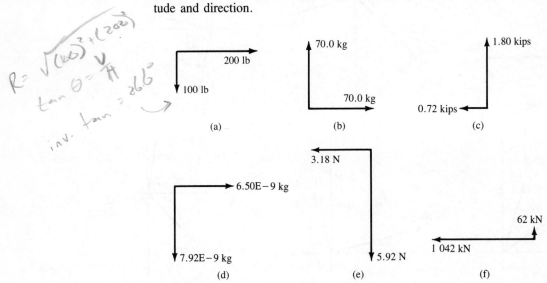

Figure 6.16 Problem 6.3.

6.4 Calculate the vertical and horizontal components of the forces shown in figure 6.17.

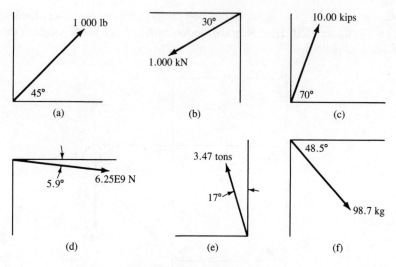

Figure 6.17 Problem 6.4.

6.5–6.7

Find the resultants of the force pairs in figure 6.18, using the method of summation of components. Give magnitude and direction.

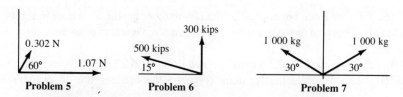

Figure 6.18 Problems 6.5 through 6.7.

6.8–6.10

Find the resultants of the force systems in figure 6.19, giving both magnitudes and directions.

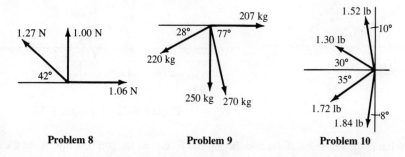

Figure 6.19 Problems 6.8 through 6.10.

6.11 What is the worst shear stress on any bolt of the pattern of four 1-in.-diameter bolts shown in figure 6.20? Hint: Resolve the force into a force and a couple at the center of the bolt pattern.

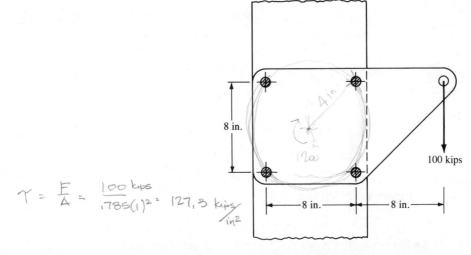

$$\tau = \frac{F}{A} = \frac{100\ \text{kips}}{.785(1)^2} = 127.3\ \frac{\text{kips}}{\text{in}^2}$$

Figure 6.20   Problem 6.11.

6.12 When 50 lb of force are applied to the lawnmower handle as shown in figure 6.21, what force tends to move the mower and what force tends to push the mower into the ground?

6.13 All of the guy wires in the antenna shown in figure 6.22 are pre-tensioned to a load of 10.0 kN. What is the resultant force on the antenna due to pre-tensioning?

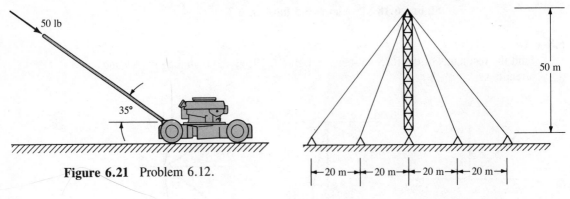

Figure 6.21   Problem 6.12.

Figure 6.22   Problem 6.13.

6.14 What is the resultant of the connecting rod forces on the rod bearing, *A*, of the crank of the rotary compressor shown in figure 6.23? The connecting rod forces for the cylinders 1 through 5 are 2 390, 7 850, 1 560, 750, and 1 030 lb compression, respectively.

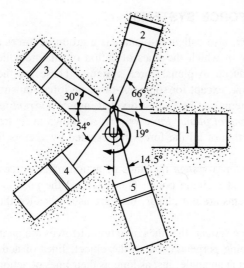

Figure 6.23 Problem 6.14.

6.15 A right-angle drive gear reducer has an input torque of 1 600 lb · in. and an output torque of 4 800 lb · in. What overturning moment must be resisted by the bolt pattern? The gear box is shown in figure 6.24. Hint: Represent the torques by moment vectors and resolve into a resultant torque.

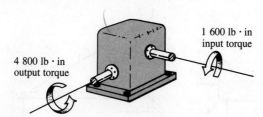

Figure 6.24 Problem 6.15.

6.16–6.17

What are the resultants of the force systems shown in figure 6.25? Use the summation-of-components method. Verify using the string polygon.

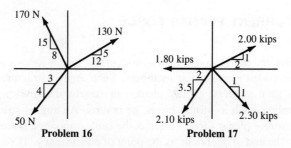

Figure 6.25 Problems 6.16 and 6.17.

6.9 EQUILIBRIUM OF PLANAR FORCE SYSTEMS

A *planar force system,* also called co-planar, is a group of forces acting together on an object or a free body in which the lines of action of all forces lie in one plane. Most structures can be reduced to planar force systems. These are the only force systems considered in this book, except for some out-of-plane forces on structures in chapter 15.

Planar force systems can be further divided into *concurrent force systems, parallel force systems,* and *general planar force systems* as shown in figure 6.26. In all of these cases, the lines of action of all the forces represented by vectors lie in the plane of the page.

In a *concurrent force system* the lines of action of all forces intersect, or can be assumed to intersect, at a single point. This is called the *point of concurrency.* When concurrent force systems are not planar, they are usually called three-dimensional force systems.

In a *parallel force system,* the lines of action of forces are parallel. Although shown in figure 6.26b as being perpendicular to the object, lines of action are not always so aligned. They may act at an angle, just as long as their lines of action are parallel. Parallel force systems may not be planar. In fact, there are many important cases where parallel forces do not share the same plane. Parallel force systems are considered in chapter 9.

The lines of action in a *general planar force system* (figure 6.26c) are not parallel, nor do they intersect at a single point. Hence, this is the general case of planar forces, as opposed to concurrent and parallel, which are special cases.

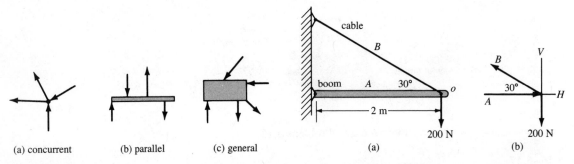

(a) concurrent	(b) parallel	(c) general	(a)	(b)

Figure 6.26 Types of planar force systems.   **Figure 6.27** Illustrative problem 6.7.

6.10 EQUILIBRIUM OF CONCURRENT PLANAR FORCE SYSTEMS

Many angled structures can be considered as concurrent planar force systems to illustrate how they distribute load among their members. Such angled structures may consist of tension members such as guys, cables, chains, or structural shapes and be fastened to compression members such as struts, props, or booms. An angled structure that can be reduced to a concurrent planar force system is the cable-supported boom shown in figure 6.27a. Point *o*, at the end of the boom, is the point of concurrency. The cable, *B*, provides a tension force to the load point, *o*. The boom, *A*, provides a compression force.

The first step is always to draw a free-body diagram which, as discussed in chapter 1, is a diagram of a body free of its supports where the supports have been replaced by the forces that the supports provide. At the time the free body is drawn, no one knows what forces the supports provide—only capability. In this case the cable, B, and the boom, A, are the supports of the point o. The *body* in a concurrent force system consists of the point, and all forces go through this point. In the free-body diagram (figure 6.27b), A and B are removed and replaced by vectors representing their forces. Since the cable pulls on the point, force B is drawn acting away from o. Because the boom pushes on the point, force A is drawn pushing on the point o.

After drawing the free-body diagram, determine how many and which equilibrium conditions apply. In previous chapters, force equilibrium along a single line was considered. But, in concurrent planar systems, two equilibrium conditions must be satisfied simultaneously. Forces in both the vertical and the horizontal directions must be in balance. Or, alternatively, forces in any two perpendicular directions, X and Y, must be in balance. Balanced forces in two mutually perpendicular directions, whether V and H or X and Y, will hold the point of concurrency at a static or still position in the plane.

The equilibrium conditions that apply are:

$$\Sigma\,V = 0 \text{ and } \Sigma\,H = 0 \quad \text{or} \quad \Sigma\,X = 0 \text{ and } \Sigma\,Y = 0 \qquad \textbf{(6.10)}$$

ILLUSTRATIVE PROBLEMS

6.7 The boom, A, and the cable, B, form a structure to suspend a 200-N load 2 m from a wall as shown in figure 6.27a. What are the axial loads in the members?

Solution:
The point of concurrency is the point o. A free-body diagram of the joint is shown in figure 6.27b.

The point is held static by members A and B. In the free-body diagram, members are replaced by equivalent forces. Member B would be expected to go into tension; therefore, it is drawn pulling on the joint. Member A would be compressed. Therefore, force A is shown pushing on the joint. Using the equilibrium conditions for a concurrent planar force system, relationship 6.10,

$\Sigma\,V\updownarrow:$

$$B_V - 200 \text{ N} = 0$$
$$B \sin 30° - 200 \text{ N} = 0$$
$$B = \frac{200 \text{ N}}{\sin 30°} = 400 \text{ N}$$

$\Sigma\,H\leftrightarrow:$

$$A - B_H = 0$$
$$A - B \cos 30° = 0$$
$$A = B \cos 30°$$

Substituting the above value of B,

$$A = 400 \text{ N} \cos 30° = 346 \text{ N}$$

The cable, B, carries 400 N of tension force. The boom, A, carries 346 N of compressive force.

6.8 A capping head for a bottling machine is actuated by a pneumatic cylinder through the mechanism shown in figure 6.28. The capping head applies a downward force of 1 500 N on the bottle when the mechanism is in its static position at the bottom of its stroke as shown. Since the mechanism is well lubricated, friction is neglected. What is the compression force in arm B?

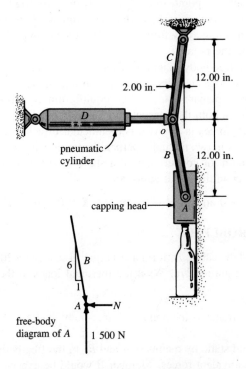

Figure 6.28   Illustrative problem 6.8.

Solution:

Mechanisms can be thought of as structures of variable configuration. When movements are significant, complicated dynamic forces must be considered. In this case, however, the mechanism is performing the static task of sealing a cap, during which force is more significant than motion.

Both the known vertical force and the compression in arm B pass through point A on the capping head. Point A is the point of concurrency and must be diagrammed as a free body as shown in figure 6.28.

Since the head pushes vertically downward upon the bottle with a force of 1 500 N, the bottle in turn pushes upward with the same amount of force, which is entered as a vertical vector of 1 500 N in the free-body diagram.

Arm B has a slope of 6 over 1. The arm pushes downward and to the right on the head and is represented on the free-body diagram by the sloping vector.

The force, N, is the force between the capping head and the slide. Since friction is being neglected, the force from the slide consists only of the normal, or perpendicular, force (as discussed in section 4.9 on friction). The force, N, is included on the free-body diagram as a horizontal force pushing upon point A from the right.

From relationship 6.8:

$$h = \sqrt{1^2 + 6^2} = 6.083$$

From the equilibrium relationships 6.9 and 6.10:

(a) $$\Sigma V \updownarrow: \quad -\frac{6}{6.083}B + 1\,500 = 0$$

Solving for B:

$$B = 1\,500 \times \frac{6.083}{6} = 1\,521 \text{ N}$$

The compression force in arm B is 1 521 N. The force, N, can be determined from the ΣH, but is not required for this problem.

Since mechanisms of this type are commonly used as force-multiplying devices, it is interesting to calculate the force required of the pneumatic cylinder. This can be done, knowing the force in arm B, with a free-body diagram of point o and solving the equilibrium equations.

PROBLEM SET 6.2

6.18–6.19

What are the forces in the members A and B of the concurrent planar structures shown in figure 6.29? Are they tension or compression forces?

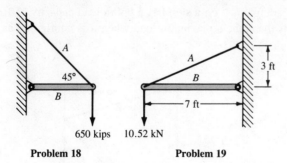

650 kips   10.52 kN

Problem 18   **Problem 19**

Figure 6.29   Problems 6.18 and 6.19.

6.20–6.21

What maximum loads, F, can the two-member structures shown in figure 6.30 support, if the booms can resist 5 000 kN before buckling in compression?

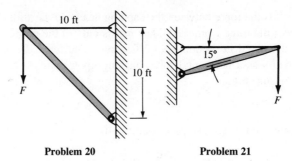

Problem 20   Problem 21

Figure 6.30   Problems 6.20 and 6.21.

6.22 A railroad spur bumper must be able to resist a horizontal force of 1.00E7 kg. What are the loads in the vertical member, B, and the diagonal prop, A, under such a load as seen in figure 6.31?

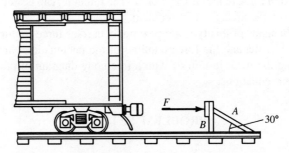

Figure 6.31   Problem 6.22.

6.23 The cargo-loading crane on a ship has a boom load capacity of 100 T in compression. What is the crane's maximum load when it is configured as shown in figure 6.32?

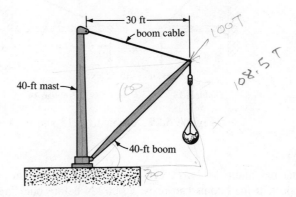

Figure 6.32   Problem 6.23.

6.24 The tripod shown in figure 6.33 supports a mass of 650 kg. What is the compressive force in newtons in the legs? Hint: The three legs share the load equally. Why?

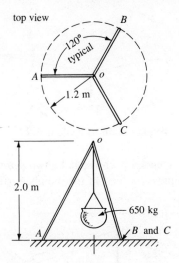

Figure 6.33   Problem 6.24.

6.25 The linkage shown in figure 6.34 carries a vertical load of 15 kips. What must be the diameters of the threaded steel rods, if NF threads are used and the allowable stress in the steel is 22 000 psi?

6.26 Write a program for a personal computer or programmable calculator that solves for the forces in any two-member structure that can be idealized as a concurrent planar force system (figure 6.35).

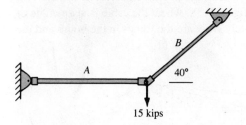

Figure 6.34   Problem 6.25.

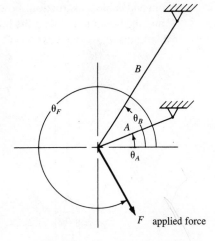

Figure 6.35   Problem 6.26.

6.27 A load of 25 T is supported at the mid-span of a cable as shown in figure 6.36. What is the tension in the cable when the angle from horizontal, θ, is (a) 15°, (b) 5°, (c) 1°, and (d) 0.1°?

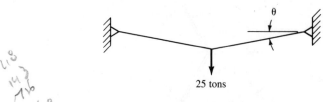

25 tons

Figure 6.36 Problem 6.27.

6.28 A load of 160 kN is suspended over a span by two symmetrical compression members. What is the compression load in the members (figure 6.37)?

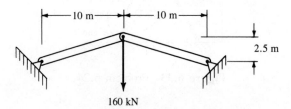

160 kN

Figure 6.37 Problem 6.28.

6.29 A parachute ride at an amusement park drops parachute and basket assemblies weighing, with passengers, 800 lb. The arms from which the lifting cable is suspended are supplied with cable brakes. The worst case of loading occurs when the cable brakes are applied in an emergency, causing a downward dynamic load in the arms four times the static weight of the parachute and assembly. The arms are of boom and cable construction shown in figure 6.38, where the cable is at an angle of 7° and the boom is 15° with the horizontal. What are the forces in the boom and the cable under the worst-case load?

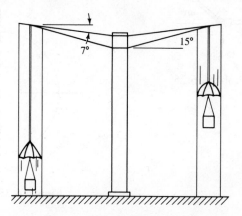

Figure 6.38 Problem 6.29.

6.30 The ski lift chair in figure 6.39 carries a vertical load of 150 kg and moves to the left. What is the tension in the cable forward of the chair and what is the tension to the rear?

15° 5°

150 kg

Figure 6.39 Problem 6.30.

6.11 THE GENERAL CASE OF PLANAR EQUILIBRIUM

Forces in a vertical plane can cause an object not only to move vertically or horizontally but also to rotate. In concurrent force systems considered in section 6.10, all forces act through a single point. When forces are concurrent, there is no rotational tendency. But when the lines of action of the forces intersect at more than one point on the object, moments of force can result that tend to make the object rotate. Therefore, to achieve static equilibrium in the general planar force system, not only must the forces in two mutually perpendicular directions (V and H or X and Y) be in balance, but also the *moments* of all forces must be balanced. This is the general case of planar equilibrium. The concurrent system discussed in the previous section is actually a special case.

$$\Sigma\, V = 0,\ \Sigma\, H = 0,\ \text{and}\ \Sigma\, M = 0 \quad \text{or} \quad \Sigma\, X = 0,\ \Sigma\, Y = 0,\ \text{and}\ \Sigma\, M = 0 \quad \textbf{(6.11)}$$

As an example, consider the sign shown supported by the angled guy wire in figure 6.40.

In the free-body diagram of the sign, the supports to be "removed" are the guy wire at B and the pin at A. The guy wire is capable of providing only a tension force at B. This

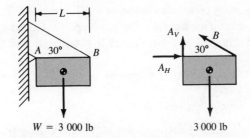

$W = 3\,000\ \text{lb}$ $3\,000\ \text{lb}$

Figure 6.40 The guyed sign is a general planar force system. Note that the lines of action of the forces in the free-body diagram do not intersect at a single point.

force will be in the direction of the wire. Consequently, a force, B, in the direction of the wire is included on the free-body diagram.

The pin at A can resist a plane force in any direction. Such a force will have vertical and horizontal components. Since neither the force nor the direction is known, the pin support at A is replaced on the free-body diagram with a pair of unknown components, A_V and A_H. These components, when solved, will give the direction and magnitude of the pin force.

No single point can be found where the lines of action of all forces intersect. In fact, five points can be found where lines of action of *pairs* of forces intersect. Therefore, this example is not a concurrent force system. It is instead a general planar force system. Therefore, all three equilibrium conditions of relationship 6.11 must be satisfied if this sign is to be in static equilibrium.

The moment condition is satisfied if equilibrium exists about *any* moment axis. The plane of the forces is the plane of the paper upon which figure 6.40 is printed. A moment axis must be perpendicular to the plane of the paper. If moment equilibrium exists about an axis through point A, it will also exist about axes through B, the center of gravity of the sign, or about any other point in the plane on or off the sign. Moment equilibrium need only be satisfied about one axis or *moment center*.

In choosing the axis or moment center about which to sum moments, it is best to minimize the number of unknowns in the resulting equilibrium equation. Forces having lines of action passing through the moment center contribute nothing to the moment about that moment center. The moment arms of those forces are zero: They do not appear in the equilibrium equation. This knowledge can be used to locate a moment center that will eliminate the largest number of forces from the moment equation.

In the free-body diagram in figure 6.40, the best moment center is point A. The unknown forces, A_V and A_H, both pass through that point. Therefore, these unknowns contribute nothing to the moment about A and do not appear in the moment equation.

ILLUSTRATIVE PROBLEM

6.9 The sign in figure 6.40 weighs 3 000 lb. What is the force in the guy wire at B? What is the shear force on the pin at A?

Solution:
A free-body diagram is constructed as shown in figure 6.40. The sign is drawn as a free body removed from its supports. The pin support is replaced with the vertical and horizontal components, A_V and A_H. The cable is replaced with force B in the known direction of the cable.

The vertical and horizontal components of the cable force are $B \sin 30°$ and $B \cos 30°$, respectively.

Now, writing the equilibrium equations,

$$\Sigma \, V \updownarrow : A_V + B \sin 30° - 3\ 000 \text{ lb} = 0$$

(a)
$$A_V + 0.500B = 3\ 000 \text{ lb}$$

$$\Sigma \, H \leftrightarrow : A_H - B \cos 30° = 0$$

(b)
$$A_H = 0.866B$$

$\Sigma\ M_A\circlearrowright$: (The symbol $\circlearrowright$ means the summation of moments about point A, taking positive to be the clockwise direction.)

$$3\ 000 \times \frac{L}{2} - B \sin 30° \times L = 0$$

Dividing through by $-L$,

$$-3\ 000/2 + 0.500B = 0$$

(c) $$B = 3\ 000\ \text{lb}$$

Substituting back into **(b)**,

$$A_H = 0.866(3\ 000) = 2\ 598\ \text{lb}$$

And substituting back into **(a)**,

$$A_V + 0.500 \times 3\ 000 = 3\ 000$$
$$A_V = 1\ 500\ \text{lb upward}$$

The shear force in the pin is the resultant of the components $A_V + A_H$.

$$A = \sqrt{1\ 500^2 + 2\ 598^2} = 3\ 000\ \text{lb}$$

The cable tension is 3 000 lb, and the shear force in the pin is 3 000 lb.

6.12 SUPPORT IDEALIZATIONS FOR PLANAR STRUCTURES

Although support idealization may appear trivial, it is a very important concept in understanding structures. By idealizing supports, one determines the load paths for the structure. Probably more structures fail due to lack of an adequate number of load paths than for any other reason.

Guys

A guy is a *pure* tension member. It may be a wire, cable, chain, rope, string, or any member with little bending resistance. Such members buckle when compression load is applied. Therefore, they are considered to have only tension capability in the direction of their axis. Compression or bending resistance is negligible.

For some purposes, long slender rods or structural shapes may be idealized as guys. When this is done, there is a deliberate decision to neglect any compressive or bending resistance that the member may be able to contribute to the support of the structure.

A guy is shown on a free-body diagram as a force pulling on the body. The arrow of the vector always points away from the body. The direction of the force is in the direction of the centerline of the guy, where it attaches to the body. Such a diagram is shown in figure 6.41.

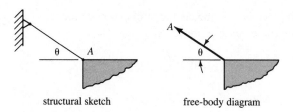

structural sketch   free-body diagram

Figure 6.41 Support idealization of a guy (chain, cable, wire, rope, or string). The direction, θ, is the same as the guy, but the magnitude, A, is unknown. A cable force may only pull on the body.

Links

A link is a member that is rigid but has freely rotatable connectors at its two points of attachment. A link has only two attachment points. A link may be a prop, strut, leg, or boom. It may be a structural shape or a machined part. Links can transmit compression as well as tension.

Links are idealized as having pinned connections at each end (figure 6.42). The pins are assumed to be frictionless. In fact, there is always friction. In some applications, the connection may be a rigid attachment such as a pattern of bolts, yet the link is assumed to have a frictionless pin connection. Such a grossly inaccurate idealization is justified when there is a deliberate decision to neglect the strengthening effect of the friction or the bending resistance of the member at the connection. Such strengthening effects may be secondary in importance and are sometimes forfeited in order to obtain a structure with more simple load paths.

A link is capable of transmitting forces (tension or compression) only in the direction of its axis. The effect of a link is shown on a free-body diagram by a force vector pushing or pulling in the direction of the link axis.

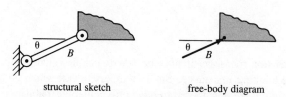

structural sketch   free-body diagram

Figure 6.42 Support idealization of a link (prop, strut, leg, or boom). The direction, θ, of the force, B, is the same as the link. The magnitude, B, is unknown. A link force may pull or push on the body.

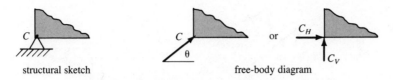

structural sketch   free-body diagram

Figure 6.43 Support idealization of a pin. A pin provides an unknown force, C, at an unknown direction, θ; or alternatively, two unknown mutually perpendicular components such as C_H and C_V.

Pins

A pin is a connection that prevents a point on an object from moving in a line, while offering no restraint to rotational motion. Linear motion in a plane can be represented by a vector having magnitude and direction or, alternatively, by a pair of mutually perpendicular components (V and H or X and Y). In order to resist linear motion, the pin must provide a force, C, at the required angle, θ, or a pair of component forces, C_V and C_H, of the required relative magnitude as shown in figure 6.43.

Pins are assumed to be frictionless and incapable of providing any moment resistance. This is an idealization. Friction is always present and provides a resisting moment. The pin idealization is often used where a considerable resisting moment is possible. However, its use represents an assumption that either the resisting moment is negligible or that its contribution to the strength of the structure will be forfeited in order to gain a more easily understood load path.

Pins are replaced on a free-body diagram with either an unknown force and angle, C and θ, or a pair of mutually perpendicular components, C_V and C_H or C_X and C_Y. Either is correct, because C is the resultant of its components. Knowing C and the angle, one can find the components, and vice versa. Pins are more complicated supports than are guys and links, because they introduce two unknowns. Both the force *and* the angle are unknown. Or, alternatively, two components are unknown.

Slides

A slide restrains motion in one direction only. It is capable of providing a force perpendicular to the slide surface. Forces parallel to the slide surface as well as moments are considered to be zero. The slide effect may be produced also by a rocker, a roller (as shown in figure 6.44), or a wear plate.

Friction is usually neglected in the slide idealization. Rollers and rockers come closest to this idealization. To represent a connection as a slide support is to assume that the friction is either negligible or that its strengthening effect upon the structure will be forfeited for simplicity.

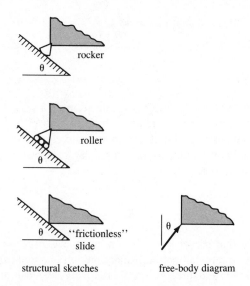

Figure 6.44 Support idealization of a frictionless slide (roller or rocker support). A slide can provide a force perpendicular to the slide surface. It does not resist parallel forces.

ILLUSTRATIVE PROBLEM

6.10 The box shown in figure 6.45 weighs 6 000 kg. What are the support reactions at *A*, *B*, and *C*?

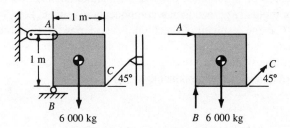

Figure 6.45 Illustrative problem 6.10.

Solution:

The free-body diagram of the box has three forces: the link force, *A*, the roller force, *B*, and the guy tension, *C*. These three unknown forces are available to satisfy the three equilibrium conditions of relationship 6.11. If more than three unknown forces appear on the free-body diagram, the problem cannot be solved by methods of statics, since three equations can be solved for only three unknowns. If less than three reaction forces appear on this free-body diagram, the structure is not stable. A general planar equilibrium such as this has three conditions of equilibrium requiring three support reactions or components.

The vertical and horizontal components of C are:

$$C_V = C \sin 45° = 0.707C$$

and

$$C_H = C \cos 45° = 0.707C$$

Setting up the equilibrium equations from relationship 6.11,

$$\Sigma V \uparrow : B + 0.707C - 6\ 000 \text{ kg} = 0$$

(a)
$$B + 0.707C = 6\ 000 \text{ kg}$$
$$\Sigma H \leftrightarrow : A + 0.707C = 0$$

(b)
$$A = -0.707C$$

Summing the moments about point B (points A or C are equally good choices),

$$\Sigma M_B \circlearrowleft : 1 \times A - 1 \times C_V + 0.5 \times 6\ 000 \text{ kg} = 0$$

The horizontal component of C contributes no moment about point B.

(c)
$$A - 0.707C = -3\ 000 \text{ kg}$$

Substituting **(b)** into **(c)**,

$$-0.707C - 0.707C = -3\ 000 \text{ kg}$$
$$1.414C = 3\ 000 \text{ kg}$$
$$C = 2\ 122 \text{ kg}$$

Substituting C into **(a)**,

$$B + 0.707 \times 2\ 122 \text{ kg} = 6\ 000 \text{ kg}$$
$$B = 4\ 500 \text{ kg}$$

Substituting C into **(b)**,

$$A = -0.707 \times 2\ 122 \text{ kg}$$
$$A = -1\ 500 \text{ kg}$$

The significance of the negative sign of A is that the link force, A, is not in the assumed direction. It is the reverse. Instead of the link being in compression, the mathematics tells us that it is in tension.

The support reactions are 1 500 kg of tension in link A; 4 500 kg of compression in the roller, B; and 2 120 kg of tension in the guy, C.

6.13 TWO-FORCE MEMBERS AND BENDING MEMBERS

When a continuous member is loaded at only two points, it is called a *two-force member*. It is capable of transmitting only a load along its axis. Its axis is the straight line between the two load points. Links, as described in section 6.12, are two-force members.

When a continuous member is loaded at three or more points, it is a *bending* member. Bending members cannot be assumed to carry only axial load. Their loading may be much more complicated than the two-force member. Bending members can transmit loads in any direction at any connecting point. Beams and levers are bending members and are

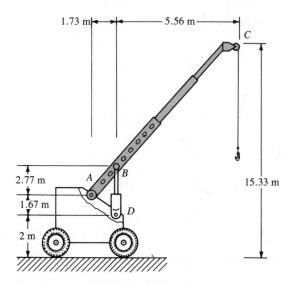

Figure 6.46 A hydraulically operated crane is a different structural system from the crane of figure 6.32. In this crane the boom is a bending member (illustrative problem 6.11).

considered fully beginning in chapter 9. For now it is sufficient to determine the reactions at guys, links, and other supports when bending members are a part of the structure.

As an example, consider the hydraulically operated crane shown in figure 6.46. The boom in this crane is raised and extended by means of hydraulic cylinders. The hook is raised by a hydraulic motor at C. The boom has the potential of being loaded at three points: the pivot, A; the cylinder, B; and the cable, C. The boom therefore is a bending member. It is erroneous to place a vector that has the direction of the axis of the boom on a free body in this case. The cylinder, B, on the other hand, is a two-force member loaded at B and D that transmits load only along its axis. The hydraulically operated crane of figure 6.46 can be understood from a free-body diagram of the boom.

ILLUSTRATIVE PROBLEMS

6.11 If the hook load on the hydraulic crane shown in figure 6.46 is 120 kN, what is the force shared by the two hydraulic cylinders on either side of the boom at B? Also, what is the pin force at A?

Solution:
Since the boom is loaded at three points, it is a bending member. The free-body diagram of the boom is shown in figure 6.47. The hydraulic cylinder constitutes a compression link and is represented by the vertical force, B. The pin force at A is represented by its vertical and horizontal components, A_V and A_H.

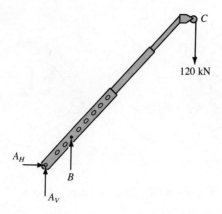

Figure 6.47 Free-body diagram of the boom from the crane of figure 6.46.

From relationship 6.11,

$$\sum V \uparrow: A_V + B - 120 \text{ kN} = 0$$
$$A_V + B = 120 \text{ kN}$$

(a)
$$\sum H \leftrightarrow: A_H = 0$$
$$\sum M_A \circlearrowleft: -1.73 \text{ m} \times B + 7.29 \text{ m} \times 120 \text{ kN} = 0$$

$$B = \frac{7.29 \text{ m}}{1.73 \text{ m}} \times 120 \text{ kN} = 506 \text{ kN}$$

Substituting back into **(a)**,

$$A_V + 506 \text{ kN} = 120 \text{ kN}$$
$$A_V = -386 \text{ kN}$$

The minus sign signifies that the pin reaction at A is down rather than up, as was initially assumed.

The cylinder force is 506 kN, and the pin force is 386 kN.

6.12 A gear rack and pinion are made to standard involute shapes that cause the angle of force between the teeth to be 20° on an effective diameter called the pitch circle. The pinion is torqued, as shown in figure 6.48a, with 100 N · m in the counterclockwise direction. The pitch diameter is 40 mm. What is the amount of vertical force tending to raise the gear shaft and the horizontal force tending to bend the gear shaft away from the fixed rack? Neglect friction.

Solution:
While the gear itself is not a structure, it is an element that can produce loads on a machine structure. The gear is drawn as a free-body diagram in figure 6.48b.

The shaft bearing can be replaced by a pair of forces, V and H. The shaft torque is entered on the diagram as a couple. The forces, V and H, represent the answers to

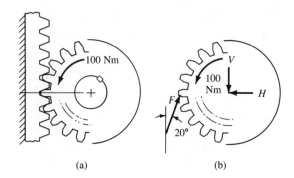

(a)   (b)

Figure 6.48 Illustrative problem 6.12.

the problem. Force F is the normal force due to the sliding of the pinion tooth on the rack tooth at the pressure angle of 20°.

The gear forces constitute a general force system. The equilibrium conditions of relationship 6.11 apply. First summing the moments about the shaft center:

$$\Sigma\, M\text{⊕}: F \sin 70° \times (0.040 \text{ m}/2) - 100 \text{ N} \cdot \text{m} = 0$$

Solving for F:

$$F = 5\ 320 \text{ N}$$

Summing the vertical and horizontal components of force:

$$\Sigma\, V\updownarrow: F \sin 70° - V = 0$$
$$\Sigma\, H\leftrightarrow: F \cos 70° - H = 0$$

Solving for V and H:

$$V = 5\ 320 \sin 70° = 5\ 000 \text{ N}$$
$$H = 5\ 329 \cos 70° = 1\ 820 \text{ N}$$

A force of 5 kN is tending to drive the pinion up the rack, since the force V must be applied downward to hold it in equilibrium. The gear shaft must provide a horizontal force of 1.820 kN to keep it in contact with the rack.

PROBLEM SET 6.3

6.31 A 1 000-kg box is supported as shown in figure 6.49. What is the pin reaction at A and the force on the roller at B?

6.32 A 50-lb boy swings on a 100-lb gate as shown in figure 6.50. What are the support reactions at A and B?

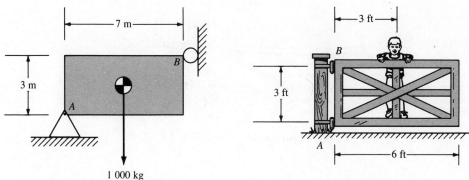

Figure 6.49   Problem 6.31.

Figure 6.50   Problem 6.32.

6.33 An A-frame structure shown in figure 6.51 supports a downward load of 75 kN at point *C*. The member *ABC* is continuous. Friction is neglected at *A*. What is the force in the link at *B*?

6.34 A light standard for a highway is shown in figure 6.52. Although it is made entirely of welded aluminum, the member *ABC* is presumed to be continuous and pin-connected at *B* and *C*. According to this conservative assumption, what is the tension force in the strut at *B*? Neglect the weight of the aluminum members.

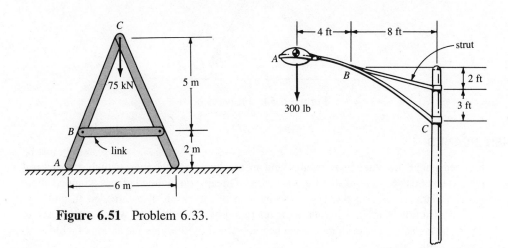

Figure 6.51   Problem 6.33.

Figure 6.52   Problem 6.34.

6.35 An airplane wing, shown in figure 6.53, carries a net vertical load due to its lifting force of 100 lb/ft of wing length. To conservatively estimate the force in the strut, the wing is assumed to be pin-connected at its root. What is the tension in the strut?

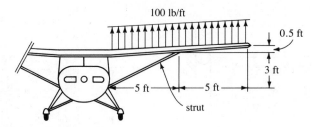

Figure 6.53   Problem 6.35.

6.36 The prop *BD* in figure 6.54 is capable of supporting 500 kN. What vertical load, *F*, can the frame support at *C*? What must be the load capacity of the pin at *A*?

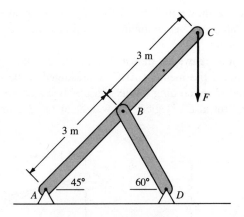

Figure 6.54   Problem 6.36.

6.14 INCLINED PLANES

Inclined panes are vital in structures and in mechanisms. Wedges, which are inclined planes, can be used to hold structures together. Threads are inclined planes wrapped about an axis. The friction on threads is essential to the working of a threaded fastener. In contrast, friction is an impediment to the operation of power threads, which can be used to drive mechanisms. These elements can be understood by studying the action of a block on a plane with friction.

With inclined planes, it is best to resolve all forces into components perpendicular to the plane and parallel to the plane. The block is assumed to be a concurrent force system with the point of concurrence at its center. In other words, the equilibrium moment is neglected. This assumption turns out to be highly accurate in most applications. Using relationship 6.10 for the concurrent force system

$$\Sigma X = 0 \quad \text{and} \quad \Sigma Y = 0$$

where *X* is the perpendicular direction and *Y* is the parallel direction. Equilibrium is used to solve for the forces required as in the following illustrative problem.

ILLUSTRATIVE PROBLEM

6.13 The block shown in figure 6.55 is at rest on the inclined plane when acted upon by the 1 000-N vertical force. The coefficient of static friction, f, between the block and the plane is 0.30. What horizontal force in the direction shown will destabilize the block?

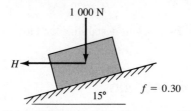

1 000 N

H

$f = 0.30$

15°

Figure 6.55 Illustrative problem 6.13.

Solution:

A free-body diagram of the block is shown in figure 6.56. The inclined plane is replaced with a normal force, N, perpendicular to the plane and a friction force, F_F, parallel to the plane. Since the motion is impending down the plane, the friction force is shown acting upward.

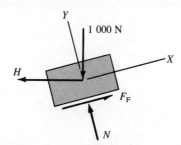

Y

1 000 N

H

X

F_F

N

Figure 6.56 Free-body diagram for illustrative problem 6.13.

Summing the components of forces in the parallel direction,

(a) $\qquad \Sigma X: F_F - H \cos 15° - 1\ 000 \sin 15° = 0$

Summing the components of forces in the normal direction:

(b) $\qquad \Sigma Y: N + H \sin 15° - 1\ 000 \cos 15° = 0$

Solving for N in terms of H, carrying 5 significant places to avoid roundoff error:

(c) $\qquad N = 965.93 - H \sin 15°$

From relationship 4.9

$$F_F = fN = 0.30N$$

(d) $\qquad F_F = 0.30(965.93 - H \sin 15°)$

Substituting into **(a)**:

(e) $0.30(965.93 - H \sin 15°) - H \cos 15° - 1\,000 \sin 15° = 0$
$289.79 - 0.077\,646H - 0.965\,92H - 258.82 = 0$

Collecting terms:

$$30.959 - 1.043\,6H = 0$$
$$H = 29.667$$

A horizontal force of 29.7 N will destabilize the block and cause it to begin sliding down the plane.

6.15 ANGLE OF FRICTION—FRICTION LOCKING

The friction force and the normal force on a block being forced to slide on a horizontal plane can be resolved into an angle called the angle of friction, as shown in figure 6.57. The angle of friction is ϕ_F.

From the figure:

(a) $$\tan \phi_F = \frac{F_F}{N}$$

From relationship 4.9 for the situation where sliding is impending:

(b) $$F_F = fN$$

Figure 6.57
ϕ_F is the angle of friction.

Substituting **(b)** into **(a)**:

$$\tan \phi_F = \frac{fN}{N} = f$$

$$\boxed{\phi_F = \arctan f}$$ **(6.12)**

The angle of friction is useful to determine whether in certain situations friction locking will occur. It is used for comparison. The proofs of these comparisons can be obtained using a block-on-plane type of idealization as in the previous illustrative problem. The comparisons follow.

Friction locking will occur if the force angle, θ, of a block on a plane as shown in figure 6.58 is less than ϕ_F. Only the angle makes the difference. The block will not slide no matter how great the magnitude of the force.

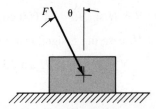

Figure 6.58   If θ is less than ϕ_F, the block will not slide no matter how great the force, F.

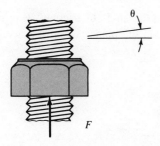

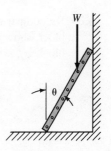

Figure 6.59 Friction locking will occur if the thread angle, θ, is less than the angle of friction.

Figure 6.60 Friction locking will be assured if the ladder angle, θ, is less than the angle of friction.

Friction locking will occur if the thread angle, θ, as shown in figure 6.59, is less than the angle of friction ϕ_F. This case illustrates the principle of threaded fasteners. As long as the force, F, is in the same direction (no vibration at the threads), the threads will remain locked.

Friction locking will occur on the ladder of figure 6.60, if the ladder angle from vertical, θ, is less than the angle of friction, ϕ_F. At this angle the load, W, may be at any location on the ladder, even right next to the wall, and remain stable. Greater ladder angles may be used if the load is nearer the floor. However, these greater angles might be dangerous because the ladder would become unstable if the load, presumably a person, were to move closer to the wall.

PROBLEM SET 6.4

6.37 What horizontal force, F, would be required to move the 1 000-lb block in figure 6.61 up the inclined plane? ($f = 0.10$)

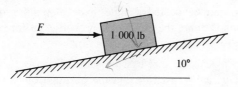

Figure 6.61 Problem 6.37.

6.38 To determine experimentally the coefficient of friction between two surfaces, a block of one material is put on a movable inclined plane of the other material. The angle is increased slowly. At 18° the block breaks away and begins to slide down the plane. What is the coefficient of friction? (See figure 6.62.)

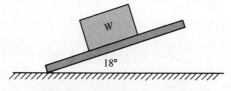

Figure 6.62 Problem 6.38.

6.39 By comparing the angle of the force, F, to the angle of friction, determine if the block shown in figure 6.63 will be stable on the plane. Ignore the weight of the block. ($f = 0.13$)

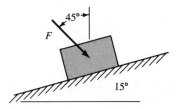

Figure 6.63   Problem 6.39.

6.40 Using the block-on-plane analogy, show that the nut in figure 6.59 will not move on the thread due to the vertical force, F, if the thread angle is less than the angle of friction.

6.41 Using a free-body diagram of a ladder with the weight, W, moved to the worst location, show that the ladder in figure 6.60 will be stable, if the angle with the vertical is less than the angle of friction. Hint: Assume, as the worst case, that the coefficient of friction at the wall is zero and that the only friction force restraining the ladder is at the floor. Take moments about the contact point with the floor along with the vertical and horizontal equilibrium equation for the ladder.

6.42 Rework illustrative problem 6.8 shown in figure 6.28 for the case where friction between the slide and the capping head is considered. ($f = 0.05$)

SUMMARY

☐ Straight-line members are connected at angles to develop more sophisticated structures. When this connection is done, the directional properties of forces must be considered. The basis for this result is the *parallelogram law*.

☐ The *resultant* is a single force that has the same effect as two or more forces. The *Pythagorean theorem* can be used to find the resultant of two perpendicular forces.
(6.1) $R = \sqrt{V^2 + H^2}$ or $\sqrt{X^2 + Y^2}$
(6.2) $\tan \theta = \dfrac{V}{H}$ or $\dfrac{Y}{X}$

☐ The *components* of a force in a plane are two forces in different directions that have the same directions as a single force. When directions are perpendicular, trigonometry can be used.
(6.3) $V = F \sin \theta$ and $H = F \cos \theta$ or $Y = F \sin \theta'$ and $X = F \cos \theta'$

☐ Components can be used to reduce nonperpendicular or multiforce systems to equivalent rectangular force systems.
(6.6) $R = \sqrt{(\Sigma\, V)^2 + (\Sigma\, H)^2}$ or $R = \sqrt{(\Sigma\, Y)^2 + (\Sigma\, X)^2}$
(6.7) $\tan \theta = \dfrac{\Sigma\, V}{\Sigma\, H}$ or $\dfrac{\Sigma\, Y}{\Sigma\, X}$

□ Directions of forces are often shown by slopes, in which case:

(6.8) $h = \sqrt{x^2 + y^2}$ then

(6.9) $V = y(F/h)$ and $H = x(F/h)$

□ There are three types of *planar force systems*. They are the *concurrent force system*, the *parallel force system*, and the *general planar force system*.

□ For equilibrium of a concurrent force system:

(6.10) $\Sigma V = 0$ and $\Sigma H = 0$ or $\Sigma X = 0$ and $\Sigma Y = 0$

□ For equilibrium of a general planar force system:

(6.11) $\Sigma V = 0, \Sigma H = 0,$ and $\Sigma M = 0$ or $\Sigma X = 0, \Sigma Y = 0,$ and $\Sigma M = 0$

□ Load paths through a structure are idealized by support reactions on the free-body diagram. Guys, links, pins, and slides are *idealized supports*. A *two-force member* can transmit only axial tension or compression. *Bending members* have three or more load points and are able to transmit forces in any direction at any connection or contact point with other members.

□ *Inclined planes* with friction may be treated as concurrent force systems. The *angle of friction* can be used to indicate whether sliding or locking might occur in a structure or a mechanism.

(6.12) $\phi_F = \arctan f$

CHAPTER SEVEN

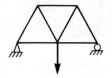 Trusses

The concepts of planar equilibrium are applied to the study of trusses. Some practicalities are included along with the historical development of these important structures. But the primary intent is to improve competency in using free-body diagrams and equilibrium relationships. The method of joints is the expansion of the equilibrium of concurrent planar force systems, and the method of sections is the expansion of the general case of planar equilibrium. Some exercises involving vehicular trusses are included.

7.1 THE PLANE TRUSS

Structures composed of straight members linked together in triangles are called *trusses*. A *plane truss* has all of its members sharing the same plane. A truss may consist of two triangles or many triangles, as in complicated bridge trusses.

Trusses generally have good rigidity and use the material of construction efficiently. The links of a truss are tension or compression members. The truss owes its strength primarily to the tension or compression strength of its members.

In visualizing the loads in a truss, consider the four-member linkage in figure 7.1a. It is unstable and incapable of carrying a load. Four-member linkages are useful in mechanisms, because every point on the linkage moves in a fixed path. But as a structure, the four-member linkage is worthless. It will collapse.

As it moves or collapses downward, points A and C move farther apart, and points B and D get nearer to one another. If A and C are prevented from moving apart by adding a diagonal member, AC, the linkage will become a truss as shown in figure 7.1b. Two triangular linkages, ABC and ACD, will be created. The linkage is stable and will support a force at B. This is an idealized truss, because all points are assumed to be connected by frictionless pins. The members of an ideal truss carry only axial loads.

The four-link mechanism of figure 7.1a could have been converted into a truss by a diagonal running from B to D instead of from A to C. But, because B and D tend to move

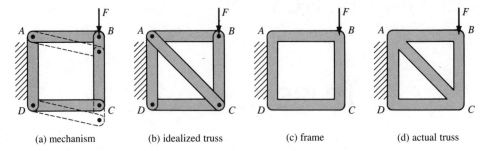

(a) mechanism   (b) idealized truss   (c) frame   (d) actual truss

Figure 7.1 The four-member mechanism (a) is unstable. Addition of a diagonal (b) makes it a stable pin-connected truss. An alternative way of stabilizing the mechanism is to add bending resistance at the joints (c). Actual trusses usually have both diagonals and bending resistance (d).

toward one another under the load shown, *BD* would be a compression member. *AC*, on the other hand, is a tension member. Usually, tension members are preferred over compression members because of the buckling tendencies of compression members. In determining whether truss diagonals carry tension or compression, it helps to envision the truss as a mechanism with the diagonal link in question removed.

The linkage of (a) could also be stabilized by making the joints at *A*, *B*, *C*, and *D* rigid instead of pin-connected, as shown in figure 7.1c. This structure is a *frame*. A frame carries its load by means of the bending resistance of its members. In a frame the axial tension or compression resistance of the members is of secondary importance.

If a diagonal, *AC*, is added to the frame as shown in figure 7.1d, the frame again becomes a truss. Such a structure is more like an actual truss in that its joints are not pin-connected. It is in fact a combination of the idealized pin-connected truss, 7.1b, and the frame, 7.1c. Part of the load resistance is due to the axial tension or compression of the members as in (a), and part is due to the bending resistance of the joints as in (b). The axial tension and compression are *primary stresses*. The stresses due to bending are *secondary stresses*. If the truss is constructed of a ductile material, the secondary stresses, like stress concentrations, tend to be self-relieving by yielding at high loads. It is conservative to ignore the secondary stresses and consider the structure to be an ideal pin-connected truss as in (b). This assumption is valid under the following conditions:

1. The truss is constructed of a ductile material.

2. All loads are applied to the joints.

3. The centerlines of all members coming together at a joint intersect at a single point.

4. The members are reasonably slender, having, for example, a length more than ten times the width.

In this chapter the pin-connected idealization will be used for all trusses, thereby considering only the primary stresses. It is a conservative idealization, because secondary stresses normally add to the strength of the truss. Secondary stresses are difficult to compute. But because computer programs that yield secondary stresses and elastic deformations in addition to primary stresses are now common, it makes little sense to do only primary stress analysis in the final design of a truss. Complete stress analysis using computer programs usually requires the services of an analytically trained structural engineer. However, much can be understood about the strength and load paths through a truss by assuming it to be pin-connected. This simplification often meets the needs of the architect or technologist, if not the structural engineer.

7.2 THE HISTORICAL PERSPECTIVE

Trusses made of bamboo or tree limbs lashed together with rope or hide are used today by primitive societies. Therefore, it can be presumed that such trusses were used in ancient times, even though they have not survived to be uncovered archeologically. Trusses are not a new idea, but their development was limited by the availability of materials.

Some very old buildings contain wooden roof trusses fastened by wooden pegs and keys. Wood was the original material used in trusses. It is lightweight and can resist both tension and compression. But the development of wooden trusses was hampered by the lack of good fasteners.

It was not until the eighteenth century that iron became generally available for fasteners. Prior to that, iron was so expensive that its use was limited to armaments and an occasional tool or fitting. But in the eighteenth century, iron fasteners in the form of bolts, pins, rivets, and gusset plates began to appear on wooden trusses. Good fastening at the joints permitted fuller use of the strength of wooden members. Wooden truss bridges were built. The bridges were covered with walls and a roof so that the joints were not exposed to the weather, which would hasten the rusting of the valuable iron fasteners.

During the industrial revolution in the nineteenth century, iron and then steel became economical as structural materials. Then not only the fasteners, but also entire members were made of metal. Steel trusses grew in complexity and size. Due to their strength and rigidity, trusses made possible rail and highway bridges of considerable span and strength.

However, steel trusses made with open sections and fastened with rivets or bolts have some undesirable features. They have a great amount of exposed surface and seem to need continual painting. Birds love to roost among their various corners and hollows, causing soiling to say nothing of the uneasy feeling of people who must pass below. And, finally, some say that from an aesthetic point of view, a truss is something that only an engineer could love.

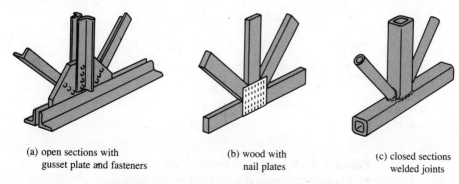

(a) open sections with   (b) wood with   (c) closed sections
gusset plate and fasteners   nail plates   welded joints

Figure 7.2 Typical truss-construction systems.

But recent developments are bringing the truss back to the forefront of modern structures. New arc welding techniques aided by computers and robotics have permitted the economical use of closed sections in the form of pipe or square tubing. These members are mitered and welded at the joints, developing their full strength. The exposed surface is minimal and the lines are clean, reducing the need for painting and providing few roosting places for birds. But, more importantly, structural engineers armed with computer programs are becoming more creative with the shapes of trusses. More and more, trusses are becoming truly three-dimensional rather than mere assemblages of plane trusses. Ingenious new methods of supporting trusses are appearing. Now, many major public buildings and structures are being built with exposed trusswork, often as a prominent architectural feature.

7.3 SUPPORTS FOR A TRUSS AS A SPAN

A *span* is a structure that gathers vertical loads and transfers them to supports on either side. Bridges, roofs, and sometimes floors are spans. The loads on a span are *live loads,* due to the weights supported, and *dead loads,* due to the weight of the structure itself. Trusses are frequently used as spans as shown in figure 7.3.

Normally, a truss (when used as a span) is *simply supported*. This means a rocker or roller support at one end restrains that end vertically, while permitting horizontal movement, as shown in figure 7.3. The other end has a vertical restraint as well as a means of horizontal restraint in order to resist any occasional horizontal forces, such as those caused by the braking of a truck on a bridge. The best idealization of this second support is a pinned connection, although the actual configuration might be quite different. Simple supports, a pin on one end and a roller on the other, are the most frequent choice for spans, because they permit thermal and elastic expansion and contraction.

A truss, when simply supported as a span under gravity loads, has the members on top in compression and members on the bottom in tension. The diagonals may be either in tension or compression, depending on how they are configured. Configurations in which diagonals are in tension are preferred. The verticals may also be in compression or tension, but they are normally compression members. Verticals should be seen as props that

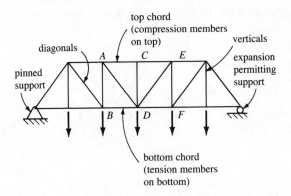

Figure 7.3 A truss as a simply supported span. In this case the span is composed of six panels.

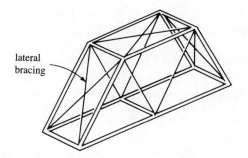

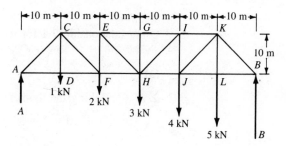

Figure 7.4 An experimental truss for student labs, showing lateral bracing.

Figure 7.5 A free-body diagram of a simply supported truss under vertical loads (illustrative problem 7.1).

hold the top and bottom members apart. Diagonals should be seen as the braces that make the mechanism a truss by forming it into a series of triangular linkages.

A truss needs to be laterally braced to resist buckling out of plane. In plane, the effective length of the compression member is the length of the link as shown in the elevation. But out of plane, the compression member is for all practicality the entire length of the top chord of the truss. The buckling strength of a member is reduced by the square of the effective length. Therefore, a truss such as shown in figure 7.3 can be weak and unstable in the lateral direction out of the plane of the truss.

To provide lateral support, trusses are usually used in pairs or groups and are connected with crossed diagonals to provide lateral bracing, as shown in figure 7.4. The vertical loads are shared by the pair of trusses, and the crossed diagonals ensure that the effective length of the compression members laterally is the same as in the plane of the truss. The lateral bracing also resists any horizontal loads due to wind. Trusses used to support roofs or floors can be laterally braced by the roof or floor panels themselves.

The truss can be drawn as a free body to calculate its support loads as shown in figure 7.5. The supports at *A* and *B* are removed and replaced with reaction forces. The pinned support at *A* can provide a horizontal as well as a vertical component. But only the vertical component is included on the diagram, because no horizontal load exists in this case to excite a horizontal reaction at *A*. It makes no difference which end, *A* or *B*, is the pinned support. All loads are parallel. Therefore, the force system is known as a *parallel planar force system*. It is a special case of a planar force system in which only the forces in the parallel direction, *V* or *Y*, and the moments about any point on the structure need to be in equilibrium.

$$\Sigma\, V = 0 \text{ or } \Sigma\, Y = 0 \quad \text{and} \quad \Sigma\, M = 0 \qquad (7.1)$$

These equilibrium conditions can be used to set up equations and solve for the support reactions.

ILLUSTRATIVE PROBLEM

7.1 What are the support reactions, *A* and *B*, for the truss shown as a free body in figure 7.5?

Solution:

Using relationship 7.1 for the parallel planar force system,

$$\Sigma V \uparrow: A - (1 + 2 + 3 + 4 + 5) \text{ kN} + B = 0$$

(a)
$$A + B = 15 \text{ kN}$$

$\Sigma M_A \circlearrowright$: Taking the moment contribution of each force in turn,

$$(10 \times 1 + 20 \times 2 + 30 \times 3 + 40 \times 4 + 50 \times 5) \text{ kN} \cdot \text{m} - 60 \text{ m} \times B = 0$$
$$550 \text{ kN} - 60B = 0$$
$$B = \frac{550 \text{ kN}}{60}$$
$$B = 9.17 \text{ kN}$$

Substituting *B* back into relationship **(a)**,

$$A + 9.17 \text{ kN} = 15 \text{ kN}$$
$$A = 5.83 \text{ kN}$$

The support, *B*, on the right carries 9.17 kN of the load and that on the left carries 5.83 kN. Since loads increase from right to left, it makes sense that the right side would be more heavily loaded.

7.4 SUPPORTS FOR CANTILEVER TRUSSES

A *cantilever* is a structure that supports loads from one side only. Cantilevers may be horizontal, vertical, or in any other direction. Trusses are frequently used as cantilevers, as shown in figure 7.6.

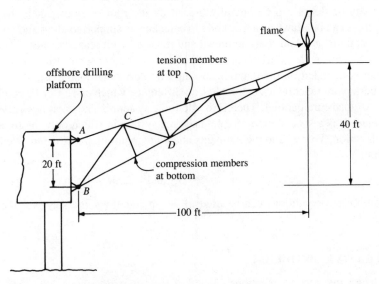

Figure 7.6 A truss as a cantilever.

When a truss is used as a cantilever under gravity loads, the upper members carry tension and the lower carry compression. This is opposite to the case of the span truss. Diagonals, as in a span, may be tension or compression depending on the configuration.

A cantilever truss is a general planar force system. It must have supports capable of providing three independent forces or components, one for each of the three equilibrium conditions:

$$\Sigma V = 0 \text{ or } \Sigma Y = 0, \ \Sigma H = 0 \text{ or } \Sigma X = 0, \ \text{ and } \ \Sigma M = 0 \qquad (7.2)$$

Figure 7.7 is a free-body diagram of the cantilever truss supporting the flare on the offshore oil rig shown in figure 7.6. Pin connections are used both at A and at B. Assuming the truss to be pin-connected, the force at A must be in the direction of the member AC. Therefore, only the magnitude is unknown. The pin connection at B can provide two unknowns, in the form of vertical and horizontal components.

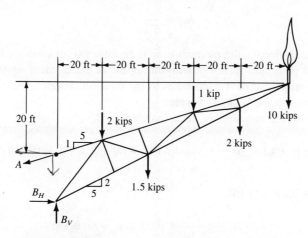

Figure 7.7 Free-body diagram of the cantilever truss shown in figure 7.6 (illustrative problem 7.2).

ILLUSTRATIVE PROBLEMS

7.2 What are the support reactions for the cantilever truss shown as a free body in figure 7.7?

Solution:
The cantilever truss is a general planar force system for which the three equilibrium conditions of relationship 7.2 apply.

First, calculate the vertical and horizontal components of A. The hypotenuse of the reference triangle is

$$h = \sqrt{1^2 + 5^2}$$
$$h = 5.10$$

From relationship 6.11,

$$A_V = \frac{1}{5.10} A \quad \text{and} \quad A_H = \frac{5}{5.10} A$$

Then, using the three equilibrium conditions of relationship 7.2:

$$\Sigma V \uparrow: \frac{-1}{5.10} A + B_V - (2 + 1.5 + 1 + 2 + 10) \text{ kips} = 0$$

(a)
$$\frac{-1}{5.10} A + B_V = 16.5 \text{ kips}$$

$$\Sigma H \leftrightarrow: \frac{-5}{5.10} A + B_H = 0$$

(b)
$$B_H = \frac{5}{5.10} A$$

B is the best choice for the moment center, because two unknowns, B_V and B_H, act through that point and therefore will not appear in the moment equation. Only the horizontal component of A contributes to the moment, because the vertical component acts through the moment center.

In summing the moments of each force about B,

$$\Sigma M_B \circlearrowleft: \frac{-5}{5.10} A \times 20 \text{ ft} + (2 \times 20 + 1.5 \times 40 + 1 \times 60 + 2 \times 80 + 10 \times 100) \text{ kip} \cdot \text{ft} = 0$$
$$A = 67.32 \text{ kips}$$

Substituting back into **(b)**,

$$B_H = \frac{5}{5.10} \times 67.3 \text{ kips} = 66.0 \text{ kips}$$

And then back into **(a)**,

$$\frac{-1}{5.10} \times 67.32 \text{ kips} + B_V = 16.5 \text{ kips}$$

$$B_V = 29.7 \text{ kips}$$

The vertical and horizontal components of the reaction at B are 29.7 and 66.0 kips upward and to the right, respectively. The reaction at A is 67.3 kips downward and to the left at a slope of 1/5.

Obviously, any connections at those supports must be capable of resisting these loads.

7.3 An 8 000-gallon water tank shown in figure 7.8a is supported on four pipe legs. The effective length of the legs is reduced by brace BE and crossed diagonal rods with turnbuckles forming a vertical cantilever truss. The truss carries 17.5 kips on each leg due to the tank and its contents and wind loads at C and B of 4.0 and 2.0 kips, respectively, idealized as concentrated forces at those points. What are the reactions at the supports, A and F?

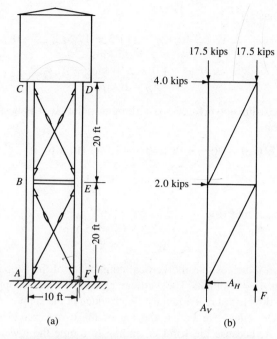

Figure 7.8 Water-tower truss with crossed diagonal turnbuckle rods.

Solution:

Many subtle considerations in this problem are important in understanding the idealization of this common structure.

Even though the legs are continuous through the joints and have rigid connections to the foundation, the diagonals make the structure a truss. It can be idealized as being pin-connected, because the bending resistance of the legs at the joints and at the supports is of secondary importance.

Second, although the wind loads act over the entire tank and support structure, it is sufficiently accurate for structural purposes to "lump" them at the nearest joint as concentrated loads.

Third, the crossed diagonals are slender. The compressive strength of the diagonals is negligible. When the wind blows from the left as shown in the free-body diagram, the mechanism *ABEF* will deform to the right, tending to stretch diagonal *AE* and compress diagonal *BF*. *BF* will buckle and relieve itself of load, leaving *AE* alone as the diagonal of consequence in that mechanism. Therefore, *BF* is removed from the free-body diagram. *BF* needs to be in the actual structure, because we have no assurance that the wind may not choose to blow from the right.

The free-body diagram can then be as shown in figure 7.8b. The support can be assumed pinned at *A*, providing horizontal as well as vertical resistance (A_V and A_H). The member *EF* can transmit only axial load when idealized as pin-connected on both ends. Therefore, at support point *F*, only a vertical reaction is shown.

From relationship 7.2,

$$\Sigma V \uparrow: A_V + F - (17.5 + 17.5) \text{ kips} = 0$$

(a)
$$A_V + F = 35 \text{ kips}$$
$$\Sigma H \mapsto: (4.0 + 2.0) \text{ kips} - A_H = 0$$
$$A_H = 6 \text{ kips}$$

And, since most unknown forces pass through point A, it is best to take moments about that point:

$$\Sigma M_A \oplus: (2.0 \times 20 + 4.0 \times 40 + 17.5 \times 10) \text{ kip} \cdot \text{ft} - F \times 10 \text{ ft} = 0$$

$$F = \frac{375}{10} \text{ kips} = 37.5 \text{ kips}$$

Substituting back into **(a)**,

$$A_V + 37.5 \text{ kips} = 35 \text{ kips}$$
$$A_V = -2.5 \text{ kips}$$

The minus sign indicates that the vertical component of the support reaction at A is not upward as first thought. The vertical reaction is in the opposite direction—downward. A downward reaction must be provided at the support to hold the truss in equilibrium. Or, otherwise stated, the leg of the tower provides an uplift on the foundation at A, because the wind is tending to topple the tower.

The reaction forces are 6 kips of horizontal load to the left, 2.5 kips of vertical load downward at A, and 37.5 kips of vertical load upward at F. The foundation "feels" a downward load of 37.5 kips at F and an uplift of 2.5 kips as well as 6 kips of horizontal load at A. The foundation at any leg must then be designed for 37.5 kips of compression and 6 kips of lateral shear. There must be enough mass at any leg to resist the uplift. This example is not the worst case, however, for uplift. The uplift is due to the overturning moment caused by the anticipated wind forces. When the tank is empty, the stabilizing effect of the weight of its contents is absent and the uplift forces are greater. To determine the maximum uplift force for the design of the foundation and the connecting bolts, the calculation would need to be repeated with the weight of the water omitted.

PROBLEM SET 7.1

7.1 What are the reactions at points A and M on the truss shown in figure 7.9, assuming it to be simply supported?

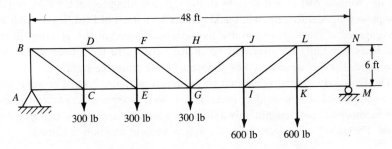

Figure 7.9 Problems 7.1, 7.22, and 7.23.

7.2 What are the reactions on the roof truss shown in figure 7.10, assumed simply supported at *A* and *G*, if a 2 000-lb weight is applied at joint *E*?

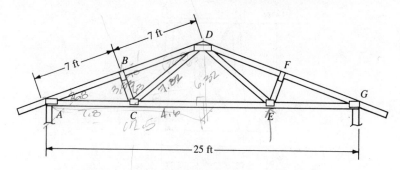

Figure 7.10 Problems 7.2, 7.14, and 7.24.

7.3 What are the support reactions on the cantilever truss shown in figure 7.11?

7.4 What are the support reactions upon the observation tower due to the wind loading shown in figure 7.12? The wind loads shown are for each of the two trusses formed by the four legs.

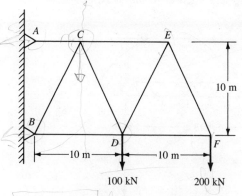

Figure 7.11 Problems 7.3 and 7.15.

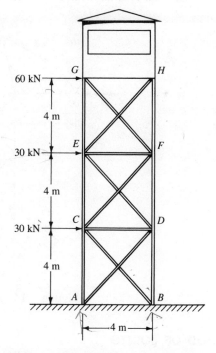

Figure 7.12 Problems 7.4 and 7.16.

7.5 The crossed diagonals on a causeway structure depicted in figure 7.13 are provided for lateral resistance to wind. What are the vertical and horizontal components of the reactions at *A* and at *D* due to the gravity and wind loads shown?

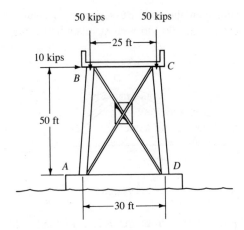

Figure 7.13   Problems 7.5 and 7.17.

7.6 What are the support reactions at *A* and *D* on the crane shown in figure 7.14 when the hook load is in the position shown? The members *AC* and *BD* are crossed diagonals. What are the support reactions when the load is removed from the hook and horizontal wind load goes to zero? There are two legs each at *A* and at *D*.

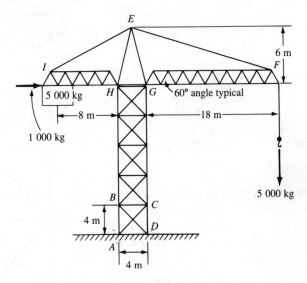

Figure 7.14   Problems 7.6 and 7.18.

7.5 THE METHOD OF JOINTS

The loads on the members of a plane truss can be determined by the *method of joints*. In this method the loads are computed by starting at some support joint on the structure and making a free-body diagram of the joint as a concurrent force system. Then, as loads are discovered, one can move successively to internal joints repeating the process.

The method of joints requires that the truss be assumed to be pin-connected. The method of joints does not consider secondary stresses. It, therefore, allows only the computation of the primary tension or compression loads. Although an assumption, pin connections quite accurately represent reality. And the errors produced by the assumption are on the conservative side.

Students who despair that they could make an entire career out of solving a complicated truss by the method of joints may be exaggerating only slightly. This method is practical only for small trusses having some symmetry. But on a small truss, such as that shown in figure 7.15, the method of joints can lead to a quick understanding of the loads involved in the structure. The method of joints is best illustrated by a problem.

ILLUSTRATIVE PROBLEM

7.4 Solve for the loads in all the members of the simply supported truss shown in figure 7.15, using the method of joints.

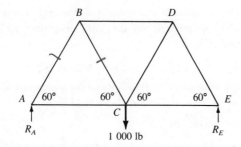

Figure 7.15 Illustrative problem 7.4. This is the same truss shown in figure 7.4.

Solution:
Before using the method of joints to find the internal forces in a truss, it is usually necessary to find all the external reactions possible. In this case the structure is symmetrical with symmetrical loads. Therefore, the support reactions should also be symmetrical, $R_A = R_E$. In order for vertical equilibrium to exist for the entire structure, the sum of the two reactions must equal 1 000 lb. Therefore, $R_A = R_E = 500$ lb.

In nonsymmetrical cases, the support reactions must be computed by methods discussed earlier in this chapter. Having the support reactions, one can then proceed to make free-body diagrams of joints. Joint A is a good place to start, because only two members intersect at that joint.

The member forces in truss analysis are best labeled with the end points of the members. For example, AB is the force in the member AB, and AC is the force in the member AC. Using this notation, the free-body diagram of joint A is such as shown in figure 7.16.

If a member is in compression, the joint will push on the member. The member in turn will push on the joint. In the free-body diagram, member AB is here assumed to be in compression and the force is shown pushing on the joint. Similarly, if a member is in tension, it will pull on the joint. Member AC is shown pulling on the joint. The

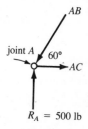

Figure 7.16 Free-body diagram of joint A of the truss shown in figure 7.15.

reaction force, R_A, is known and is shown as a 500-lb force pushing upward on the joint, A.

The free-body diagram is a concurrent force system, because the lines of action of all forces lie in a plane and intersect at a single point, A. In a concurrent force system, the sum of forces in two mutually perpendicular directions, ΣV and ΣH or ΣX and ΣY, must be equal to zero, as discussed in the previous chapter.

$$\Sigma V \uparrow: 500 - AB \sin 60° = 0$$

$$AB = \frac{500}{\sin 60°} = 577 \text{ lb}$$

$$\Sigma H \leftrightarrow: -AB \cos 60° + AC = 0$$
$$AC = AB \cos 60°$$
$$AC = 577 \cos 60° = 289 \text{ lb}$$

The loads in members AB and AC, respectively, are 577 lb compression and 289 lb tension.

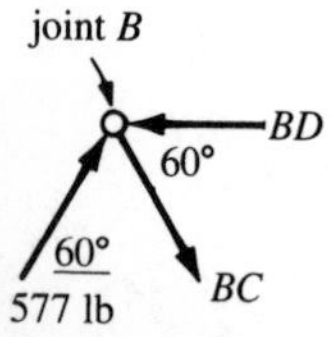

Now, when moving to point B and making a free-body diagram, AB is entered as a known force of 577 lb pushing on the joint as shown in figure 7.17. For the free-body diagram of joint B,

$$\Sigma V \uparrow: 577 \sin 60° - BC \sin 60° = 0$$
$$BC = 577 \text{ lb}$$
$$\Sigma H \leftrightarrow: 577 \cos 60° + BC \cos 60° - BD = 0$$

Substituting the now known BC,

$$577 \cos 60° + 577 \cos 60° - BD = 0$$
$$BD = 577 \text{ lb}$$

By symmetry, the rest of the member loads can be deduced:

$$DE = AB$$
$$CD = BC$$
$$CE = AC$$

Therefore, all the member forces are now known, and the solution is complete. It is necessary to indicate whether the members are in tension or compression when giving the solution of a truss.

$$AB = 577 \text{ lb compression}$$
$$AC = 289 \text{ lb tension}$$
$$BC = 577 \text{ lb tension}$$
$$BD = 577 \text{ lb compression}$$
$$CD = 577 \text{ lb tension}$$
$$CE = 289 \text{ lb tension}$$
$$DE = 577 \text{ lb compression}$$

joint B

Figure 7.17
The free-body diagram of joint B of the truss shown in figure 7.15.

PROBLEM SET 7.2

7.7 What are the forces in the members of the truss shown in figure 7.18?

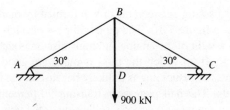

Figure 7.18 Problem 7.7.

7.8 What are the forces in the members of the truss shown in figure 7.19?

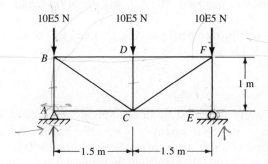

Figure 7.19 Problem 7.8.

7.9 A main feed pipe must span a street in an oil refinery. In order to decrease the sag, the pipe is formed into a truss by the attachment of a strut, *DB*, and tension rods, *AB* and *BC*, as shown in figure 7.20. If the rods are to be made of structural steel with an allowable stress of 22 000 psi, what must be their diameter? The pipe and contents weigh 75 lb/ft of length.

7.10 What are the forces in the members of the truss shown in figure 7.21?

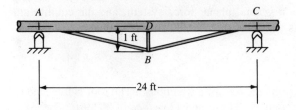

Figure 7.20 Problem 7.9.

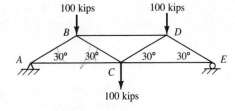

Figure 7.21 Problem 7.10.

7.11 What are the forces in members *BK*, *BL*, and *JL* of the bridge truss shown in figure 7.5? See illustrative problem 7.1.

7.12 What are the forces in members *BC* and *BD* in the cantilever truss for the gas flare shown in figure 7.6? See illustrative problem 7.2.

7.13 What are the forces in all members of the water tower truss shown in figure 7.8? See illustrative problem 7.3.

7.14 A young mechanic plans to lift an engine and transmission assembly from a car by suspending a hoist from the roof trusses of his parents' garage. The hoist will be attached to a pipe passing between the crotch formed by members *DE* and *EF* of the roof truss shown in figure 7.10. The pipe will pass through two adjacent trusses so that the 1 000-lb weight of the engine and transmission is shared by two trusses. The douglas fir 2 × 4's from which the truss is made are good for 1 200 psi stress in tension or compression. Buckling is unlikely because of the lateral support provided by the roof panels. The nail plates can transmit 90 percent of the strength of the wood. Is there any danger of the roof's collapsing as the young man lifts his machinery? Support your opinion with calculations. Use data from Appendix VI.

7.15 What are the forces in members *AC*, *BC*, and *BD* of the cantilever truss shown in figure 7.11?

7.16 What are the forces in members *AC*, *AD*, and *BD* of the cantilever truss shown in figure 7.12?

7.17 What are the forces in members *AB*, *AC*, *BD*, and *CD* of the causeway bracing shown in figure 7.13?

7.6 THE METHOD OF SECTIONS

In a static structure, every part must be in equilibrium. Every joint of a truss, every member, the entire truss, and every assembly of members must be in equilibrium. In the method of joints, successive joints are put in equilibrium. In calculating support reactions, the entire structure is set into equilibrium. Now, in the method of sections, assemblies of members will be put into equilibrium.

The method of sections is particularly valuable in understanding span trusses. In spans, the most heavily loaded tension members and compression members normally occur at the mid-span, if the loading is symmetrical. If the method of joints were used, one would have to work joint by joint from the supports to the mid-span in order to get to the most heavily loaded members. But with the method of sections, it is possible to calculate the forces in the mid-span members using only one free-body diagram.

The free body in this method is a *section* of a truss. A section is any group of triangular linkages. The choice of the section to draw as a free body is the key. Once the free body is drawn, it is treated as a general planar force system, and the equilibrium equations are written and solved as shown in the next illustrative problem.

ILLUSTRATIVE PROBLEM

7.5 The bar joist in figure 7.22 supports the roof of a warehouse. The roof is intended to support a *dead load* due to its weight of 10 lb/ft² and a *live load* of 30 lb/ft². The joists are spaced on 2-ft centers. What would be the maximum forces in the members of the joist?

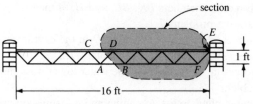

Figure 7.22 Illustrative problem 7.5.

Solution:

A bar joist is a truss made of a steel bar bent to form diagonals and welded between angles at the top and the bottom. They are very popular for floor and roof support in light construction, because they are economical and allow plenty of room for the routing of utility lines.

To solve the problem, it is first necessary to determine all the loads on the joist. The total of the live and dead loads is 40 lb/ft². Since the joists are spaced on 2-ft centers, each one can be assumed to carry a 2-ft-wide strip of roof area extending one foot to either side. Since the span is 16 ft, the total roof area supported by the joist is 32 ft², and the total load is 1 280 lb. Because of symmetry, the reaction at either end of the joist is 640 lb.

To assume the truss is pin-connected, the distributed load must be considered to be concentrated at the joints. The most obvious concentrations are those at the top of the truss. These joints are spaced 2 ft apart along the length of the truss. Each joint can be considered to carry the load on all the area 1 ft to either side. Since each truss carries a strip of area 2 ft wide, each joint supports 4 ft² of roof area. At 40 lb/ft², the joint loads are 160 lb.

The highest tension and compression member loads will be at mid-span at members *AB* and *CD*. *AB* is truly at mid-span. *CD* is just to the left of mid-span, because joint *D* is at mid-span. The member going to the right of joint *D* will have the same load as *CD* because of symmetry.

In applying the method of sections, it is necessary to mentally cut a section out of the truss that will cut *AB* and *CD*, because these are the member loads desired. The section *BDEF* shown in figure 7.23 can be isolated by cutting members *CD*, *AD*, and *AB*, as well as the supporting point, *E*. The section should be chosen so that no more than three unknown members are cut.

When making the free-body diagram of the section, the members "cut" must be replaced by forces. *AB* and *CD* were intended. But, the diagonal *AD* must also be cut in order to isolate the sections. A force must be provided on the free-body diagram for the member *AD*. These are the unknown forces.

The known loads must be put on the free-body diagram. The support reaction is 640 lb. The joint loads are 160 lb each, except at the end joint (which supports only 2 ft² of roof rather than 4 ft²). When all loads have been added to this section, the equilibrium conditions can be applied.

$$\Sigma \, V \updownarrow: \; -AD \sin 45° - 160 - 160 - 160 - 160 - 80 + 640 = 0$$
$$AD \sin 45° = -80$$
$$AD = \frac{-80}{\sin 45°} = -113 \text{ lb}$$

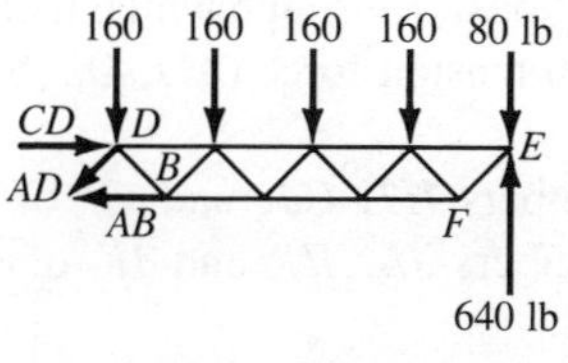

Figure 7.23 Free-body diagram of a section of the truss in figure 7.22.

$$\Sigma H \rightarrow: CD - AD \cos 45° - AB = 0$$

(a)
$$CD = AB - 80$$
$$\Sigma M_D \circlearrowleft: AB(1) + 160(2) + 160(4) + 160(6) + 80(8) - 640(8) = 0$$
$$AB = 2\,560 \text{ lb}$$

Substituting back into **(a)**,

$$CD = 2\,560 - 80$$
$$CD = 2\,480 \text{ lb}$$

The maximum force in the members on the bottom of the bar joist is that of *AB* and is 2 560 lb tension. The maximum compression force on the top members is that of *CD* and is 2 480 lb compression. The force in the diagonal *AD* is 80 lb compression. It was assumed to be tension, but the calculations yielded a negative sign. *AD* is not the maximum diagonal force in the truss. Forces on the diagonal are greater near the supports, whereas the chords are more heavily loaded at mid-span. The maximum diagonal force can be obtained quickly by applying the method of joints at the support point, *E*. One section free-body diagram and one joint free-body diagram can normally give the extremes of loading on the members of a simply supported truss symmetrically loaded as a span.

PROBLEM SET 7.3

7.18 What are the tensions in members *EF* and *EI* of the crane shown in figure 7.14?

7.19 What are the forces in members *AB*, *CD*, and *BC* of the truss in figure 7.24?

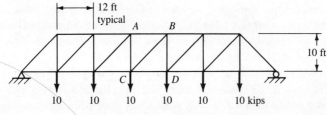

Figure 7.24 Problem 7.19.

7.20 In figure 7.3, the forces shown at the joints are 15 kN. The joints are separated vertically and horizontally by a distance of 9 m. What are the forces in members *AD*, *AC*, *BD*, and *CD*?

7.21 Using the method of sections, determine which member of the truss shown in figure 7.5 carries the maximum tension force (*AD*, *DF*, *FH*, *HJ*, *JL*, or *LB*). What is the maximum tension force?

7.22 Find the forces in members *HJ*, *GJ*, and *GI* of the truss shown in figure 7.9.

7.23 Find the forces in members *JL*, *IL*, and *IK* of the truss shown in figure 7.9.

7.24 The roof truss shown in figure 7.10 is used in a building designed for a snow load of 30 lb/ft². The dead load is 5 lb/ft². These loads are based upon the horizontally projected area of the roof. If the trusses are placed on 24-in. centers, what is the stress in the wooden member *CE*, which has a cross section 3.5 × 1.5 in.?

7.25 Find the forces in members *EG*, *FG*, and *FH* of the truss shown in figure 7.25.

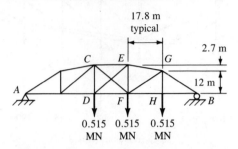

Figure 7.25   Problems 7.25 and 7.26.

7.26 What are the forces in members *CE*, *DF*, *CF*, and *DE* of the truss shown in figure 7.25? To solve the problem, assume that one of the crossed diagonals carries zero load. Does the opposite diagonal carry zero load? Is this reasonable? Can either of the crossed diagonals be removed from the structure without ill effect?

7.27 Under the dynamic conditions caused by striking a bump, the pair of trusses forming the primary structure of a racing car is subjected to the loads shown in figure 7.26. What are the loads in the pairs of members *AB*, *AD*, *BC*, *CD*, and *DE*?

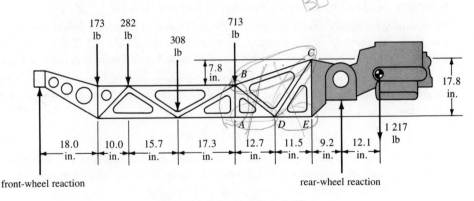

Figure 7.26   Problem 7.27.

7.28 A certain maneuver produces the loads on the aluminum tubular truss of an aircraft shown in figure 7.27. What are the loads in the pairs of members *EG*, *FG*, and *FH*? (There are two trusses.)

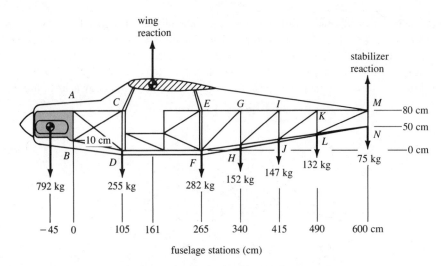

Figure 7.27   Problems 7.28 and 7.29.

7.29 What are the loads in members *AC*, *BC*, *AD*, and *BD* in the aircraft truss shown in figure 7.27? See Problem 7.28.

SUMMARY

☐ *Trusses* are structures composed of straight-line members linked together in triangles. The stresses due to the axial loads are called *primary stresses;* those due to bending are called *secondary stresses*. Primary stresses are usually the most important and can be accurately estimated by assuming the truss to be *pin-connected*.

☐ Trusses may be *simply supported spans* or *cantilevered*. The loads applied are called *live loads*. The weight of the structure itself is called the *dead load*. Trusses may be treated as parallel planar force systems in order to determine their support reactions.

☐ In the *method of joints*, the axial forces in the members of a truss are determined by making free-body diagrams of successive joints of the truss. Each joint becomes a concurrent planar force system:

$$\Sigma V = 0 \quad \text{and} \quad \Sigma H = 0$$

☐ In the *method of sections* the axial forces in the most critical members are determined from a free-body diagram of a selected group of triangular elements of a truss, called a section. The section becomes a general planar force system for which:

$$\Sigma V = 0, \Sigma H = 0, \quad \text{and} \quad \Sigma M = 0$$

CHAPTER EIGHT

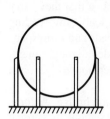

Shaped Structures: Cables, Suspension Bridges, Arches, and Pressure Vessels

The study of tension and compression structures now takes us to the gracefully curved shapes that have been contrived to replace less efficient bending members. The past is again revisited for a perspective on the evolution of such structures. The subject of suspension cables, which take their parabolic shape naturally, is pursued first. Then the reverse, the parabolic arch, is examined. Uniformly weighted catenary cables and the weighted catenary arch are studied.

For mechanical engineering technology students, especially, the subject of pressure vessels is introduced. First, the case of membrane stresses in thin-walled spherical and cylindrical vessels is discussed. Then the design rationalizations from the ASME *Boiler and Pressure Vessel Code* are presented, along with the code formulas for spherical and cylindrical vessels and elliptical, torispherical, and conical heads.

8.1 HISTORICAL PERSPECTIVE

Shaped structures, as well as being the world's most efficient structures, have captured the imagination of mankind from ancient times to the present. Tradition says that Czarist Russia was converted to Eastern Orthodox religion because envoys of the Czar were awestruck by the enormous domed church called Hagia Sophia, which still stands in Istanbul. The envoys reasoned that if the Byzantines could envelope such a large free space within a single building, they must surely have a close tie to the Almighty.

231

Even today in modern Istanbul, layers of civilization—modern upon Turkish upon Byzantine upon Roman upon ancient Greek—inspire the art of structures. Many are examples of shaped structures, which derive their strength from their special shapes. As well as domes such as Hagia Sophia, these structures include more recent Islamic mosques and the cables in the modern suspension bridge that spans the Bosporus, linking the two continents of Europe and Asia (figure 8.1).

In this chapter we consider structures which are carefully formed so that they carry only tension or compression, rather than bending, when they are under load. Any bending is small and of secondary importance. These shaped structures carry loads and form spans that would result in enormous bending stresses were the structures made of straight members or elements. The structures in this family are domes, arches, suspension bridges, and pressure vessels. All these structures are related by their shape.

An arch is a compression structure that is the reverse of a suspension bridge. A dome is a three-dimensional arch and can be thought of as an arch revolved about its vertical axis, sweeping out a shape as it is revolved. A pressure vessel is the opposite of a dome, in that it is a tension structure, although the pressure loads are different from gravity loads and require some different shape considerations.

Figure 8.1 The old and new in Istanbul.
(Photo by Näzän Alkan. Reprinted by
permission of Barbara Wolf.)

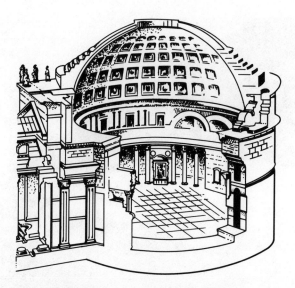

Figure 8.2   The Pantheon in Rome.

The dome and arch, which are compression structures, were present in ancient civilizations. This is because ancient builders had at their disposal only brittle materials, such as stone, which has good compression strength but is of little value in tension. The Romans are sometimes credited with inventing the dome and arch, although no one really knows if they first conceived of them. But certainly the Romans built many domes and arches, some of which still stand. Perhaps the most outstanding example of a Roman dome is the Pantheon, which means "temple to all gods." The Pantheon (figure 8.2) is a brooding bulk from the outside that is lofty and airy from within. A beam of sunlight from the central hole in its apex moves around the space, featuring one by one the niches where statues of ancient gods once stood.[1]

Most bridges from ancient times until the industrial revolution were stone arches. When iron and steel became available for construction, they were used to form arches for bridges, even though arches are compression structures, and iron and steel are used to greatest effect in tension. This design was probably used because people were accustomed to seeing bridges as arches. Captain James Buchanan Eads built his famous steel bridge across the Mississippi River at St. Louis in the shape of an arch. The Eads Bridge (shown in figure 8.3 in front of the modern Jefferson National Expansion Memorial Arch) is one of the world's great structures because of its advanced use of steel and the development of underwater "caisson" techniques, which were used to build the foundations for the piers between the 500-ft arch spans. The Eads Bridge has a graceful beauty and is remarkably functional. It provides a straight, wide roadway on a deck above the railway deck. The motorist is greeted by unobstructed views of the river. The bridge was an act of great vision for one who lived his entire life before the age of the motor car.

However, it was John Roebling, together with his son Washington, who fully realized the potential of steel in a shaped structure. John proved the concept of the suspension bridge with what is now called the Roebling Suspension Bridge, which links Covington, Kentucky, and Cincinnati, Ohio. It has a span of 1 057 ft and was completed in 1866. John and Washington Roebling then designed and built the first major suspension bridge.

Figure 8.3 St. Louis Riverfront with Eads
Bridge and Saarinen's Arch. (Photo courtesy of
Bill Stover.)

The Brooklyn Bridge was completed in 1883 in New York City (figure 8.4). This bridge,
with a span of 1 595 ft, made people aware of the enormous spans possible with steel.
John Roebling died of tetanus from an injury on the job site before construction began.
Washington, at age 32, took over the project but he was seriously injured by emerging too
rapidly from a pressurized caisson during the construction of the underwater foundations
for the bridge. Confined to his sickroom and with impaired vision, Roebling relied on his
wife Emily to finish the project. She visited the site daily, interpreted drawings and
calculations, checked materials, let contracts, and conferred with Washington. As project
managers, women technologists follow in the great tradition of Emily Roebling, without
whom the Brooklyn Bridge could not have been completed.[2]

Some credit the Roeblings with making possible not only a kind of bridge, but also
urban civilization itself. While there may be much romanticism in this notion, the fact
remains that most major urban areas face natural barriers that must be surmounted. Great
cities such as New York and San Francisco would be much less without their suspension
bridges. The Roeblings' bridge led to the remarkably beautiful Golden Gate Bridge in San
Francisco (4 200-ft span), which was the longest span in the world, until the Verrazano
Narrows Bridge (4 260 ft) in New York City was completed in 1964.

Suspension bridge building was stopped for some years by the discovery in 1941 at
Tacoma Narrows in Washington that wind-induced vibration can destroy a suspension

bridge. So it took much courage and confidence to build the Mackinac Bridge in the 1950s, which links the upper and lower peninsulas of Michigan. Continents were first linked by a suspension bridge at Istanbul in 1973. And now, a great suspension bridge on floating piers is envisioned linking Europe with North Africa at Gibraltar. Yet even these grand plans are dwarfed by some of the projects envisioned for the unhampered environment and exotic new materials of space.

8.2 PARABOLIC CABLES

Although in the lore of long-span structures, the arch came before the suspension bridge, it is best to start the study of shaped structures with suspension bridges. The suspension bridge cable falls naturally into an optimum structural shape—the parabola. To prove that the mathematical shape of a suspension cable is a parabola, one makes the assumption that the load upon the cable, w, (in such units as kips/ft or kg/m) is distributed uniformly over the horizontal span of the cable, as shown in figure 8.5. The coordinates of the x and y axes are drawn at the point of maximum sag of the cable. A free-body diagram is formed by cutting a section of the cable at the origin, o, and at any arbitrary point, a. At the origin the cable force, H, is horizontal, because a cable can only transmit force in the direction tangent to its curve, which is horizontal at the origin. At any arbitrary distance, x, the cable force is T (which is not necessarily equal in magnitude to H). This study will show that T is equal to H only at the origin.

Figure 8.4 John, Washington, and Emily Roebling's Brooklyn Bridge. (Photo courtesy of AP/Wide World Photos.)

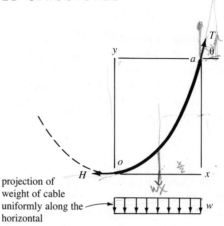

projection of
weight of cable
uniformly along the
horizontal

Figure 8.5 Free-body diagram of a section of
a uniformly loaded cable.

For equilibrium, the summation of moments must be equal to zero. Taking moments
about point a, where x and y are the coordinates of point a,

$$\sum M_a \circlearrowright: \quad Hy - wx\left(\frac{x}{2}\right) = 0$$

$$Hy = \frac{wx^2}{2}$$

and

$$\boxed{y = \left(\frac{w}{2H}\right)x^2} \tag{8.1}$$

Since w and H are constants, the quantity in parentheses, $w/2H$, is a constant. Therefore,
the equation has the form

$$y = Cx^2$$

which is the equation of a parabola having an axis of symmetry on the y axis. It can be
concluded, therefore, that the uniformly weighted cable has a parabolic shape.

For vertical equilibrium,

$$\sum V \uparrow: \quad T \sin \theta - wx = 0$$

$$T \sin \theta = wx \tag{8.2}$$

For horizontal equilibrium,

$$\sum H \leftrightarrow: \quad T \cos \theta - H = 0$$

$$T \cos \theta = H \tag{8.3}$$

handwritten: 1, 3, 8, 16, 20

Since $T \sin \theta$ is the vertical component of T, and $T \cos \theta$ is the horizontal component of T, the magnitude of T can be expressed using the Pythagorean theorem and relationships 8.2 and 8.3,

$$T = \sqrt{(wx)^2 + H^2} \qquad (8.4)$$

Relationships 8.1 and 8.4 can be used to find the forces in a uniformly loaded cable as shown in illustrative problem 8.1.

ILLUSTRATIVE PROBLEM

8.1 A pipeline spanning a river is supported by two suspension cables as shown in figure 8.6. The pipeline and contents weigh 25 lb/ft. The cables span 200 ft and have a sag of 32 ft. What is the tension in the main span of the cables at mid-span and the tension at the support tower?

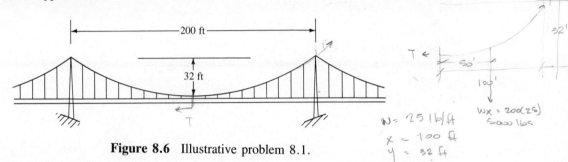

Figure 8.6 Illustrative problem 8.1.

Solution:
Solving relationship 8.1 for H where x equals one-half the span and y equals the sag,

$$H = \left(\frac{w}{2y}\right) x^2$$

$$H = \frac{25 \text{ lb/ft}}{2 \times 32 \text{ ft}} (100 \text{ ft})^2$$

$$H = 3\ 906 \text{ lb}$$

From relationship 8.4,

$$T = \sqrt{(wx)^2 + H^2}$$
$$T = \sqrt{(25 \text{ lb/ft} \times 100 \text{ ft})^2 + (3\ 906 \text{ lb})^2}$$
$$T = 4\ 640 \text{ lb}$$

Since there are two cables, the tension at mid-span is 1 953 lb, and the tension at the support tower is 2 320 lb for each cable.

8.3 TAUT CABLES

A cable that carries only its own weight, such as an electrical transmission line, does *not* take true parabolic shape, because its weight is not distributed uniformly along the horizontal, as shown in figure 8.5. As the slope of the cable increases towards the supports, a greater length of cable (and hence weight of cable) is associated with each unit of horizontal distance. Therefore, relationships 8.1 through 8.4 are not precise for a free-hanging cable. However, if the cable is taut (that is, has only a small amount of sag), these parabolic relationships are close enough to make a satisfactory approximation. If the sag of the cable is less than 10 percent of the span, the error in using relationships 8.1 through 8.4 on a freely suspended cable is within acceptable limits for most purposes.

To illustrate the variation of tension with sag in a shallow or taut cable, relationship 8.1 can be substituted into relationship 8.4 as follows:

From 8.1,

$$H = \frac{wx^2}{2y} \tag{8.5}$$

Substituting into 8.4,

$$T = \sqrt{(wx)^2 + \left(\frac{wx^2}{2y}\right)^2}$$

$$T = wx \sqrt{1 + \left(\frac{x}{2y}\right)^2}$$

Note from relationship 8.6 that if the cable is tightened so that the sag approaches zero, y approaches zero and the tension, T, approaches infinity. Since cable materials do not have infinite strength, it is impossible to tighten a cable such that it has zero sag. As the tension increases, the cable will stretch either elastically or plastically so that it sags (or ruptures).

ILLUSTRATIVE PROBLEM

8.2 A utility line is stretched across a span of 90 m. The insulated line has a mass of 0.32 kg/m. What is the minimum sag that the cable may have if the allowable tension in the cable is 15 000 N?

Solution:
From relationship 8.6,

(a) $$\frac{T}{wx} = \sqrt{1 + \left(\frac{x}{2y}\right)^2}$$

Squaring both sides,

(b)
$$\left(\frac{T}{wx}\right)^2 = 1 + \left(\frac{x}{2y}\right)^2$$

Substituting the values $T = 15\ 000$ N, $w = 0.32$ kg/m, and $x = 45$ m into the left side of **(b)**,

$$\left(\frac{T}{wx}\right)^2 = \left(\frac{15\ 000\ \text{N}}{0.32\ \text{kg/m} \times 45\ \text{m}} \times \frac{1\ \text{kg}}{9.8\ \text{N}}\right)^2 = 11\ 298$$

Therefore, relationship 8.6 **(b)** becomes

$$11\ 298 = 1 + \left(\frac{x}{2y}\right)^2$$

Solving for y,

$$y = \frac{x}{2\sqrt{11\ 297}}$$

$$y = 0.004\ 704x = 0.004\ 704 \times 45\ \text{m} = 0.211\ 68\ \text{m}$$

The sag must be greater than 0.212 m.

8.4 THE CATENARY CABLE (OPTIONAL)

When a cable has a sag of more than 10 percent of its span, its shape begins to deviate noticeably from a parabola. The deviation increases as the sag increases. The true shape of the curve is called a *catenary*. In fact, the definition of a catenary is the shape that is taken by a suspended chain, cable, or any strand that offers little resistance to bending. The shape of the catenary is defined by the relationship

$$y = C\left(\cosh\frac{x}{C} - 1\right) \tag{8.7}$$

where y and x are the vertical and horizontal distances, respectively, from the origin; C is again the constant; and cosh is a mathematical function called the hyperbolic cosine.

The tension forces in a catenary cable are given by the relationships

$$H = wC \quad \text{and} \quad T = \sqrt{(ws)^2 + H^2} \tag{8.8}$$

This is similar to relationship 8.4 for the parabolic cable, but in this case the weight is distributed along the cable length, s, rather than the horizontal span, x. The cable length becomes significantly different from the span when the sag of the cable gets greater than 10 percent. The cable length is related to x by the following expression:

$$s = C\sinh\frac{x}{C} \tag{8.9}$$

where C is the same constant as in relationship 8.7 and sinh is the hyperbolic sine.

For a complete derivation of relationships 8.7 through 8.9, see reference 3 at the end of this chapter.

ILLUSTRATIVE PROBLEM

8.3 A 90-m cable weighing 2.00 kg/m is suspended between two supports at the same elevation but 60 m apart. What is the amount of sag and the maximum tension in the cable?

Solution:

The constant, C, must be determined by trial and error (either from relationship 8.9 if the cable length is known or by relationship 8.7 if the sag is known). In this case the cable length, s, is known to be 45 m (half the total length). Therefore, relationship 8.7 can be used as follows:

$$s = C \sinh \frac{x}{C}$$

Letting $s = 45$ and $x = 30$, and dividing through by C:

(a) $$\frac{45}{C} - \sinh \frac{30}{C} = 0$$

Now different values of C must be tried until equation **(a)** is satisfied. To do this, let f be the value of the function for the different values of C. Equation **(a)** is satisfied if a value of C can be found such that f, defined as follows, equals zero:

$$f = \frac{45}{C} - \sinh \frac{30}{C}$$

As a first approximation, let $C = 10$.

$$f = \frac{45}{10} - \sinh \frac{30}{10} = 4.5 - 10.017\ 87 = -5.517\ 87$$

The proper value of C is the one that would make f equal to zero. Obviously, 10 is not the proper value of C.

Trying $C = 11$,

$$f = \frac{45}{11} - \sinh \frac{30}{10} = -3.521\ 95$$

which is closer to zero. So, we are moving in the right direction, but not far enough.

Next, we try some more values to get f closer to zero as follows:

C	f
15	−0.626 855
16	−0.371 227
17	−0.187 246
18	−0.052 803 8
18.3	−0.019 821 9
18.4	−9.469 27E−3
18.5	+5.788 8E−4

The value of f when C is 18.4 or 18.5 is very small, but more significantly, one is negative and the other is positive. Therefore, f goes through zero between 18.4 and 18.5. The solution for C is between 18.4 and 18.5. The answer is thus precise to within one-tenth of a meter, which is sufficient here. If greater accuracy is required, one can search further for the $f = 0$ point between $C = 18.4$ and 18.5 with an additional decimal place of accuracy, and then still another. It is easy to write a small computer program to find the solution for C to any degree of accuracy required. This means of solving problems is classified as a *method of successive approximations*.

Here we accept $C = 18.5$ as the answer, because it is closest to $f = 0$. The units for C are the same as for s and y.

Using relationship 8.7,

$$y = 18.5 \text{ m}\left(\cosh\frac{30 \text{ m}}{18.5 \text{ m}} - 1\right) = 30.1 \text{ m}$$

Therefore, the sag in the cable is 30.1 m.

From relationship 8.8,

$$H = wC = 2 \text{ kg/m} \times 18.5 \text{ m} = 37.0 \text{ kg}$$
$$T = \sqrt{(2 \times 45.0)^2 + (37.0)^2}$$
$$T = 97.3 \text{ kg}$$

The mid-span tension is 37 kg, and the maximum tension, at the supports, is 97.3 kg.

8.5 ARCHES

Arches are constructed in many shapes. Since they are made of materials that can resist bending, arches do not find their own shapes as do cables, but must be built to shape. Bending is undesirable in unreinforced, brittle materials, because bending can cause tension in a member. As discussed in chapter 3, most brittle materials are very weak in tension. Brittleness aside, pure compression load without bending makes more efficient use of the material.

A clever way to construct an arch so that it has all compression and no bending is to make it in the shape of a sagging cable, but upside down. (See figure 8.7.) Cables hang

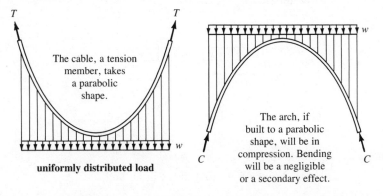

Figure 8.7 Deriving the arch from the cable.

naturally in a shape such that tension is the dominant internal force. If one constructs the identical shape with a rigid material, compression forces rather than tension forces would be expected when the shape is inverted to form an arch. If bending is of negligible effect with respect to tension in the cable, bending should also be of negligible or at least secondary effect with respect to compression in the arch.

The Parabolic Arch

Relationships 8.1 and 8.5 for parabolic cables apply also for parabolic arches, such as those of the Niagara River Bridge of figure 8.8, if one uses the coordinate system shown in figure 8.9. The origin is at the high point of an arch rather than at a low point of a cable. Therefore the vertical distances, y, have a negative sign. For negative y, H will be negative, indicating compression rather than tension.

In using relationships 8.4 or 8.6 for a parabolic arch, note that, because the term containing y is under the radical, T may be mathematically either positive or negative. If

Figure 8.8 The Niagara River Bridge. (Photo courtesy of Frank Seed, Niagara Falls Bridge Commission.)

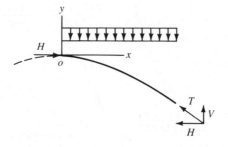

Figure 8.9 End thrust in an arch.

H is negative (compressive), T will also be negative, indicating compression. This rule must be true for horizontal equilibrium. In the case of arches, T can be thought of as indicating "thrust," which is compressive, rather than indicating "tension" as it does in the case of cables.

8.6 END THRUST

As a load pushes down on an arch, the arch pushes outward on its supports. This outward push is called *end thrust*. The more shallow the arch, the greater the end thrust. Ancient builders probably learned the hard way about end thrust, but evidence shows that they learned effectively. Arches were given firm abutments such as hillsides or mountainsides. If such natural abutments were not so considerate as to place themselves where people wanted to build arches, other means of abutment had to be provided. One means of doing this is to balance the end thrust of one arch with that of another. Some of the most remarkable structures of the ancient Romans are the aqueducts and bridges with cascades of arches, each balancing the end thrusts of its neighbors.

ILLUSTRATIVE PROBLEM

(Optional: The difference in elevation of the supports is a complexity that some teachers may prefer to bypass with introductory students.)

8.4 A gas pipeline bridge (shown in figure 8.10) over a creek is made by forming a pair of steel pipes, with outside diameters of 12 in., into a parabolic arch so that the pipes are simultaneously the conduits of the gas and the supporting structure. The span of the pipe bridge is 90 ft. One support has an elevation 2.5 ft above the other. The apex of the arch is 10 ft above the lowest support. Assuming that the load due to the weight of the two $\frac{1}{4}$-in.-thick pipes and their contents is 90 lb/ft, what is the maximum compressive stress in the pipe arches? What is the maximum horizontal end thrust?

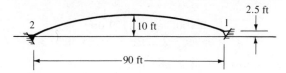

Figure 8.10 Pipe bridge of illustrative problem 8.4.

Solution:
Parabolic cables and arch problems are easy when the two supports are on the same horizontal level. In actuality, this is often not the case. This problem is complicated by the fact that the parabolic pipe arch has unequal supports. The horizontal location of the highest point must be determined. Therefore, it is necessary to first do some geometry to find the origin, o. Referring to figure 8.11, $y_1 = -7.5$, $y_2 = -10$, and $x_1 - x_2 = 90$ ft.

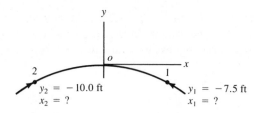

Figure 8.11 Solution of illustrative problem 8.4.

In relationship 8.1,

$$y = \left(\frac{w}{2H}\right) x^2$$

$w/2H$ can be treated as a constant, C. Therefore, at points 1 and 2, respectively,

(a) $$y_1 = Cx_1^2 \quad \text{or} \quad x_1 = \pm\sqrt{y_1/C}$$

and

$$y_2 = Cx_2^2 \quad \text{or} \quad x_2 = \pm\sqrt{y_2/C}$$

Squaring both sides of the geometrical expression $x_1 - x_2 = 90$ yields:

$$x_1^2 - 2x_1x_2 + x_2^2 = 8\ 100$$

Substituting for x_1 and x_2, and multiplying through by C yields:

$$y_1 \pm 2\sqrt{y_1y_2} + y_2 = 8\ 100C$$

Substituting -7.5 for y_1 and -10 for y_2,

$$-7.5 \pm 17.32 - 10 = 8\ 100C$$

Therefore,

$$8\ 100C = -0.179 \text{ or } -34.82$$
$$C = \text{either } -2.22\text{E}{-}5 \text{ or } -4.30\text{E}{-}3 \text{ ft}^{-1}$$

Using relationship (a) to find x_1 and x_2,

	For $C = -2.22$E-5		For $C = -4.30$E-3
$x_1 = \pm\sqrt{-7.5/-2.22\text{E}{-}5}$		or	$\pm\sqrt{-7.5/-4.30\text{E}{-}3}$
$x_2 = \pm\sqrt{-10/-2.22\text{E}{-}5}$		or	$\pm\sqrt{-10/-4.30\text{E}{-}3}$
$x_1 = \pm580$		or	±41.76 ft
$x_2 = \pm671$		or	±48.24 ft

From these choices of answers, clearly those in the first column are not acceptable. The absolute values of x_1 and x_2 must be less than the total span of 90. Of the answers in the second column, only $x_1 = +41.76$ and $x_2 = -48.24$ ft are acceptable. Therefore, the proper value of C is -4.30E-3 ft^{-1}.

Since

$$C = \frac{w}{2H}$$

$$H = \frac{w}{2C} = \frac{90 \text{ lb/ft}}{2 \, (-4.30\text{E}-3 \text{ ft}^{-1})}$$
$$H = -10\ 465 \text{ lb}$$

The maximum thrust will occur at x_2, because it is the farthest from the origin. Using relationship 8.4 to find the thrust at x_2,

$$T = \sqrt{(90 \text{ lb/ft} \times 48.24 \text{ ft})^2 + (-10\ 465 \text{ lb})^2}$$
$$T = 11\ 330 \text{ lb}$$

T must be compressive for an arch. The stress is obtained when dividing the thrust by the area of the two pipes.

$$\sigma = \frac{T}{A} = \frac{11\ 330 \text{ lb}}{2(12 \text{ in.} \times \pi \times 0.25 \text{ in.})} = 601 \text{ psi compression}$$

The end thrust is the horizontal compressive force of 10 470 lb. The maximum compressive stress is 601 psi.

PROBLEM SET 8.1

8.1 A suspension bridge has a main span of 1 000 m. The two cables have a sag of 90 m. If the bridge is to be designed for a uniformly distributed load of 100 kN/m and the allowable stress in the wire strands is 200 MPa, how many 2-mm-diameter strands are required in each of the two cables?

8.2 The footbridge in figure 8.12 has a parabolic cable with the geometry shown. What is the span of the bridge?

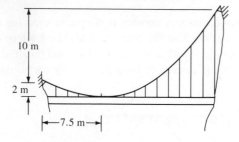

Figure 8.12 Problem 8.2.

8.3 A telephone cable weighs 0.85 lb/ft and spans 90 ft with a sag of 18 in. What is the maximum tension force in the cable?

8.4 A high-voltage power transmission line has towers spaced 100 yards apart. The cable sags 5 yards on level ground. The line is routed up a hill with a slope of 4 percent and strung so that the lower end of the cable is 3 yards above the lowest point of sag. What is the percentage increase in the cable stress caused by the hill? Hint: Solve for even supports, then for uneven supports. See illustrative problem 8.4.

8.5 A stainless steel guy wire is 0.120 in. in diameter. Its anchor is 10 ft lower than the support point, which is 150 ft away horizontally. The wire is tightened until it is horizontal at the anchor point. What is the stress in the wire?

8.6 A chain hangs in a catenary. It weighs 1.75 lb/ft of length and sags 4 ft over a span of 6 ft. What is the tension at the chain supports?

8.7 Using a personal computer, plot the points of a catenary curve that has a sag equal to its span. Compare the curve to a parabola of equal dimensions.

8.8 What is the horizontal end thrust on a parabolic arch with a 90-m span, a 60-m rise, and a 100-kN/m uniform load?

8.9 A shallow parabolic arch supporting a walkway between buildings has a rise of 5 ft and a span of 100 ft. What is the maximum compressive load in the arch if the uniform load is 1.2 kips/ft?

8.10 A concrete barrel-vaulted roof has parabolic spans of 20 ft with rises of 2 ft as shown in figure 8.13. The concrete roof is 2 in. thick and has a density of 150 lb/ft^3. Neglecting the strengthening effect of reinforcement, what is the allowable load per square foot of roof surface, if the allowable concrete stress is 1 000 psi?

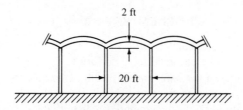

Figure 8.13   Problem 8.10.

8.11 Figure 8.14 shows a masonry construction system used for floors of multistoried structures found in some Mediterranean and Middle Eastern cities. It is called the "jack-arch" system. From what you know about end thrust, explain why a building constructed this way is extremely hazardous under earthquake conditions.

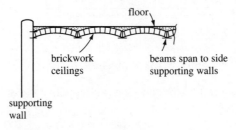

Figure 8.14   Problem 8.11.

8.12 How many feet of electrical transmission line would be necessary to string between two supports 200 m apart horizontally in order to have a sag equal to 20 percent of the span?

8.13 Write a computer program to calculate the constant C within 0.1 percent and the cable length for a uniform catenary cable, given the span and the sag.

8.14 Write a computer program giving the constant C (within 0.1 percent), the sag, and the cable tensions, if you know the cable length, the span, and the weight of the uniform cable.

8.7 CATENARY ARCHES (OPTIONAL)

Arches can be built in a variety of natural compression shapes called *weighted catenaries*. Arches rarely have uniform cross sections as do cables. Therefore, the catenary curve is different from that described by relationship 8.7. A remarkable example of a weighted catenary is the Jefferson National Expansion Memorial in St. Louis, shown in figure 8.3 along with the Eads Bridge.

The arch was an entry by architect Eero Saarinen in a design competition for a monument commemorating the Louisiana Purchase by Thomas Jefferson and the subsequent westward expansion of the United States. Saarinen's concept was an enormous free-standing arch of elegant structural design signifying the "Gateway to the West." The cross section of the arch is an equilateral triangle at all points, but the triangle increases in size from 15 ft on a side at the apex to 45 ft at the base. The increase is linear with y, the distance downward from the apex as shown in figure 8.15. The increase in cross-sectional size is intended to carry the increased compression near the base, but it also causes an increasing weight distribution as one approaches the supports. Thus arises the need for a weighted catenary shape if this arch is to be entirely in compression.

Saarinen proposed that the arch be built so that the centroid of its cross section fall on the weighted catenary curve,

$$y = 68.767\,2 \left(\cosh \frac{x}{99.668\,2} - 1 \right) \tag{8.10}$$

with a base of 598.44 ft and an elevation of 625.09 ft, making the outside dimensions 630 ft high and wide.* Its construction, completed in 1965, required that the two legs be jacked apart before the key section at the apex could be put into place so that the required H force was developed. When the required pushing force was applied, all loads were changed such that all sections carry pure compression.

The arch has $\frac{1}{4}$-in. stainless steel as a skin over a hollow carbon-steel framework above the 300-ft level and a concrete-filled steel framework below 300 ft. Although Saarinen's equation did not allow for the sharp discontinuity in weight distribution due to the concrete below 300 feet, the arch is still primarily a compression structure.

Though parabolic in appearance, the arch is in reality two levels of sophistication beyond the parabola. The parabola is shaped for a load uniform in the horizontal direction. The catenary is for a load uniformly distributed along the arc length. The Saarinen Arch is

---

*Source: Sverdrup and Parcel Engineers, St. Louis, Missouri.

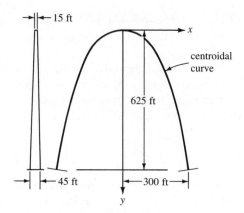

Figure 8.15 Dimensions of the Saarinen Arch.

designed for a load uniformly increasing in the y direction. Figure 8.16 provides a comparison of the three shapes.

During construction, the skin of the arch wrinkled in the lateral direction. Some think it was due to the differential thermal expansion between the stainless and carbon steels as the structure responded to the heat from the sun. Other than that cosmetic difficulty, the arch has been structurally sound and widely acclaimed for its aesthetics.

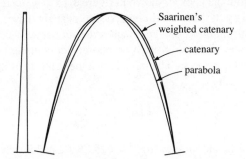

Figure 8.16 Comparison of actual shape of the Jefferson National Expansion Memorial Arch, a weighted catenary, to a parabola and a catenary curve of equal base and height.

PROBLEM SET 8.2

8.15 Using a personal computer or a calculator, calculate the x and y components of the centroid curve for the Saarinen Arch at 15-ft intervals of elevation from top to bottom.

8.8 PRESSURE VESSELS

The dominant features of petroleum refineries and food and chemical processing plants are their pressure vessels, which include storage tanks or silos, fractionating towers,

Figure 8.17 Pressure vessels in a petroleum refinery. (Photo courtesy of Atlantic Richfield Corp.).

boilers, cookers, digesters, heat exchangers, and catalytic cracking units. As shown in figure 8.17, such pressure-containing units are essential to industry. They must withstand enormous forces, often in a high-temperature environment, while containing materials that may be corrosive, abrasive, viscous, toxic, flammable, or radioactive. Failure of the vessels could mean the release of hazardous materials and/or explosive amounts of energy with resulting destruction of property, illness, injury, or loss of life.

The design of pressure vessels is governed by the American Society of Mechanical Engineers (ASME) *Boiler and Pressure Vessel Code*.[4] The ASME Code, an international standard, has been highly effective in reducing failures from the frequent boiler explosions that occurred after the turn of the century. Now such a failure is rare, usually caused by fire, accident, or failure of some part of the system other than the pressure vessel. The ASME Code provides rules, formulas, and allowable stresses for the safe design and manufacture of pressure-containing elements.

Most pressure vessels carry their loads by shell or membrane action. Such vessels are constructed in precise shapes so that material is used most efficiently. The stresses in the vessel walls are tension for internal pressure or compression for vacuum or external pressure. Bending is a secondary effect. These conditions can be achieved if the shapes

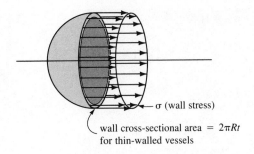

σ (wall stress)

wall cross-sectional area = $2\pi Rt$
for thin-walled vessels

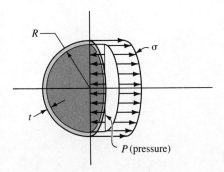

Figure 8.18 Section drawing of free-body
diagram of a sphere cut in half.

used are natural shapes of inflatable membranes such as bubbles or balloons. Hence,
pressure vessels are made in the shapes of spheres, cylinders, cones, or most any surface
of revolution. *Surfaces of revolution* are surfaces that have a circular intersection with any
plane perpendicular to the axis. The most common shapes are spheres and cylinders.

The stress in a sphere can be determined by membrane analysis, as shown in the
free-body diagram of figure 8.18.

Assuming that the weight of the vessel and its contents are negligible, the free body
is acted upon by pressure and stress at the cutting plane. The pressure and stress act over
their respective areas causing horizontal forces. Assuming that the wall thickness, t, is
small with respect to the radius of the vessel, the stress area is $2\pi Rt$. The pressure acts on
the area πR^2. For horizontal equilibrium,

$$\sum H \leftrightarrow: \quad -2\pi Rt\sigma + \pi R^2 P = 0$$

Solving for stress,

$$\sigma = \frac{PR}{2t} \qquad\qquad (8.11)$$

This is the fundamental relationship for membrane stresses in a spherical pressure vessel. The stress in the wall of a spherical pressure vessel is uniform in all directions. As will be seen later, this is not the case for other vessel shapes.

ILLUSTRATIVE PROBLEM

8.5 An aluminum storage tank is spherical in shape with a diameter of 20 ft. It is to be constructed for a pressure of 120 psi. What must be the minimum thickness if the membrane stress may not exceed 18 000 psi?

Solution:
From relationship 8.11,

$$t = \frac{PR}{2\sigma}$$

$$t = \frac{120 \text{ psi} \times 10 \text{ ft}}{2 \times 18\,000 \text{ psi}} \times \frac{12 \text{ in.}}{1 \text{ ft}}$$

$$t = 0.40 \text{ in.}$$

The vessel must have a wall thickness of at least 0.40 in.

Cylindrical Shells

The membrane stresses in a cylindrical shell depend upon the directions. Circumferential stress, σ_θ, is the stress that acts in a plane perpendicular to the shell axis. Boilermakers call this *hoop* stress. Meridional (or longitudinal) stress, σ_ϕ, acts in a plane that includes the shell axis. Longitudinal stress is also called *girth* stress. Hoop and girth stress can be determined from the free-body diagrams shown in figure 8.19.

Circumfer = hoop = σ_θ
Longitudinal = girth = σ_ϕ

For free-body diagram 8.19a, horizontal equilibrium requires that the pressure and stress forces be equal.

$$\Sigma H \leftrightarrow: \quad 2\pi Rt\sigma_\phi - \pi R^2 P = 0$$

$$\boxed{\sigma_\phi = \frac{PR}{2t}} \tag{8.12}$$

This is the relationship for meridional stress in a cylindrical pressure vessel. Note that it is the same as the stress in a spherical shell.

For free-body diagram 8.19b, the vertical forces due to pressure and stress must also be in equilibrium,

$$\Sigma V \updownarrow: \quad 2Lt\sigma_\theta - 2RLP = 0$$

Note that the arbitrary length, L, can be divided out of the expression. Therefore,

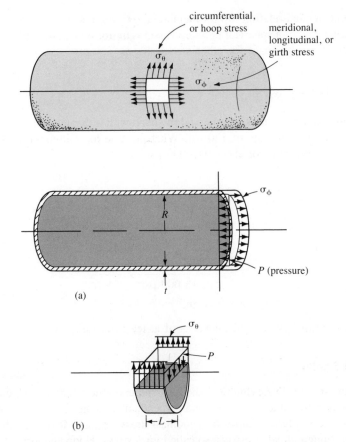

Figure 8.19 Free-body diagrams of cylindrical shell portions.

$$\sigma_\theta = \frac{PR}{t}$$

$\sigma_\theta = 2\sigma_\phi$ (8.13)

This is the relationship for circumferential stress in a cylindrical pressure vessel. It is exactly twice the meridional stress. Therefore, a cylindrical pressure vessel is most likely to fail by rupture in the direction of the circumferential stress, which could be along a longitudinal seam, provided that the vessel and seams are equally strong in all directions.

Since the circumferential stress normally governs, it is accurate to say that, by comparing relationships 8.11 and 8.13, the stress in a cylindrical shell is twice that of a spherical pressure vessel of the same radius and thickness. Thus, the sphere is the most efficient shape of all for pressure containment. But spheres are generally more expensive to manufacture, and for equal volumes they take more land or building space than do cylindrical pressure vessels.

ILLUSTRATIVE PROBLEM

8.6 Estimate the burst pressure of a welded cylindrical pressure vessel with spherical heads made of 155A titanium alloy in the annealed condition, if its diameter is 200 mm and the wall thickness is 5.00 mm. Assume the welds to be as strong as the vessel wall.

Solution:
The vessel is likely to burst in the circumferential direction in the cylindrical portion. Solving relationship 8.13 for P,

(a)
$$P = \frac{\sigma_\theta t}{R}$$

Bursting is a tension failure of the vessel wall. It is a full rupture failure. Therefore, the ultimate strength should be used here for stress. From Appendix I, the ultimate strength of 155A titanium alloy is 1 033 MPa.

$$P = \frac{1\ 033\ \text{MPa} \times 5.00\ \text{mm}}{100\ \text{mm}}$$

$$P = 51.7\ \text{MPa}$$

The result is the likely burst pressure for the pressure vessel.

Unfortunately, pressure in these units is not familiar to either the American or the European designer. The American thinks in terms of psi, and the European in atmospheres. Since technology is an international activity, first convert to psi.

$$P = 51.7\ \text{MPa} \times \frac{1\ \text{psi}}{6.89\text{E}{-}3\ \text{MPa}} = 7\ 500\ \text{psi}$$

For the European, one standard atmosphere is taken to be 14.7 psi. Therefore,

$$P = 7\ 500\ \text{psi} \times \frac{1\ \text{atmosphere}}{14.7\ \text{psi}} = 510\ \text{atmospheres}$$

Europeans doing pressure work usually call the unit of atmosphere the "bar." Think of "barometer," an instrument used to measure atmospheric pressure. To aid in conversion,

$$1\ \text{bar} = 14.7\ \text{psi} \times \frac{6.89\text{E}{-}3\ \text{MPa}}{1\ \text{psi}}$$

$$1\ \text{bar} = 0.101\ 3\ \text{MPa}$$

$$\boxed{1\ \text{MPa} \approx 10\ \text{bars}}$$

This result is accurate within 1.3 percent, and for common pressure vessel work is taken to be equal.

Going further,

$$1 \text{ bar} = 0.101\ 3 \text{ MPa} = 0.101\ 3\text{E6}\ \frac{\cancel{N}}{\cancel{m^2}} \times \frac{\cancel{m^2}}{1 \times \text{E4 cm}^2} \times \frac{1 \text{ kg}}{9.8\ \cancel{N}} = 1.033 \text{ kg/cm}^2$$

In pressure work, Europeans usually take

$$\boxed{1 \text{ bar} \approx 1 \text{ kg/cm}^2}$$

The European result is within 3 percent of the American, which considers an atmosphere to be 14.7 psi. European pressure gages are normally calibrated in bars or kg/mm^2 units.

PROBLEM SET 8.3

8.16 Estimate the burst pressure for a spherical pressure vessel made of AISI 1018 hot-rolled steel $\frac{1}{4}$ in. thick to a diameter of 60 in.

8.17 What are the hoop and girth stresses in a cylindrical pressure vessel, 350 cm in diameter and 12 mm thick, under a pressure of 23 bars?

8.18 A cold-drawn copper pipe is $\frac{1}{2}$ in. in diameter with a wall thickness of 0.035 in. Estimate its burst pressure.

8.19 What must be the thickness of an oxygen tank for a spacecraft if it is to be made of Type 316 stainless steel and must have a burst pressure of at least 600 psi? The tank is to be a fully annealed sphere with a 22.75-in. inside diameter.

8.20 Calculate the hoop and girth stresses in a cylindrical pressure vessel 10 ft in diameter and 90 ft long, having a wall thickness of 0.187 5 in. The pressure is 100 psi, and the material of construction is SA283 carbon-steel plate.

8.21 What is the wall stress in a Lucite® plastic sphere, 2.20 m in diameter and 18 mm in wall thickness, submerged to a depth of 300 m in sea water? It is used for undersea television work.

8.9 THE CODE DESIGN OF PRESSURE VESSELS FOR INTERNAL PRESSURE

Technologists may be involved in the design, construction, operation, or inspection of pressure vessels. For this reason, a brief introduction to the ASME *Boiler and Pressure Vessel Code* is included here.

The Code, consisting of eleven sections, comprises a collection of books that requires more than a shelf of a good-sized bookcase. The Code is continually updated with addenda from the ASME. Periodically, on multi-year intervals, the Code is completely replaced. The price of the Code makes it too expensive for most individuals. Therefore, the Code is normally a set of reference books owned by an industrial organization or library.

Of the eleven sections, the rules of Section VIII, Pressure Vessels, govern the majority of vessels. The other sections are for power boilers, nuclear vessels, reinforced plastic vessels, and material and welding control. Section VIII consists of two books, Division 1 and Division 2. Division 2 rules require a thorough stress analysis beyond the scope of this text, but offer the user lower safety factors that can be of economic advantage for unusually expensive or special vessels. Division 1 rules, discussed herein, require a fundamental knowledge of structures and materials. Hereinafter, the word "Code" shall mean ASME, Section VIII, Division 1.[4,5]

Paragraphs UG-27 and UG-32 of the Code give the thickness of shells for internal pressure as follows:

Spherical Shells:

$$t = \frac{PR}{2Sj - 0.2P} \quad \text{or} \quad P = \frac{2Sjt}{R + 0.2t} \qquad (8.14)$$

Longitudinal Stress in Cylindrical Shells (Circumferential Joints):

$$t = \frac{PR}{2Sj + 0.4P} \quad \text{or} \quad P = \frac{2Sjt}{R - 0.4t} \qquad (8.15)$$

Circumferential Stress in Cylindrical Shells (Longitudinal Joints):

$$t = \frac{PR}{Sj - 0.6P} \quad \text{or} \quad P = \frac{Sjt}{R + 0.6t} \qquad (8.16)$$

where t is minimum required thickness, P is the design or working pressure, R is the inside radius, S is the allowable stress at design temperature, and j is the joint efficiency.

If there were no joints, j would be equal to 1.00. When the pressure, P, is low with respect to the allowable stress, S, the second term in the denominators of these relationships is negligible. Under such conditions, Code formulas 8.14, 8.15, and 8.16 compare algebraically to the membrane stress relationships 8.11, 8.12, and 8.13, respectively.

Note: External pressure or vacuum conditions cause compression stresses in pressure vessels. Shell buckling is a likely mode of failure in such cases. A special part of the Code, not included here, must be used for external pressure. These formulas are for internal pressure conditions only.

Minimum Thickness

Material thickness is usually increased over the Code minimum. For carbon-steel vessels, a corrosion allowance of $\frac{1}{16}$ to $\frac{1}{8}$ in. in addition to the Code minimum thickness is normally added. This corrosion allowance is specified by the process engineer and is monitored through the life of the vessel either by ultrasonic inspection or by visual inspection during a shutdown of the unit. Coatings, liners, or weld-deposited cladding may also be used in addition to the required thickness. The vessel wall around openings or nozzles in the vessel must be carefully reinforced with extra thickness as specified in detail in the Code.

Allowable Stress

The allowable stress for Section VIII, Division 1, is essentially one-fourth the ultimate strength. Ultimate strength rather than yield strength is used because the catastrophic bursting of the vessel is more feared than the dimensional instability of yielding. The conservative safety factor of four takes into consideration variations in material properties, construction inaccuracies, the fact that stress concentrations and secondary stresses are ignored, and the dire consequences of a pressure vessel explosion. However, yield strength is not completely ignored. Two-thirds of the yield strength may be the allowable stress if it is lower than one-fourth the ultimate strength.

The effects of temperature are also considered. Values of the allowable stresses for various materials at temperatures are included in the Code. One must use the allowable stress for the operating temperature specified. At elevated temperatures, the material may begin to creep. *Creep* is a slow and progressive elongation under sustained loads. In the creep range, the allowable stresses will be 100 percent of the stress for a creep rate of 0.01 percent in 1 000 hours, or 80 percent of the stress for creep rupture at the end of 100 000 hours, if these stress values are lower than those based upon the yield or ultimate strengths.

Table 8.1 is an excerpt of some of the allowable stress values from the Code.

Table 8.1 Section VIII, Division 1, Code-allowable stresses for low-carbon steel in tension (ksi) at temperatures in Fahrenheit

		−20° to 650°	*700°*	*750°*	*800°*	*850°*	*900°*	*950°*
SA36	plate	12.7	—	—	—	—	—	—
SA285C	plate	13.8	13.3	12.1	10.2	8.4	6.5	—
SA53	pipe	12.0	11.7	10.7	9.3	7.9	6.5	—

Joint Efficiency

Pressure vessel joints are critical. Most pressure vessel joints are butt welds. They may be multipass welds applied to plates with edges previously machined smooth and shaped. Most often, welds are machine-deposited using the submerged arc or gas-shielded arc processes. When a weld contains a crack, too many voids, or too many inclusions of dirt or slag, the weld metal must be ground out and redeposited. Radiographic inspection is the best way to discover weld defects. Therefore, for full x-ray, the joint efficiency, *j*, of 1.00 may be used. If the welded joint is spot x-rayed, a joint efficiency of 0.85 is used. A joint efficiency of 0.70 must be used for butt-welded joints, if not x-rayed.

ILLUSTRATIVE PROBLEM

8.7 What must be the thickness of the pressure vessel shown in figure 8.20 to be made of SA285C low-carbon steel plate and which must operate at a temperature of 850° F and a pressure of 1 000 psi? The inside diameter of the vessel is to be 30 in., and it is to have a length of 120 in. with spherical heads. The corrosion allowance is $\frac{1}{16}$ in. Longitudinal welds will have full x-ray. Circumfreitudinal welds and those in the sphere will have no x-ray.

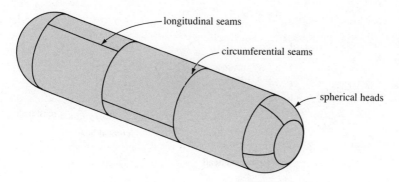

Figure 8.20   Illustrative problem 8.7.

Solution:

The allowable stress, *S*, is 8 400 psi. Using Code formula 8.14 for the heads,

$$t = \frac{PR}{2Sj - 0.2P} = \frac{1\ 000\ \cancel{psi} \times 15\ \text{in.}}{(2 \times 8\ 400\ \cancel{psi} \times 0.7) - (0.2 \times 1\ 000\ \cancel{psi})}$$

$$t = 1.298\ \text{in.}$$

Using Code formula 8.15 for the longitudinal stresses in the cylinder,

$$t = \frac{PR}{2Sj + 0.4P} = \frac{1\ 000 \times 15}{(2 \times 8\ 400 \times 0.7) + (0.4 \times 1\ 000)}$$

$$t = 1.234\ \text{in.}$$

Using Code formula 8.16 for the circumferential stress in the cylinder,

$$t = \frac{PR}{Sj - 0.6P} = \frac{1\ 000 \times 15}{(8\ 400 \times 1) - (0.6 \times 1\ 000)}$$

$$t = 1.923\ \text{in.}$$

The circumferential stress governs the thickness of the cylinder. Adding the corrosion allowance and going up in size to the next likely stock thickness of plates,

$$t = 1\tfrac{3}{8}\ \text{in. for the heads}$$

and $\qquad t = 2$ in. for the cylindrical walls of the vessel

8.10 HEADS FOR PRESSURE VESSELS

Spherical heads are highly efficient, but they are expensive to manufacture. For large diameters or thicknesses, they must be made in pieces and welded together as shown in figure 8.20. Elliptical heads and torispherical heads can be formed in one piece, either by spinning or pressing. Some other types of pressure vessel heads other than spherical are shown in figure 8.21.

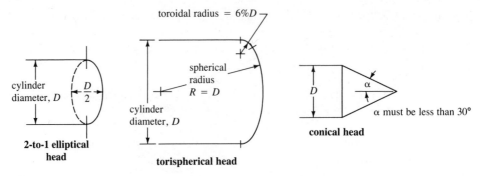

Figure 8.21 Other types of pressure vessel heads.

The Code formulas governing the design of these special heads for pressure on the concave side are:

2-to-1 Elliptical Head:

$$t = \frac{PD}{2Sj - 0.2P} \quad \text{or} \quad P = \frac{2Sjt}{D + 0.2t} \tag{8.17}$$

6-Percent Torispherical Head:

$$t = \frac{0.885PD}{Sj - 0.1P} \quad \text{or} \quad P = \frac{Sjt}{0.885D + 0.1t} \tag{8.18}$$

Conical Head:

$$t = \frac{PD}{2\cos\alpha(Sj - 0.6P)} \quad \text{or} \quad P = \frac{2Sjt\cos\alpha}{D + 1.2t\cos\alpha} \tag{8.19}$$

PROBLEM SET 8.4

8.22 A spherical pressure vessel was found in storage and measured. It has a thickness of 0.234 in. and an outside diameter of 27.35 in. It is known to be made of SA36 carbon-steel plate. What is its pressure rating for air compressor service in the condition as found? No radiographic inspection is intended.

8.23 What is the pressure rating for 12-in. schedule 40 seamless pipe of SA53 carbon steel at a temperature of 800° F?

8.24 What should be the thickness of a 2-to-1 elliptical head made of SA285 grade C material for a 24-in.-diameter heat exchanger shell where the operating conditions are 785° F and 324 psi?

8.25 A digester for a paper pulp mill is a vertical cylindrical shell 75 ft high and 25 ft in diameter. It is filled with a material with a density of 60 lb/ft³. What must be the thickness of the conical bottom, which has a 30° slope from the vertical, if it is made of SA53 plate and is to be operated at 215° F? There will be no x-ray, and the corrosion allowance is $\frac{1}{16}$ in.

8.26 A 6-percent torispherical head is to close a 70-in.-diameter vessel made of 18-8 stainless steel, having an allowable stress of 18 000 psi. What must be the thickness of the head if it is to have no corrosion allowance and must contain a pressure of 215 psi?

8.27 A vertical cylindrical storage tank is to be constructed of SA36 plate for an oil refinery tank farm. The tank will have a height of 75 ft and a diameter of 200 ft. The vessel is to have a spot x-ray on the vertical welds. The corrosion allowance is to be $\frac{1}{16}$ in. What must be the thickness of the lowest course of plates for the tank (which is to be designed for water, although it will be put into light hydrocarbon service)?

8.28 A high-pressure ammonia reactor is a cylinder that must operate at 6 000 psi and 700° F. The inside diameter is 30 in. The material is to be a 16 Cr–12 Ni–2 Mo high-alloy steel forging with an allowable stress of 12 900 psi at operating pressure. What must be the wall thickness according to Section VIII, Division 1 rules?

SUMMARY

☐ The shape of a *parabolic cable* is:

$$(8.1) \quad y = \left(\frac{w}{2H}\right)x^2$$

☐ The tension in a parabolic cable is given by:

$$(8.5) \quad H = \frac{wx^2}{2y}$$

$$(8.6) \quad T = wx\sqrt{1 + \left(\frac{x}{2y}\right)^2}$$

☐ The *catenary cable* has the shape:

$$(8.7) \quad y = C\left(\cosh\frac{x}{C} - 1\right)$$

☐ The tension in the catenary cable is given by:

$$(8.8) \quad H = wC \quad \text{and} \quad T = \sqrt{(ws)^2 + H^2} \quad \text{where} \quad s = C\sinh\frac{x}{C}$$

☐ Arches are the opposite of free-hanging cables. The relationships for cables may be used for similar arches, but the tension forces become compression. *End thrust* is the compression force at the supports.

☐ The tensile stress in a thin-walled spherical pressure vessel is:

$$(8.11) \quad \sigma = \frac{PR}{2t}$$

☐ The *meridional* or *girth stress* in a cylindrical pressure vessel is:

$$(8.12) \quad \sigma_\phi = \frac{PR}{2t}$$

☐ The *circumferential* or *hoop stress* is:

$$(8.13) \quad \sigma_\theta = \frac{PR}{t}$$

☐ In the metric system, one atmosphere of pressure is called a *bar*. One bar is approximately equal to 1 kg/cm^2.

☐ The ASME Code formulas are extended from the above stress formulas for thin-walled pressure vessels.

REFERENCES

1. *Curvilinear Forms in Architecture*. Portland Cement Association. Chicago, 1966.

2. Billington, D. P. *The Tower and the Bridge*. New York: Basic Books, 1983.

3. Beer, F. P. and Johnston, E. R. *Mechanics for Engineers: Statics and Dynamics*. New York: McGraw-Hill, 1972.

4. *Boiler and Pressure Vessel Code,* Section VIII, Division I. American Society of Mechanical Engineers, 1980.

5. Jawad, M. H. and Farr, J. R. *Structural Analysis and Design of Process Equipment*. New York: John Wiley & Sons, 1984.

CHAPTER NINE

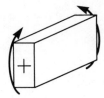

Bending Members

The parallel planar force system is now treated in detail as a means of solving for reactions that are external forces. The subject is continued to include shearing and bending moments, which are internal forces. Shearing force and bending moments are immediately applied to the study of stresses in beams. Having thoroughly considered tension and compression, we move on in this chapter to the study of bending. Bending is approached as Galileo approached it from the point of view of rectangular beams. Both normal bending stresses and shearing stresses are considered for the special case of rectangular beams. Wooden beams, which are usually rectangular, are treated as a practical application of the theory.

9.1 BEAMS AND LEVERS: PARALLEL FORCE SYSTEMS

Beams and levers are actually the same kind of structural members. They are *bending* members, also called *flexure* members, and they rely on bending resistance to carry loads. A lever is used to change the direction of a force or to gain a mechanical advantage. A beam is used to support a load either by spanning two supports or by suspending a load from one side as a cantilever. Yet, as shown in figure 9.1, the free-body diagrams of beams and levers are identical to each other; they can also be thought of as inverses of one another.

As can be seen from their free-body diagrams, beams and levers are examples of parallel planar force systems. The lines of action of all the forces are parallel. In this case the parallel direction is vertical. No horizontal forces are applied. Therefore, there are no forces that will cause a horizontal reaction.

The forces and/or reactions on either beams or levers can be determined from two equilibrium conditions. The sum of vertical forces on the free-body diagram and the sum of moments about any point on the free-body diagram must equal zero.

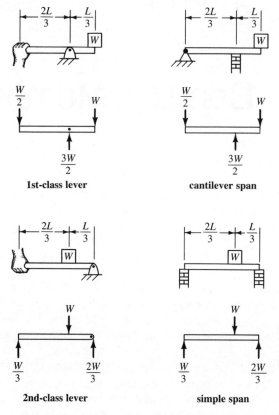

Figure 9.1 Free-body diagrams show that levers and beams are identical structural members.

9.2 DISTRIBUTED LOADS

As discussed in chapter 3, all loads are in fact distributed. Even so-called concentrated loads are distributed over a small, perhaps microscopic, area. However, many loads are distributed over a large section of a beam or a lever. In fact, the job of a beam is often to gather up a lot of distributed load and transfer that load to supports.

Distributed loads are represented by diagrams as shown in figure 9.2. Uniformly distributed and uniformly increasing loads are common. The height of the diagram represents the load per length in units of kips/ft, lb/in., N/m, or kg/cm.

For the purpose of calculating external reactions, distributed loads can be reduced to equivalent concentrated forces. In the case of a uniform load it is intuitive that the equivalent concentrated load must be equal to the weight per unit length, w, times the length. The location of the equivalent concentrated load must be in the center of the load package that it replaces.

In the case of a uniformly increasing load, the magnitude of the equivalent concentrated load can be rationalized as follows. The load goes from zero to w_o over the length of

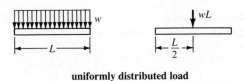

uniformly distributed load

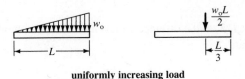

uniformly increasing load

Figure 9.2 Commonly encountered distributed load packages and their equivalent concentrated loads.

the load package. The average load must, therefore, be $w_o/2$. The total effect is then $w_oL/2$.

However, the location of that equivalent concentrated load is not obvious. The location can be determined from the fact that the concentrated load must produce the same moment as the distributed load. Referring to figure 9.3a, the distributed load at any horizontal distance x is given by the relationship,

(a)
$$w = \frac{w_o}{L}x$$

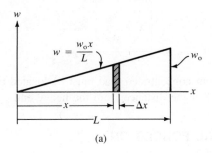

(a)

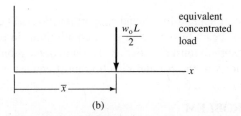

(b)

Figure 9.3 Determining the location of the equivalent concentrated load for a uniformly increasing distributed load.

The incremental moment, ΔM, due to the distributed load acting over an incremental length of the beam, Δx, has a moment arm of x.

(b) $$\Delta M = w \times \Delta x$$

Substituting w from relationship (a),

(c) $$\Delta M = \frac{w_\mathrm{o}x^2}{L}\Delta x$$

The resultant force, $w_\mathrm{o}L/2$, at the distance $\bar{x}$ must have the same moment as the sum of all the ΔM along the beam.

(d) $$\frac{w_\mathrm{o}L}{2}\bar{x} = \Sigma\,\Delta M$$

Substituting ΔM from relationship (c),

(e) $$\frac{w_\mathrm{o}L}{2}\bar{x} = \Sigma\,\frac{w_\mathrm{o}}{L}x^2\,\Delta x$$

Expressing the summation as an integral,

(f) $$\frac{w_\mathrm{o}L}{2}\bar{x} = \frac{w_\mathrm{o}}{L}\int_0^L x^2dx = \frac{w_\mathrm{o}}{L}\left.\frac{x^3}{3}\right]_0^L$$

$$\frac{w_\mathrm{o}L\bar{x}}{2} = \frac{w_\mathrm{o}L^2}{3}$$

Therefore,

$$\boxed{\bar{x} = \frac{2}{3}L}$$ (9.1)

In other words, the equivalent concentrated load must be located two-thirds of the way from the origin *or* one-third of the distance from the heavy end.

9.3 SUPPORT REACTIONS—EXTERNAL FORCES ON BENDING MEMBERS

External forces consist of applied forces and support reactions. The reactions must be of such magnitude that the beam or lever is in static equilibrium. In parallel force systems, two conditions of equilibrium must be satisfied. The sum of the moments must equal zero, and the sum of the forces in the parallel direction must equal zero.

ILLUSTRATIVE PROBLEM

9.1 A beam 20 ft long is simply supported at its ends. It carries a distributed load which increases uniformly from 10 kips/ft at end A to 30 kips/ft at end B. What are the support reactions?

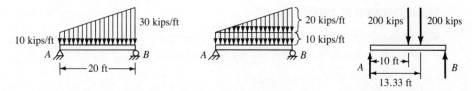

(a) the simply supported beam under a uniformly increasing distributed load

(b) the load divided into rectangular and triangular packages

(c) a free-body diagram with distributed loads replaced by equivalent concentrated loads

Figure 9.4 Illustrative problem 9.1.

Solution:

The beam and load are shown in figure 9.4a. First the load is divided into a triangular load package with a maximum value of 20 kips/ft and a rectangular load package of 10 kips/ft over the entire length of the beam (figure 9.4b).

The rectangular load package can be replaced by an equivalent concentrated load of 10 kips/ft × 20 ft = 200 kips at mid-span as shown in figure 9.4c.

The triangular load package can be replaced by a concentrated load of 20 kips/ft × 20 ft/2 = 200 kips two-thirds of the way from the light end of the load package per relationship 9.1.

The supports at A and at B are removed and replaced by unknown forces, making figure 9.4c a complete free-body diagram. The reactions are then solved from the equilibrium conditions $\Sigma M = 0$ and $\Sigma V = 0$.

$$\Sigma \, M_A \circlearrowleft : 200 \times 10 + 200 \times 13.33 - 20B = 0$$
$$B = 233 \text{ kips}$$
$$\Sigma \, V \updownarrow : \ A + B - 200 - 200 = 0$$
$$A = 400 - B = 400 - 233$$
$$A = 167 \text{ kips}$$

The reactions at supports A and B are 167 and 233 kips upward, respectively.

A support reaction might also be considered to be a concentrated moment as in the beam built into a wall at one end, as shown in figure 9.5.

This is a *cantilever* beam. Actually, the beam must extend on into the wall, where it experiences a complicated set of distributed loads over the small length that is within the wall. But the net result of the complex distributed support loads within the wall can accurately be assumed to be a concentrated moment just where the beam emerges from the wall. For free-body diagrams, the beam is cut at that point of emergence.

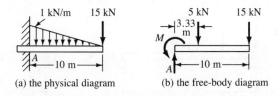

(a) the physical diagram

(b) the free-body diagram

Figure 9.5 Illustrative problem 9.2.

ILLUSTRATIVE PROBLEM

9.2 A cantilever beam carries a uniformly decreasing load and a concentrated load as shown in figure 9.5. What are the support reactions?

Solution:

The uniformly decreasing load can be replaced by an equivalent concentrated load one-third of the way from the heavy end of the load package. A free-body diagram can be constructed by cutting the beam at the wall and representing the support capability of the wall with a concentrated moment, M, and the force, A.

Applying the two equilibrium conditions:

$$\Sigma M_A \circlearrowleft : -M + 5 \times 3.33 + 15 \times 10 = 0$$
$$M = 166.7 \text{ kN} \cdot \text{m}$$
$$\Sigma V \updownarrow : A - 5 - 15 = 0$$
$$A = 20 \text{ kN}$$

The reactions at the support are a moment of 166.7 kN · m counterclockwise and a vertical force of 20 kN.

PROBLEM SET 9.1

9.1–9.2

What is the amount of force, F, required to raise the weight using the levers of figures 9.6 and 9.7?

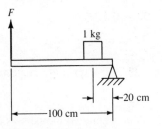

Figure 9.6   Problem 9.1.

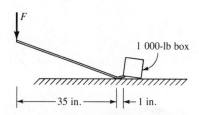

Figure 9.7   Problem 9.2.

9.3–9.12

What are the reactions at supports A and B for the simply supported beams shown in figures 9.8 through 9.17?

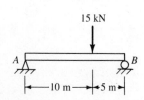

Figure 9.8   Problem 9.3.

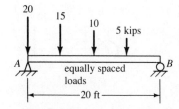

Figure 9.9   Problem 9.4.

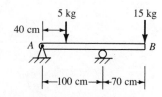

Figure 9.10   Problem 9.5.

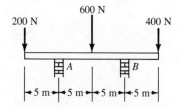

Figure 9.11 Problem 9.6.

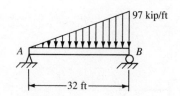

Figure 9.12 Problem 9.7.

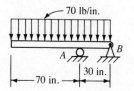

Figure 9.13 Problem 9.8.

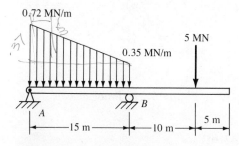

Figure 9.14 Problem 9.9.

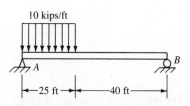

Figure 9.15 Problem 9.10.

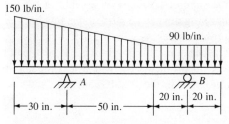

Figure 9.16 Problem 9.11.

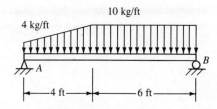

Figure 9.17 Problem 9.12.

9.13–9.14

What are the support reactions for the cantilever beams shown in figures 9.18 through 9.19?

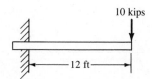

Figure 9.18 Problem 9.13.

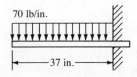

Figure 9.19 Problem 9.14.

9.15–9.16

What are the required operating forces and the fulcrum reactions on the levers shown in figures 9.20 and 9.21?

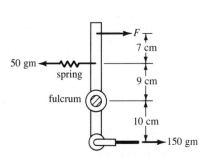

Figure 9.20   Problem 9.15.   **Figure 9.21**   Problem 9.16.

9.4 USING A COMPUTER TO SOLVE PARALLEL PLANAR FORCE SYSTEMS (OPTIONAL)

While a computer is not needed to solve for the support reactions on a beam or for the external forces on a lever, the necessary computer program can be written in a few lines of code. Since the solution of many subsequent problems begins with the determination of reactions, such a program will be useful later. Furthermore, this program is a fundamental building block of solutions to more complex beam problems for which a computer is needed. Illustrative problem 9.3 gives such a program in the BASIC language and discusses its use.

ILLUSTRATIVE PROBLEM

9.3 Write a program in BASIC to solve for the support reactions for a beam on two supports that may have overhangs on either side and may be loaded with any number of forces. Use the program to determine the reactions on the beam loaded as shown in figure 9.22.

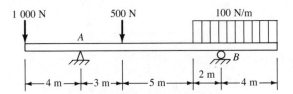

Figure 9.22   Illustrative problem 9.3.

Solution:
The loads in figure 9.22 can be described as three downward forces, each with a magnitude F and a location X to the right of support A, as shown in figure 9.23. Reactions A and B are upward and separated by the span.

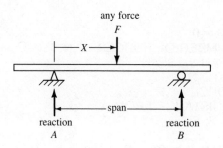

Figure 9.23   Illustrative problem 9.3.

In the program, $\Sigma\, V$ and $\Sigma\, M$ are indicated by SF and SM, respectively. SF and SM are set initially to zero as shown in the flowchart, figure 9.24. Then the number of applied forces, N, in this case 3, is entered. For each force in turn, the program asks for the F and X. It increments SF by F and SM by F*X. When all the forces have been entered it calculates A and B from the equilibrium relationships:

$$B = \Sigma\, M/\text{span}$$
$$A = \Sigma\, V - B$$

The resulting reactions A and B are then printed.

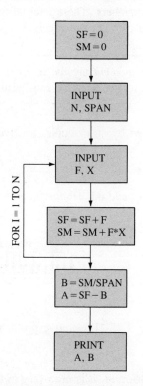

Figure 9.24   Illustrative problem 9.3.

Program:

```
100   SF = 0: SM = 0
110   INPUT "NUMBER OF FORCES =",N
120   INPUT "SPAN =",SPAN
130   FOR I = 1 TO N
140   INPUT "FORCE =",F
150   INPUT "X DISTANCE FROM A = ",X
160   SF = SF + F: SM = SM + F*X
170   NEXT I
180   B = SM/SPAN: A = SF − B
190   PRINT "REACTIONS AT A AND B ARE",A,B
200   GO TO 100
210   END
```

Illustrative problem 9.3 is solved by inputting 3 for N, 10 for SPAN, and 1000, −4, 500, 3, 600, and 11 for the three forces and their respective moment arms. The answers received are 1 750 and 350, respectively, for A and B. The reactions are 1 750 N and 350 N, both upward, at A and B.

9.5 INTERNAL FORCES—SHEARING FORCE AND BENDING MOMENT

External forces, including applied loads and reactions, cause internal forces that are variable along the axes of bending members. These internal forces are called *shearing force* and *bending moment*. They must be defined together.

> *Shearing force* and *bending moment* are the internal forces at a particular cross section on the axis of a beam necessary for equilibrium.

This definition is best illustrated by an example, as given in illustrative problem 9.4.

ILLUSTRATIVE PROBLEM

9.4 What are the shearing forces and the bending moments at cross sections C and D of the beam shown in figure 9.25?

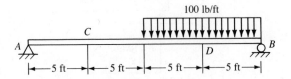

Figure 9.25 Illustrative problem 9.4.

Solution:
Before finding the internal forces, it is necessary to first find the support reactions. By the methods of section 9.3, the reaction at A can be determined to be 250 lb, and

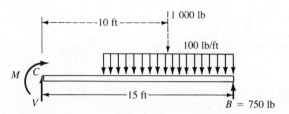

Figure 9.26 Free-body diagram of portion of the beam to the right of cross section C.

B to be 750 lb, both upward. To find the bending moment and shearing forces at section C, it is necessary to make a free-body diagram of a portion of the beam from section C to the end of the beam. Either end will do. Here the portion of the beam to the right of section C is shown in figure 9.26. The effect of the portion on the left is represented by the moment, M, and the shearing force, V, placed on the diagram as shown.

Not only must the entire beam be in equilibrium, but every portion of the beam must be in equilibrium as well. Therefore, equilibrium relationships can be written for the portion of the beam as follows:

$$\Sigma M_C \circlearrowleft : M + 10 \times 1\ 000 - 15 \times 750 = 0$$
$$M = -10\ 000 + 11\ 250 = 1\ 250\ \text{lb} \cdot \text{ft}$$
$$\Sigma V \uparrow : V - 1\ 000 + 750 = 0$$
$$V = 250\ \text{lb}$$

At cross section C the bending moment, M, is $1\ 250\ \text{lb} \cdot \text{ft}$ and the shearing force, V, is 250 lb.

These internal forces are necessary at C to hold the beam in equilibrium. The same values are obtained if the portion of the beam to the left of section C is used as the free-body diagram.

At cross section D, the bending moment and shearing force can be determined similarly by making a free-body diagram of the section of the beam to the right of D, as shown in figure 9.27. Carefully note that the distributed load is included in these free-body diagrams for shearing force and bending moment, even though the distributed load may have been replaced earlier with an equivalent concentrated load when calculating the support reactions. *We bring back the distributed loads when we calculate shear and moment.*

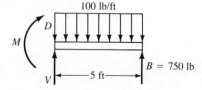

Figure 9.27 Free-body diagram of portion of beam to the right of cross section D.

Writing the equilibrium equations for this new free-body diagram:

$$\Sigma\, M_D \circlearrowleft : M + 100\ \text{lb/ft} \times 5\ \text{ft} \times 5\ \text{ft/2} - 750\ \text{lb} \times 5\ \text{ft} = 0$$
$$M = -1\ 250 + 3\ 750 = 2\ 500\ \text{lb} \cdot \text{ft}$$
$$\Sigma\, V \updownarrow : V - 100\ \text{lb/ft} \times 5\ \text{ft} + 750\ \text{lb} = 0$$
$$V = 500\ \text{lb} - 750\ \text{lb} = -250\ \text{lb}$$

At cross section D, the bending moment, M, is 2 500 lb · ft and the shearing force, V, is −250 lb. Note that the bending moment is greater at D than at C. The shearing force is negative at D, whereas it was positive at C. Both bending moment and shearing force are highly variable along the length of a beam. Chapter 11 is devoted to techniques for drawing diagrams of bending moment and shearing force.

9.6 SIGN CONVENTION FOR SHEARING FORCE AND BENDING MOMENT

As with other types of problems, one can choose one's own sign convention and follow it through. But, a standard sign convention for shear force and bending moment is used consistently throughout engineering and technology. It takes some time to become accustomed to the standard way signs are applied to these internal forces. Therefore, it is best not to change from one sign convention to another. The positive directions are shown in figure 9.28.

At section 1, the left part of the beam has been removed. Therefore, the positive direction of the bending moment is clockwise. The positive direction for the shearing force is upward. This is the case when the material to the left is removed in making the free-body diagram.

At section 2, the material to the right has been removed. Therefore, the positive direction for bending moment is counterclockwise and the positive direction for shearing force is downward.

Alternatively, think of it as shown in figure 9.29. If the two pieces are put back together, the M's and V's are in opposite directions. Therefore, they cancel one another and disappear. These internal forces only become exposed when the beam is taken apart for a free-body diagram.

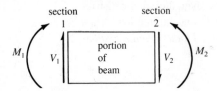

Figure 9.28 Positive directions for shearing forces and bending moments.

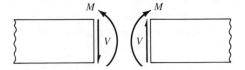

Figure 9.29 An alternative method of representing the positive directions for shearing force and bending moment.

9.7 LOCATION OF MAXIMUM BENDING MOMENT

The tendency of a loading system to bend a beam (or lever) at any cross section is called *bending moment*. The bending moment is variable along the length of the bending member. Usually the location of the maximum bending moment is determined. So a free-body diagram can be constructed to find the value of the maximum bending moment. If the

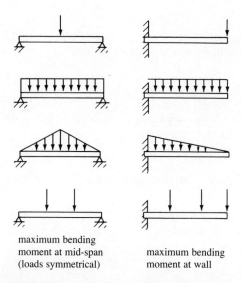

maximum bending
moment at mid-span
(loads symmetrical)

maximum bending
moment at wall

Figure 9.30 The general cases where the
location of the maximum bending moment is
known at the outset.

maximum bending moment is known, a beam or lever of constant cross section for it can
then be designed or given a load rating. Most bending members are of constant cross
section throughout.

In some loading systems, the location of the maximum bending moment can be
drawn from personal experiences such as breaking sticks and wood. For example, the
maximum bending for a simply supported beam with a single concentrated load at the
center will occur at mid-span. Another obvious case is a concentrated load on a cantilever
beam. The maximum bending moment would be expected to occur at the support. This
intuition can be carried further, as shown in figure 9.30. In general, these three rules
apply:

☐ For a simply supported beam carrying a symmetrical set of downward loads, the maximum bending moment will be at mid-span and the maximum shear force will be at the
supports.

☐ A cantilever beam with any pattern of downward loads will *always* have the maximum
bending moment and the maximum shearing force at the built-in end.

☐ The maximum bending moment in a simply supported beam carrying several concentrated loads will always occur at the location of one of the loads.

When none of these three conditions applies, the cross section having maximum bending
moment will need to be located, using the methods discussed in chapter 11.

PROBLEM SET 9.2

9.17–9.20

What are the values and locations of the maximum bending moments in the
simply supported beams shown in figures 9.31 through 9.34? All spans are
30 m.

Figure 9.31
Problem 9.17.

Figure 9.32
Problem 9.18.

Figure 9.33
Problem 9.19.

Figure 9.34
Problem 9.20.

9.21–9.23

What are the maximum bending moments and shearing forces in the cantilever beams shown in figures 9.35 through 9.37?

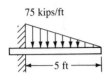

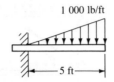

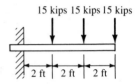

Figure 9.35 Problem 9.21.   **Figure 9.36** Problem 9.22.   **Figure 9.37** Problem 9.23.

9.24 In the aircraft bell crank shown in figure 9.38, what is the bending moment at section 1? What is it at section 2?

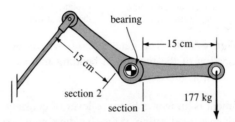

Figure 9.38 Problem 9.24.

9.25–9.28

What are the bending moments and shearing forces at sections C and D of the beams shown in figures 9.39 through 9.42?

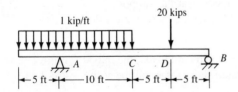

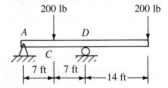

Figure 9.39 Problem 9.25. Note: Point D is immediately to the left of the 20-kip load.

Figure 9.40 Problem 9.26. Note: Point C is immediately to the left of the 200-lb load.

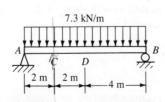

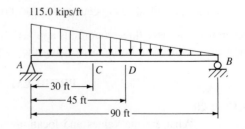

Figure 9.41 Problem 9.27.

Figure 9.42 Problem 9.28.

9.29 What are the maximum bending moments and shearing forces in the beam shown in figure 9.43?

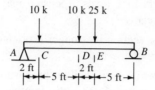

Figure 9.43   Problem 9.29.

9.8 GALILEO'S WORK WITH BEAMS

Galileo Galilei (1564–1642) is thought to be the first to study beams and document his conclusions. His book *Two New Sciences* was published in 1638.[1] The work was done during the last eight years of Galileo's life. During this time he was under house arrest for having defied the Church in teaching and publishing astronomical theories in which the earth is portrayed as one of several planets rather than the center of the universe. Galileo's fascinating biography has many implications. Perhaps first among them is the invention of the scientific method, in which experiment is held at least equal in importance to abstract philosophical reasoning. Furthermore, Galileo was the first to begin the quantitative and experimental studies of mechanics, materials, and structures. Before Galileo, structures were designed using tradition and intuition. Galileo is the intellectual father of the rational design of structures. Figure 9.44 shows a sketch by Galileo of the test of the bending of a wooden plank.

Two New Sciences, the source of this figure, is written with humor as a dialogue between a teacher and two students. One of the students always raises incorrect explanations for experimental observations based upon the accepted reasoning of the time. He is sarcastically given the name *Simplicio*.

Figure 9.44   Galileo's sketch of the rectangular beam problem.

Galileo's work was not completely correct in that his calculations of stresses in a beam were about one-third lower than actuality, but he correctly observed the following:

1. Members that can resist very high loads in tension may be able to support only very small loads as cantilever beams.

2. A beam of rectangular cross section is stronger if oriented so that the long side of the rectangle is in the direction of the load.

3. The strength of a beam under its own weight is proportional to the inverse square of its length, but only to the first power of its height. In other words, the height of the cross section of a rectangular beam must increase by the square of the span length in order for it to have the same stress under its own weight.

4. Tests of small models of a large structure of the identical material will not predict the success of the large structure of the same proportions.

5. Every material seems to have a certain "absolute" resistance to fracture that does not increase with size. This property means that there are practical limits to the sizes of structures. As Galileo stated,

> "You can plainly see the impossibility of increasing the size of structures to vast dimensions either in art or in nature; likewise the impossibility of building ships, palaces, or temples of enormous size in such a way that their oars, yards, beams, iron-bolts, and, in short, all their other parts will hold together; nor can nature produce trees of extraordinary size because the branches would break down under their own weight; so also it would be impossible to build up the bony structures of men, horses, or other animals so as to hold together and perform their normal functions if these animals were to be increased enormously in height; for this increase in height can be accomplished only by employing a material which is harder and stronger than usual, or by enlarging the size of the bones, thus changing their shape until the form and appearance of the animals suggest a monstrosity. . . . If the size of a body be diminished, the strength of that body is not diminished in the same proportion; indeed the smaller the body the greater its relative strength. Thus a small dog could probably carry on his back two or three dogs of his own size; but I believe that a horse could not carry even one of his own size."[1]

Today Galileo would be delighted with the seemingly unlimited prospects for the sizes of structures in space, a territory that he knew better than most of his contemporaries.

9.9 BEAMS OF RECTANGULAR CROSS SECTION

It is most convenient to begin the study of beams as Galileo did, with the straight beam of rectangular cross section. If such a beam were made of foam rubber and marked with a pair of lines perpendicular to its axis, as shown in figure 9.45a, the beam would deform as shown in figure 9.45b, when bending moment is applied as shown.

The following observations can be made from figure 9.45.

1. Points a and b move closer together, indicating compression at the top of the section.

2. Points c and d move apart the same amount as a and b moved together, indicating tension at the bottom of the section.

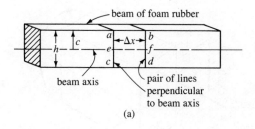

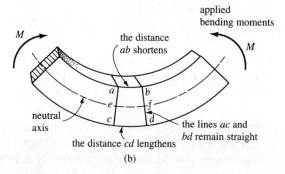

Figure 9.45 The local deformation of a beam under an applied bending moment.

3. Points e and f remain the same distance apart on the arc, indicating no strain at the beam axis. This is called the *neutral* axis.

4. Lines ac and bd remain straight, showing that sections remain plane during bending.

These observations are true whether the material of the beam is elastic or plastic. If the material is elastic, however, the stress distribution over the section is related to strain distribution. Since the strain distribution is in turn related to the deformation, one would expect the stress distribution to be as shown in figure 9.46.

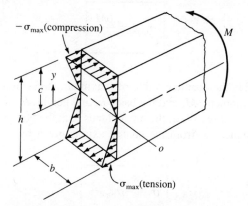

Figure 9.46 Stress distribution under elastic conditions for a rectangular beam under a bending moment.

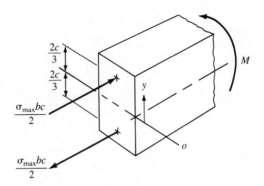

Figure 9.47 Replacement of the two triangular stress blocks in figure 9.46 with equivalent concentrated forces.

The maximum stress varies linearly from maximum compression, $-\sigma_{max}$, at the top to zero at the neutral axis and then to maximum tension of the same value, σ_{max}, at the bottom. These two triangular stress blocks may be treated exactly as uniformly increasing distributed loads and replaced by equivalent concentrated loads of $\sigma_{max}bc/2$ at distances of $2c/3$ from the neutral axis, as shown in figure 9.47.

Writing an equilibrium equation for the moments about the axis, o, which intersects the neutral axis of the beam,

$$\Sigma M\oplus: \quad \frac{\sigma_{max}bc}{2} \times \frac{2c}{3} + \frac{\sigma_{max}bc}{2} \times \frac{2c}{3} - M = 0$$

$$\sigma_{max} = \frac{3M}{2bc^2}$$

Since $c = h/2$,

$$\boxed{\sigma_{max} = \frac{6M}{bh^2}}$$ **(9.2)**

This relationship yields the absolute value of maximum stress in a rectangular beam in terms of the bending moment, M, and the base, b, and height, h, of the rectangular cross section. The stress is compressive at the top and tensile at the bottom of the cross section. The stress at any distance, y, from the neutral axis is given by

$$\boxed{\sigma = -\sigma_{max}\frac{y}{c}}$$ **(9.3)**

where compression is taken to be of negative sign.

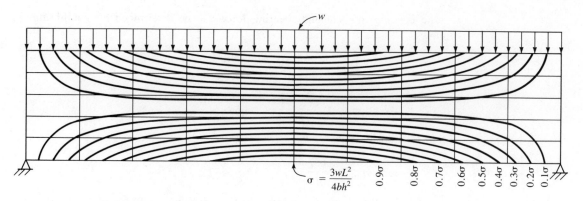

Figure 9.48 Lines of constant bending stress in beams of rectangular cross section due to a distributed load, w.

Most often relationship 9.2 is written in the form

$$\sigma_{max} = \frac{Mc}{I} \quad \text{or} \quad \frac{M}{Z} \tag{9.4}$$

where $I = bh^3/12$ and $Z = bh^2/6$ for a rectangular cross section. Equation 9.4 is a more general form of the expression. The quantities I and Z will be discussed further in the next chapter.

The bending stresses vary with the vertical and horizontal location within the beam. Figure 9.48 shows this variation. The curved lines are constant stress lines and should be read like a topographical map.

ILLUSTRATIVE PROBLEM

9.5 If the stone in Galileo's drawing (figure 9.44) weighs 300 kg and the beam is 3 m from hook to wall, what is the maximum bending stress? The beam cross section is 30 cm high and 60 cm across. What would be the stress if the beam were placed so that the 30-cm dimension were horizontal and the 60-cm were vertical?

Solution:
The maximum bending moment will occur at the cross section where the beam emerges from the wall, which is also the location of maximum stress. A free body of the beam cut at that section is as shown in figure 9.49.

Figure 9.49 Free-body diagram of Galileo's beam.

The bending moment and shearing force can be determined for equilibrium as follows:

$$\sum M_o \circlearrowright:\ M + 300\ \text{kg} \times 3\ \text{m} = 0$$
$$M = -900\ \text{kg} \cdot \text{m}$$

$$\sum V \uparrow:\ V - 300\ \text{kg} = 0$$
$$V = 300\ \text{kg}$$

From relationship 9.2,

$$\sigma_{max} = \frac{6M}{bh^2}$$

$$\sigma_{max} = \frac{6(-900\ \text{kg} \cdot \text{m})}{60\ \text{cm} \times (30\ \text{cm})^2} \times \frac{9.8\ \text{N}}{1\ \text{kg}} \times \frac{1.0\text{E}6\ \text{cm}^3}{1\ \text{m}^3}$$

$$\sigma_{max} = -0.980\ \text{MPa}$$

Therefore, the stress will be 0.980 MPa tension at the top of the beam at the wall and 0.980 MPa compression at the bottom.

If the beam were oriented so that the strong direction carries the load,

$$\sigma_{max} = \frac{6(-900)}{30 \times 60^2} \times 9.8$$

$$\sigma_{max} = -0.490\ \text{MPa}$$

In other words, the beam would be twice as strong if the 60-cm dimension were vertical.

9.10 STRESS CONCENTRATIONS

Bending stresses also can be increased by discontinuities in cross section, such as shoulders and notches. The values of the stress-concentration factors for flexure are not identical to those for axial loads. However, they do depend similarly upon the root radii in comparison to the other dimensions. Figure 9.50 gives the stress concentrations for bending of a notched beam and for a stepped beam.[2] The stress-concentration factor is applied to the stress as computed from relationship 9.2. Holes on the neutral axis are not important stress raisers in the case of bending stress.

9.11 SHEAR STRESS IN RECTANGULAR BEAMS

A variation of bending moment along the length of the beam causes shearing stress to develop in beams in the horizontal direction. This case can be shown as in figure 9.51. A piece is cut from a beam between planes 1 and 2, which are only a slight distance, Δx, apart (figure 9.51b). Since there is no vertical load on this small part of the beam, the shearing forces, V, and both sections 1 and 2 must be equal.

The bending moment at section 1 is taken to be M. At section 2 it is $M + \Delta M$ to show that it changes by a slight bit, ΔM, as one goes from 1 to 2. For the free-body

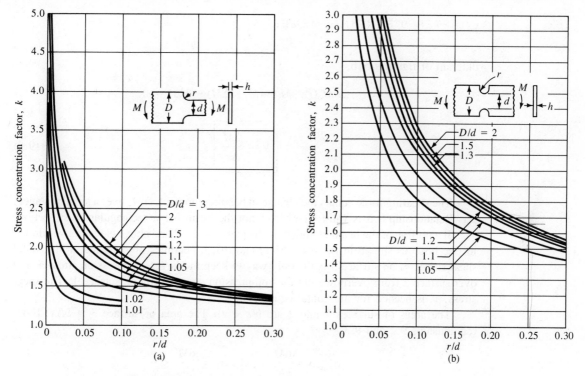

Figure 9.50 Stress concentrations due to bending in rectangular beam shapes. (a) shoulder, (b) notch.

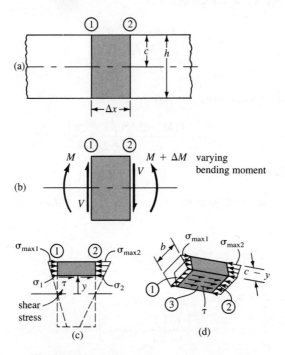

Figure 9.51 Shear stress, τ, required for horizontal equilibrium.

diagram of figure 9.51b,

$$\sum M_o \circlearrowleft: \quad M - (M + \Delta M) + V \Delta x = 0$$

$$\boxed{V = \frac{\Delta M}{\Delta x}} \tag{9.5}$$

The shearing force equals the change in bending moment, ΔM, over a small distance, Δx. Relationship 9.5 is valid whether or not the beam has a rectangular cross section.

If a portion of the beam bounded by the planes 1 and 2 and above elevation y is drawn in free body as shown in figures 9.51(c) and (d), it becomes obvious that a horizontal shear stress, τ, on surface 3 is the only way to keep the larger compression stress acting over surface 2 from overpowering the compression stress acting on surface 1. The shear stress, τ, is needed for horizontal equilibrium.

The areas of surfaces 1 and 2 are $b(c - y)$. The area of surface 3 is $b\Delta x$. From relationships 9.2 and 9.3,

$$\sigma_{max1} = \frac{6M}{bh^2} \quad \text{and} \quad \sigma_1 = \frac{6M}{bh^2} \times \frac{y}{c}$$

For this derivation, compression is taken to be positive as shown in figure 9.51d. Since $c = h/2$, the average stress on surface 1 is

$$\overline{\sigma}_1 = (\sigma_{max1} + \sigma_1)/2$$

$$\overline{\sigma}_1 = \frac{6M}{bh^2}\left(\frac{1}{2} + \frac{y}{h}\right)$$

Similarly, the average stress on surface 2 is

$$\overline{\sigma}_2 = \frac{6(M + \Delta M)}{bh^2}\left(\frac{1}{2} + \frac{y}{h}\right)$$

For horizontal equilibrium,

$$\sum H \mapsto: \quad \overline{\sigma}_1 b(c - y) - \overline{\sigma}_2 b(c - y) + \tau b \Delta x = 0$$

Solving for τ,

$$\tau = \frac{1}{\Delta x}(\overline{\sigma}_2 - \overline{\sigma}_1)(c - y) \quad \text{or} \quad \frac{h}{\Delta x}(\overline{\sigma}_2 - \overline{\sigma}_1)\left(\frac{1}{2} - \frac{y}{h}\right)$$

Substituting for $\overline{\sigma}_2$ and $\overline{\sigma}_1$,

$$\tau = \frac{6}{bh}\left[\frac{1}{4} - \left(\frac{y}{h}\right)^2\right]\frac{\Delta M}{\Delta x}$$

From relationship 9.5,

$$\tau = \frac{6V}{bh}\left[\frac{1}{4} - \left(\frac{y}{h}\right)^2\right] \tag{9.6}$$

Relationship 9.6 give the shearing stress, τ, at any elevation, y, in a beam of rectangular cross section. At the top of the beam, $y = h/2$. Therefore, τ is equal to zero. At the bottom of the beam, $y = -h/2$ and τ is again equal to zero.

The maximum shear stress at any given section along a rectangular beam is at the neutral axis where $y = 0$. Then,

$$\tau_{max} = \frac{3}{2}\frac{V}{bh} \tag{9.7}$$

Notice that the maximum shearing stress is 50 percent higher than the vertical shear force, V, divided by the cross-sectional area, bh. This difference is because the shear stress is not uniform from top to bottom. It is distributed in rectangular beams as shown in figure 9.52.

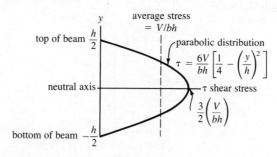

Figure 9.52 Distribution of the shear stress through the cross section of a rectangular beam.

The horizontal shear stress, τ_h, and the vertical shear stress, τ_v, are shown in figure 9.53 on an infinitesimally small square element of the beam. On such a small area, τ_v must be equal to τ_h in order to have moment equilibrium. Therefore, the average value of the shearing stress distribution in the cross section must be equal to the vertical shearing force divided by the beam cross-sectional area, V/bh.

Horizontal shearing force can be visualized using a deck of cards. When the deck is flexed, the cards will move horizontally against each other at the ends of the deck, showing the tendency to develop shearing stress. Because the cards are free to shear against one another, as shown in figure 9.54, it requires only 52 times the force to bend the deck as to bend a single card.

Figure 9.53 On an infinitesimally small square element, τ_v must be equal to τ_h for the element to be in moment equilibrium.

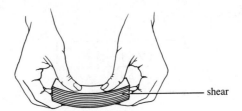

Figure 9.54 Cards move horizontally against
each other when the deck is flexed,
demonstrating the tendency for horizontal shear.

If the cards were cemented together so as to transmit shear stress from one to the other instead of sliding freely upon each other, the deck would act like a monolithic block and be difficult to bend by hand.

ILLUSTRATIVE PROBLEM

9.6 Four pieces of wood are exactly 1 in. by 2 in. in cross section and 48 in. long. They are stacked as shown in figure 9.55 and simply supported. A 200-lb man stands at mid-span. (a) What is the maximum bending stress? (b) What is the maximum bending stress if the pieces of wood are glued together? (c) What is the maximum shear stress in the glued joints?

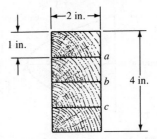

Figure 9.55 Illustrative problem 9.6.

Due to symmetry, the reaction loads on the beam are 100 lb at each support. Taking the weight of the man to be a concentrated load at mid-span, the maximum bending moment will be at mid-span. A free-body diagram of one-half the beam is shown in figure 9.56.

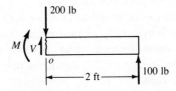

Figure 9.56 Free-body diagram for illustrative
problem 9.6.

For equilibrium,

$$\Sigma M_o \circlearrowleft: \quad M - 100 \text{ lb} \times 2 \text{ ft} = 0$$
$$M = 200 \text{ lb} \cdot \text{ft or 2 400 lb} \cdot \text{in.}$$
$$\Sigma V \uparrow: \quad V - 200 + 100 = 0$$
$$V = 100 \text{ lb}$$

If the beam is cut anywhere to the left of the 200-lb load, $V = 100$ lb. To the right of the load, $V = -100$ lb. Therefore, $V = 100$ lb maximum.

(a) If the beams are free to shear against each other, the four pieces of wood share the load equally. Therefore, the maximum M and V on each beam is 600 lb · in. and 25 lb, respectively. From relationship 9.2,

$$\sigma_{max} = \frac{6 \times 600 \text{ lb} \cdot \text{in.}}{2 \text{ in.} \times 1 \text{ in.}^2} = 1\ 800 \text{ psi}$$

The maximum bending stress is 1 800 psi.

(b) If the pieces are glued together, they act as one beam. Therefore,

$$\sigma_{max} = \frac{6 \times 2\ 400 \text{ lb} \cdot \text{in.}}{2 \text{ in.} \times (4 \text{ in.})^2} = 450 \text{ psi}$$

The maximum bending stress is 450 psi.

(c) In the glued beam, the shear stress at glue line a is given by relationship 9.6,

$$\tau = \frac{6V}{bh}\left[\frac{1}{4} - \left(\frac{y}{h}\right)^2\right]$$

$$\tau = \frac{6 \times 100 \text{ lb}}{2 \text{ in.} \times 4 \text{ in.}}\left[\frac{1}{4} - \left(\frac{1}{4}\right)^2\right]$$

$$\tau = 14.06 \text{ psi}$$

The maximum shear stress is at the neutral axis at glue line b. The maximum neutral axis stress is given by relationship 9.7,

$$\tau_{max} = \frac{3}{2}\frac{V}{bh}$$

$$\tau_{max} = 18.75 \text{ psi}$$

The shearing stress at glue line c is the same as at a. Shear stresses of 14.06 psi and 18.75 psi are produced in the glue by laminating the four pieces together. But, the advantage in doing so is the factor-of-four reduction, from 1 800 psi to 450 psi, in the normal bending stresses.

Figure 9.57
Problem 9.30.

PROBLEM SET 9.3

9.30 What are the maximum bending and shearing stresses in the beam of figure 9.57, which has a cross section 5 cm wide by 13 cm high?

9.31 A rectangular beam with a cross section 0.93 m high by 0.27 m wide carries the load shown in figure 9.58. What are the maximum bending and shearing stresses?

Figure 9.58
Problem 9.31.

5 equally spaced
loads of 83.5 N

Figure 9.59
Problem 9.32.

9.32 The rectangular beam in figure 9.59 has a cross section 15 cm wide by 50 cm high. What is the maximum bending stress and shearing stress?

9.33 If the bending stress may not be greater than 1 200 psi and the shearing stress may not be greater than 600 psi, what must be the height of a rectangular beam where h equals $4b$ if it is to carry the load in figure 9.60?

9.34 What is the distributed load capacity, w, for the beam shown in figure 9.61? The cross section is a rectangle 8 in. high by 2.5 in. wide. The maximum stress is 3 500 psi in flexure and 1 000 psi in shear.

9.35 What is the maximum flexural stress in a plastic beam of rectangular cross section 0.50 cm × 3 cm, if it is simply supported on a 50-cm span and carries a uniformly distributed load of 92 N/cm and if the cross section is oriented against the load (a) the strong way? (b) the weak way?

9.36 In a mechanism a lever shown in figure 9.62 is made of 0.375-in. flat metal stock. What is the maximum bending stress at the inside of the shoulder? All dimensions are in inches.

Figure 9.60
Problem 9.33.

Figure 9.61
Problem 9.34.

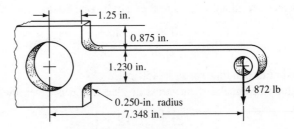

Figure 9.62   Problem 9.36.

9.37 A test specimen of flat plastic 6 mm thick and 20 mm high carries a bending moment of 132 N · cm at a point where it is notched from each side as shown in figure 9.63. What is the bending stress at the notch?

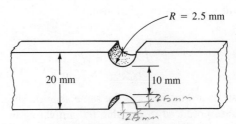

Figure 9.63   Problem 9.37.

9.12 WOODEN BEAMS

Most wooden beams are rectangular in shape. Some standard sizes of timber beams are given in Appendix VI, along with some properties of sections that will be explained in chapter 10.

The materials considered thus far in this book are isotropic. *Isotropic* means that properties are the same in all directions. However, wood is highly *anisotropic*, which means that the properties are different in different directions. Wood is a complex biological material with cellulous fibers bonded in a material called lignin. Varying growth rates in spring and summer produce the annual rings in the wood. The tree grows faster in spring, producing a less dense wood lighter in color. Summer wood is more dense and darker in color. These variations in density and color form the grain of the wood. The best structural properties of the wood are in the direction of the grain. These properties depend upon the species, moisture content, and grade of the lumber. Timber is graded according to grain quality and the presence of knots and checks. The allowable unit stresses for some softwood lumbers from the American Institute of Timber Construction are given in table 9.1. These allowable stresses include safety factors.

Table 9.1 Allowable stresses in psi for some stress-graded lumber[3]

Species and Grade	Extreme Fiber in Bending and Tension Parallel to Grain	Horizontal Shear	Compression Perpendicular to Grain	Compression Parallel to Grain	Elastic Modulus 10^6 psi
Douglas fir:					
1500 industrial	1 500	120	390	1 200	1.76
1200 industrial	1 200	95	390	1 000	1.76
Dense select structural	2 300	125	455	1 825	1.76
Construction	1 200	120	390	1 200	1.76
2-in.-thick southern pine:					
Dense structural	2 900	150	455	2 200	1.76
No. 1 dense	1 750	120	455	1 550	1.76
No.2	1 200	105	390	900	1.76
No. 2 decking	1 200	105	390	900	1.76
Redwood:					
Dense structural	1 700	110	320	1 450	1.32
Heart structural	1 300	95	320	1 100	1.32

Timber beams are susceptible to failure by shearing stress, because wood's shear strength parallel to the grain is very low compared to its bending strength. In other words, wood tends to shear along the grain. The shearing stress is highest at the neutral axis. In simply supported beams, the maximum shear force is at the end of the beam. Therefore, a shearing stress failure is most likely to appear as a horizontal crack at the end of a wood beam at the neutral axis, as shown in figure 9.64d.

It is not unusual to find shear cracks in floor joists. However, such cracks are partially caused by drying and shrinking of the wood. The consequence of a shear crack at

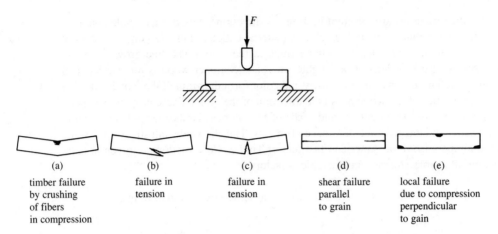

Figure 9.64 Types of failure in timber beams.

Table 9.2 Minimum uniformly distributed live loads (lb/ft^2)

Corridors	100
Dining rooms and restaurants	100
Garages (passenger cars)	100
(Also, 150 percent of the maximum wheel load as a concentrated load anywhere on the floor)	
Library reading rooms	60
Library stock rooms	150
Manufacturing	125
Office building	80
Lobbies	100
Multi-family apartments	40
Public rooms	100
Single family	40
second floors and habitable attics	30
uninhabitable attics	20
Schools	
classrooms	40
corridors	100
Stores, retail	
first floors	100
upper floors	75

the end of a beam is that the beam begins to behave as two beams weaker than the original single beam. If the load is increased, the beam will ultimately fail in bending where the bending moment is maximum. The greater the height of the beam with respect to the span, the more susceptible it is to shear failure.

Timber structures are designed according to building codes. Such codes usually reference the American Standard Association for design loads. Table 9.2 gives selected values for design loads from those standards. These are *live loads,* which do not include the weight of the structure itself. The weight of the structure is called *dead load*. The ASA standards also give loads for snow, wind, and seismic effects. These loads depend upon the geographical region in the United States and are expressed in terms of maps such as in figure 9.65 on page 290.

ILLUSTRATIVE PROBLEM

9.7 A single-story residential dwelling is to be custom-built with No. 2 southern pine 2 × 6 tongue-and-groove decking. What must be the size of the 2-in. construction grade douglas fir floor joists on 24-in. centers in order to have 16-ft spans?

Solution:
Assuming that the joists are simply supported and that 2 ft^2 of decking is supported by each foot of joist, the maximum bending moment is at mid-span and is obtained from the free-body diagram of figure 9.66.

Using the information in table 9.2, the distributed load, w, is 40 × 2 lb/ft.

$$\Sigma\, M_o\circlearrowleft :\ M + 80 \times 8 \times 4 - 8 \times 640 = 0$$
$$M = 2\,560 \text{ lb} \cdot \text{ft}$$

Solving relationship 9.2 for h,

$$h = \sqrt{\frac{6M}{b\sigma_{max}}}$$

From Appendix VI, b equals 1.50 in. for 2-in. lumber. The allowable stress is 1 200 psi. Therefore,

$$h = \sqrt{\frac{6 \times 2\,560 \text{ lb} \cdot \text{ft}}{1.50 \text{ in.} \times 1\,200 \text{ lb/in.}^2} \times \frac{12 \text{ in.}}{1 \text{ ft}}}$$
$$h = 10.12 \text{ in.}$$

A 2 × 10 is only 9.25 in. high. Therefore, the Code would require the use of the next size. A 2 × 12 should be specified. It has a height of 11.25 in.

The shear stress must also be calculated to see if it is within the 120 psi allowable. The maximum shear is at the end of the beam, $V = 8w$ or 640 lb. From relationship 9.7,

$$\tau_{max} = \frac{3}{2}\frac{V}{bh}$$

$$\tau_{max} = \frac{3}{2} \times \frac{640}{1.50 \times 11.25} = 56.9 \text{ psi}$$

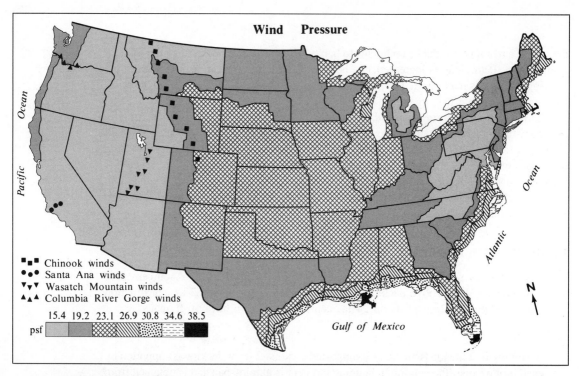

Wind Pressure

- ■·■ Chinook winds
- ●·● Santa Ana winds
- ▼·▼ Wasatch Mountain winds
- ▲·▲ Columbia River Gorge winds

psf 15.4 19.2 23.1 26.9 30.8 34.6 38.5

N

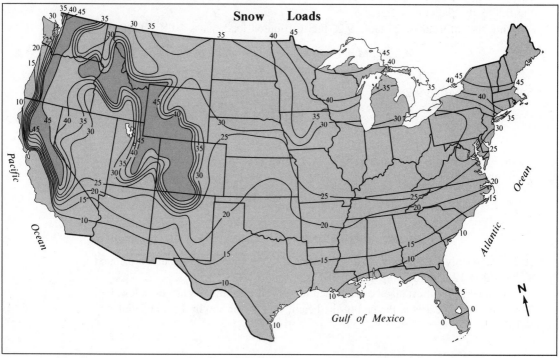

Snow Loads

N

Figure 9.65 Snow loads for roof slopes up to 20° and wind pressures at 30 ft of elevation in psf, from the American Standards Association.

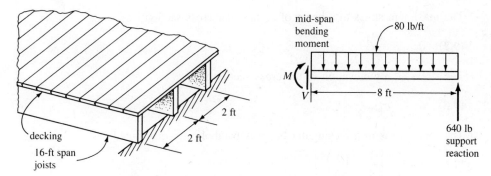

Figure 9.66 Free-body diagram for illustrative problem 9.7.

The shearing stress is within the allowable range.

The calculations should be redone adding the weight of the floor and joist to check that the stresses are also within the allowable for live and dead loads together.

PROBLEM SET 9.4

9.38 What is the load capacity of a simply supported 4×6 dense structural redwood beam 20 ft long under a uniformly distributed load considering both bending and horizontal shear?

9.39 A 2×12 of dense select structural douglas fir is 10 ft long. What is the allowable distributed load if the beam is simply supported?

9.40 Choose a standard size timber beam of construction grade douglas fir to span 25 ft if it is to be simply supported and carry a load of 100 lb/ft.

9.41 Design a roof system spanning 20 ft $\times$ 26 ft for a snow shelter using construction-grade douglas fir.

SUMMARY

☐ In determining reactions, a *distributed load* can be replaced by concentrated force equal in magnitude to the total amount of the load. The concentrated force should be located at the middle of a *uniform load* package and one-third of the way from the heavy end of a *uniformly increasing load* package.

☐ A beam or a lever may be a *parallel planar force system*. Reactions can be found from the equilibrium conditions $\Sigma F = 0$ and $\Sigma M = 0$.

☐ *Shearing force* and *bending moment* are the internal forces necessary for equilibrium in a beam or lever. They can be determined at any point along the length of a bending member by making a free-body diagram of the member cut at the point of interest and then applying the conditions of equilibrium.

☐ Galileo began the study of strength of materials by experimenting with rectangular beams.

☐ The maximum stress in a beam of rectangular cross section is:

(9.2) $\sigma_{max} = \dfrac{6M}{bh^2}$

☐ The stress at any elevation in the cross section is:

(9.3) $\sigma = -\sigma_{max}\dfrac{y}{c}$

☐ The shear stress in a rectangular beam is parabolically distributed.

(9.6) $\tau = \dfrac{6V}{bh}\left[\dfrac{1}{4} - \left(\dfrac{y}{h}\right)^2\right]$

☐ The maximum shear stress is:

(9.7) $\tau_{max} = \dfrac{3}{2}\dfrac{V}{bh}$

☐ Wood is an *anisotropic* material. It has different Code-allowable stresses for different grain orientations.

☐ Building codes for construction are usually based upon floor, wind, and snow loads, which are assumed to be uniform pressures.

REFERENCES

1. Galilei, Galileo. *Two New Sciences*. Translated by Stillman Drake. Madison, WI: University of Wisconsin Press, 1974.

2. Peterson, R. E. *Stress Concentration Design Factors*. New York: John Wiley & Sons, 1966.

3. *Timber Construction Manual*. American Institute of Timber Construction. New York: John Wiley & Sons, 1966.

CHAPTER TEN

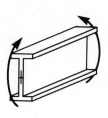

Beams of Non-Rectangular Cross Section

Having thoroughly explored stresses in rectangular beams, we will now address bending in non-rectangular beams. The properties of non-rectangular sections including area, centroids, and moment of inertia are derived using calculus. Built-up sections are considered using the parallel-axis rule. The general bending stress and the shear stress relationships are introduced and applied. Cross sections unsymmetrical about the neutral axis are considered. The chapter closes with the subject of equivalent cross sections for beams of two materials.

10.1 THE PURPOSES OF NON-RECTANGULAR CROSS SECTIONS

Beams of rectangular cross section do not make the most effective use of material. Bending produces tension at the bottom of the beam and compression at the top as discussed in chapter 9 (or vice versa, in the case of a cantilever). The material near the neutral axis is very lightly stressed. For this reason, beam shapes have been contrived so that most of the material is in flanges, where the stresses are high and where the material can be most useful. Such shapes (shown in figure 10.1) are made by rolling or extruding and have only a small amount of material in the web, near the neutral axis, to transmit the shear between the top and the bottom flanges. Shear stress, as discussed in chapter 9, is the minor form of stress associated with bending and necessitates only a small amount of material in a structural member.

Sometimes, non-rectangular shapes are used for reasons other than support. Some examples are shafts, pipes, and functional parts such as turbine blades. These elements are intended to transmit torque, transfer fluids, or control motion. Their shape is determined by their primary purpose. However, in performing their functions, shafts, pipes, and many other mechanical parts must function as bending members carrying flexural loads.

293

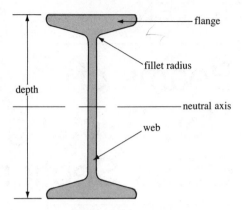

Figure 10.1 A structural shape.

For all these reasons, it is essential that the various properties of cross-sectional areas, which are important for structures, be understood.

10.2 CROSS SECTIONS AND PLANE AREAS

Plane areas are additive. This quality means that area can be divided into smaller, more easily determinable portions that can be added together to yield total area. Mathematically, this addition can be expressed by the relationship,

$$A = \Sigma \, \Delta A$$ (10.1)

where A is the total area of the cross section and ΔA represents areas of the smaller portions of the cross section.

ILLUSTRATIVE PROBLEMS

10.1 What is the cross-sectional area of the tee section shown in figure 10.2a?

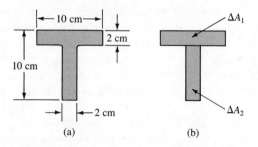

Figure 10.2 Illustrative problem 10.1. Dividing an area into easily determinable component areas.

Solution:
The area can be considered as two rectangles, ΔA_1 and ΔA_2, as shown in figure 10.2b.

$$\Delta A_1 = 2 \text{ cm} \times 10 \text{ cm} = 20 \text{ cm}^2$$
$$\Delta A_2 = 2 \text{ cm} \times 8 \text{ cm} = 16 \text{ cm}^2$$

Using relationship 10.1,

$$A = \Sigma \, \Delta A = 20 \text{ cm}^2 + 16 \text{ cm}^2 = 36 \text{ cm}^2$$

The area of the total cross section is 36 cm².

For more sophisticated areas, the ΔA's can be reduced to many small increments and summed using a computer. Or, if the shape can be expressed in mathematical form, the methods of calculus can be used.

10.2 Use calculus to determine the area of a parabola of base b and height h as shown in figure 10.3.

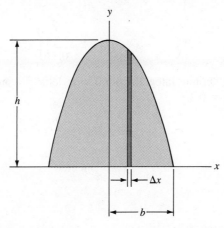

Figure 10.3 Illustrative problems 10.2 and 10.5.

Solution:
The equation of the parabola shown in figure 10.3 is

(a)
$$y = h\left(1 - \frac{x^2}{b^2}\right)$$

When $x = \pm b$, $y = 0$. When $x = 0$, $y = h$. Δx is taken to be an infinitesimally small distance along the x axis. The value of y changes little over this increment. Therefore, the incremental area, ΔA, is essentially equal to a rectangular slice of area, $y\Delta x$. From relationship 10.1,

(b)
$$A = \Sigma \, y\Delta x$$

Using calculus, the summation sign is replaced by an integral and the infinitesimal sign by the differential.

$$A = \int y \, dx$$

Putting this equation in the form of a definite integral,

(c)
$$A = 2 \int_0^b y \, dx$$

Integrating the expression in the x direction from $x = 0$ to b gives the area of one-half the parabola. Therefore, the factor of 2 needs to be included for the total area.

To perform the integration, it is necessary to replace y with an expression in terms of x from expression **(a)**,

(d)
$$A = 2 \int_0^b h \left(1 - \frac{x^2}{b^2} \right) dx$$

Integrating,

(e)
$$A = 2h \left(x - \frac{x^3}{3b^2} \right) \Big]_0^b$$

To evaluate the definite integral from 0 to b, let $x = b$ and then subtract the same terms, letting $x = 0$.

(f)
$$A = 2h \left(b - \frac{b}{3} \right)$$

$$A = \frac{4}{3} bh$$

10.3 POLAR COORDINATES

Some areas can be determined most directly by using polar coordinates. This method can best be demonstrated by a problem.

ILLUSTRATIVE PROBLEM

10.3 Solve for the area of a circle of radius R using polar coordinates.

Solution:
The ΔA can be an annular area as shown in figure 10.4. The area of an incremental annulus is

$$\Delta A = 2\pi r \Delta r$$

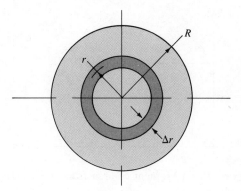

Figure 10.4 Illustrative problem 10.3.

From relationship 10.1,

$$A = \sum \Delta A = \sum 2\pi r \Delta r$$

Writing this in integral form and integrating in the direction of r from 0 to R,

$$A = 2\pi \int_0^R r \, dr$$

$$A = 2\pi \left[\frac{r^2}{2} \right]_0^R$$

$$A = \pi R^2$$

The example of illustrative problem 10.3 yields the formula for the area of a circle, πR^2. Calculus need not be used when the formula for area is known. The formulas for most areas whose boundaries can be described with a mathematical relationship can be obtained from tables such as that in Appendix III. Generally, it is more expedient to use formulas from a table of areas than to find the necessary integral from a table of integrals. However, calculus is useful in the rare cases where an area formula cannot be found in a table. Remember, formulas for areas found in tables were originally derived using calculus.

10.4 MOMENT OF AREA

The *moment of area* is an area times the perpendicular distance from a coordinate axis. The axis lies in the plane of the area. The moment arm is the perpendicular distance from the coordinate axis to the centroid of the area, as shown in figure 10.5.

The area shown has two moments, Q_x and Q_y; one from the x axis and one from the y axis. The two mutually perpendicular axes are arbitrarily defined in position and orientation in each case.

$$Q_x = yA \quad \text{and} \quad Q_y = xA \tag{10.2}$$

where A is the area and x and y are the moment arms.

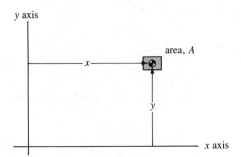

Figure 10.5 Moment of area from the x and y coordinate axes.

The Q values are subscripted according to the axis from which the moment arm is taken rather than in the direction of the moment arm. This subscript convention is used to be consistent with subsequent notation.

The notion of a moment of area evolves from the moment of force, which is force times a moment arm. Moment of force has a physical meaning, whereas moment of area is merely a mathematical property of an area. However, the moment of an area is useful in some computations later in this text. Moment of area has units of length cubed such as in.[3] or cm[3].

Moments of area defined from the same set of axes are additive algebraically. The word *algebraically* means that the sign of the moment is taken into consideration. It is possible to have a "negative" area, which is a cutout or hole in an area. When an area is divided into several components, the moments can be written in terms of increments of area, ΔA.

$$Q_x = \sum y \Delta A \quad \text{and} \quad Q_y = \sum x \Delta A \tag{10.3}$$

If the area is bounded by a curve that can be defined mathematically, it is sometimes possible to use calculus to solve relationship 10.3.

ILLUSTRATIVE PROBLEMS

10.4 What are the moments of area of the tee shown in figure 10.6a about vertical and horizontal axes through point o?

Solution:

The area can be divided into two rectangular areas, A_1 and A_2, as shown in figure 10.6b. Since these are rectangular areas, the moment arms go to centers of the rectangles. It is best to work in tabular form as follows:

Area	ΔA (in.^{2})	x (in.)	y (in.)	$x\Delta A$ (in.^{3})	$y\Delta A$ (in.^{3})
1	6.00	3.00	0.500	18.00	3.00
2	5.00	3.00	3.500	15.00	17.50
Σ	11.00			33.00	20.50

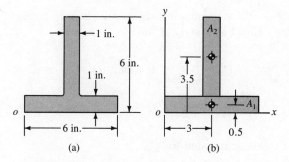

Figure 10.6 Illustrative problems 10.4 and 10.6.

Columns $x\Delta A$ and $y\Delta A$ were obtained by multiplying the ΔA values by the x and the y values, respectively, across the rows. The summation is obtained by adding the appropriate columns.

From relationships 10.1 and 10.3,

$$A = \Sigma\, \Delta A = 11.00 \text{ in.}^2$$
$$Q_x = \Sigma\, y\Delta A = 33.00 \text{ in.}^3$$
$$Q_y = \Sigma\, x\Delta A = 20.50 \text{ in.}^3$$

10.5 What are the moments of area from the x and y axes for one-half of the parabola shown in figure 10.3?

Solution:
For the parabola,

$$y = h\left(1 - \frac{x^2}{b^2}\right)$$

From relationship 10.3,

$$Q_x = \Sigma\, y\Delta A \quad \text{and} \quad Q_y = \Sigma\, x\Delta A$$

The area can be divided into incremental vertical slices of height, y, and width, dx, as shown in figure 10.7.

$$\Delta A = y\, dx$$

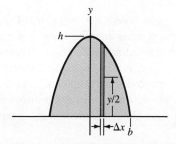

Figure 10.7 Illustrative problems 10.5 and 10.7.

In the case of Q_x, the moment arm must actually be half the height of the incremental area when defined as a vertical slide as done here.

$$Q_x = \int_0^b \frac{1}{2}y^2 \, dx \quad \text{and} \quad Q_y = \int_0^b xy \, dx$$

(a) $\qquad Q_x = \int_0^b \frac{h^2}{2}\left(1 - \frac{x^2}{b^2}\right)^2 dx \quad \text{and} \quad Q_y = \int_0^b xh\left(1 - \frac{x^2}{b^2}\right) dx$

Integrating first the expression for Q_x,

$$Q_x = \frac{h^2}{2}\int_0^b \left(1 - \frac{2x^2}{b^2} + \frac{x^4}{b^4}\right) dx$$

$$Q_x = \frac{h^2}{2}\left(x - \frac{2x^3}{3b^2} + \frac{x^5}{5b^4}\right)\Bigg]_0^b$$

$$Q_x = \frac{h^2}{2}\left(b - \frac{2}{3}b + \frac{1}{5}b\right)$$

(b) $\qquad Q_x = \frac{4}{15}bh^2$

Integrating then the expression for Q_y from (a),

$$Q_y = h\int_0^b \left(x - \frac{x^3}{b^3}\right) dx$$

$$Q_y = h\left(\frac{x^2}{2} - \frac{x^4}{4b^2}\right)\Bigg]_0^b$$

$$Q_y = \frac{h}{2}\left(b^2 - \frac{b^2}{2}\right)$$

(c) $\qquad Q_y = \frac{hb^2}{4}$

The relationships for the moments of area are (b) and (c).

10.5 CENTROIDS

The centroid of an area is a point where the moments of area about any two mutually perpendicular axes through the point is equal to zero. Some characterize the centroid as the point where area can be thought of as concentrated. This notion is permissible as long as one remembers that area is always distributed and cannot be concentrated.

Others maintain that the centroid is the point where there is as much area on one side as there is on the other, but this is only true for symmetrical areas.

In figure 10.8, the moments about the axes x and y are, respectively,

$$\boxed{Q_x = \bar{y}A \quad \text{and} \quad Q_y = \bar{x}A}$$ (10.4)

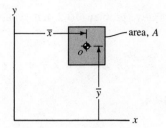

Figure 10.8 $\bar{y}$ and $\bar{x}$ can be found by dividing the appropriate moment of area by the area.

where $\bar{x}$ and $\bar{y}$ are the distances from the x and y axes, respectively, to the centroid. ($\bar{x}$ and $\bar{y}$ are called "x bar" and "y bar".)

These relationships suggest that one can find the distance to the centroid of any area by dividing the moment of area by the area.

ILLUSTRATIVE PROBLEMS

10.6 What are the distances to the centroid of the tee section shown in figure 10.6?

Solution:
From relationships 10.4,

$$\bar{x} = \frac{Q_y}{A} \quad \text{and} \quad \bar{y} = \frac{Q_x}{A}$$

Using the results of illustrative problem 10.4,

$$\bar{x} = \frac{33.00 \text{ in.}^3}{11 \text{ in.}^2} \quad \text{and} \quad \bar{y} = \frac{20.50 \text{ in.}^3}{11 \text{ in.}^2}$$

$$\bar{x} = 3.00 \text{ in.} \quad \text{and} \quad \bar{y} = 1.86 \text{ in.}$$

The centroid is in the middle of the stem of the tee and 1.86 in. above its base.

10.7 What are the x and y distances to the centroid of the half parabola shown in figure 10.7, where $h = 7$ cm and $b = 4$ cm?

Solution:
The Q_x and Q_y from illustrative problem 10.5 can be used.

$$Q_x = \frac{4}{15} bh^2$$

$$Q_x = \frac{4}{15} \times 4 \times 7^2 = 52.3 \text{ cm}^3$$

$$Q_y = \frac{hb^2}{4}$$

$$Q_y = \frac{7 \times 4^2}{4} = 28 \text{ cm}^3$$

From the results of illustrative problem 10.2, half a parabola as shown in figure 10.7 would be

$$A = \frac{2}{3}bh = \frac{2}{3} \times 4 \times 7 = 18.67 \text{ cm}^2$$

From relationships 10.4,

$$\bar{x} = \frac{Q_y}{A} \quad \text{and} \quad \bar{y} = \frac{Q_x}{A}$$

$$\bar{y} = \frac{52.3 \text{ cm}^3}{18.67 \text{ cm}^2} = 2.80 \text{ cm}$$

$$\bar{x} = \frac{28 \text{ cm}^3}{18.67 \text{ cm}^2} = 1.50 \text{ cm}$$

The centroid is 2.80 cm above and 1.50 cm to the right of the origin.

10.6 SYMMETRY

If a cross section, or area, is symmetrical, the centroid will lie on the axis of symmetry. This relationship was shown in illustrative problem 10.6, where the centroid of the tee section was located in the x direction exactly at the middle of the stem of the tee (which is the axis of symmetry). For a singly symmetric area, one needs to calculate only the position of the centroid along the axis of symmetry. If the area is doubly symmetric (having two axes of symmetry, as shown in figure 10.9), the centroid must be located at the intersection of the two axes, and no calculations are required.

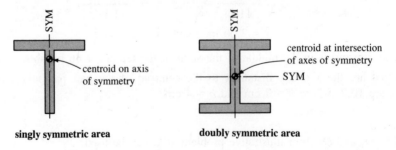

Figure 10.9   Using symmetry to help locate the centroid.

ILLUSTRATIVE PROBLEM

10.8 What is the location of the centroid of the extruded aluminum cross section shown in figure 10.10a?

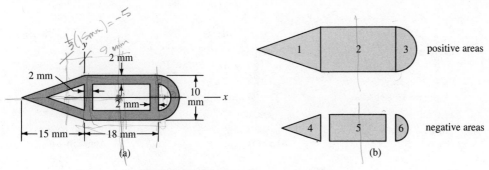

Figure 10.10 Illustrative problem 10.8.

Solution:

Due to its symmetry about the x axis, the centroid of the area must lie on the x axis. The problem is to find its position on the x axis. First, the cross section is divided into simple geometric shapes. The hollows are taken to be negative areas. The formulas in Appendix III are used to calculate areas and moment arms. From Appendix III, the moment arm of a semicircle is $4R/3\pi$ from the base; that of a triangle is $h/3$ from the base. The information is entered in tabular form as follows:

Shape	ΔA Area (mm^2)	x Moment Arm (mm)	$x\Delta A$ (mm^3)
1	75	−5	−375
2	180	+9	1 620
3	39.27	20.12	790
4	−27	−3	81
5	−84	+9	−756
6	−14.14	19.27	−272.5
Σ	169.13		1 087.5

From relationships 10.4,

$$\bar{x} = \frac{Q_y}{A} = \frac{1\ 087.5 \text{ mm}^3}{169.13 \text{ mm}^2}$$

$$\bar{x} = 6.43 \text{ mm}$$

The centroid is located at coordinates 6.43 and 0.

10.7 STRUCTURAL SHAPES

Shapes rolled of steel or extruded of aluminum or some other non-ferrous metals are called *structural shapes*. They can also be forged or cast. Figure 10.11 gives the nomenclature for structural shapes.

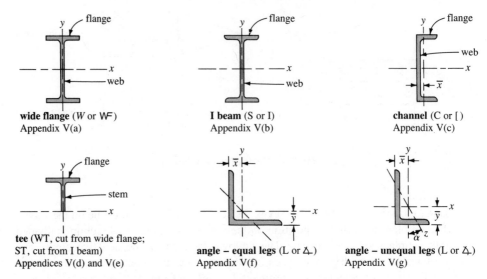

Figure 10.11 Structural shapes (refer to
Appendix V for dimensions).

Normally, the areas and locations of the centroids can be found from references. Appendix V is a partial table of properties of American standard shapes for steel, which can be used as examples. Properties of cross-sectional areas, other than area and centroid locations, given in the tables are explained later in this chapter. Structural shapes are indicated by type, size, and weight per unit length.

10.8 BUILT-UP SECTIONS

Sometimes beams are made by welding standard shapes together to form a combined cross section. In earlier times, these shapes may have been riveted together. These are called *built-up* sections. When determining the properties of a built-up section, it is best to use whatever data can be found from tables on the components of the section.

ILLUSTRATIVE PROBLEM

10.9 A built-up beam is constructed of a tee (ST12 × 60) and a channel (C10 × 30) welded together as shown in figure 10.12. What is the area and where is the centroid of the beam located?

Solution:

By putting the axis at the base of the section and ignoring the contribution of the weld to the area, we can list the properties of the two sections from Appendix V as

Component	Area (in.^{2})	y Location of Centroid (in.)	$y\Delta A$ (in.^{3})
ST12 × 60	17.6	3.52	61.95
C10 × 30	8.82	12.024	106.05
Σ	26.42		168.00

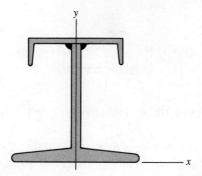

Figure 10.12   Illustrative problems 10.9 and 10.13.

The y location of the centroid of the channel is the height of the tee plus the thickness of the channel web minus the distance from the channel back to the channel centroid $(12 + 0.673 - 0.649)$.

Using relationships 10.4,

$$\bar{y} = \frac{Q_x}{A} = \frac{168.00 \text{ in.}^3}{26.42 \text{ in.}^2}$$

$$\bar{y} = 6.35 \text{ in.}$$

The cross-sectional area of the built-up section is 26.4 in.^{2} and the centroid is located at coordinates 0 and 6.35 in.

PROBLEM SET 10.1

In this set of problems, ignore all fillet radii when calculating section properties.

10.1 Calculate the area and location of the centroid of a structural tee section having the dimensions shown in figure 10.13. Compare the answer to the data for the nearest standard structural shape from Appendix V.

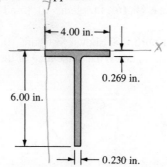

Figure 10.13   Problems 10.1 and 10.17.

10.2 Calculate the area and the centroidal location of the extruded aluminum channel section shown in figure 10.14.

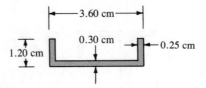

Figure 10.14   Problems 10.2 and 10.18.

10.3 Calculate the area and location of the centroid of the structural steel angle shown in figure 10.15. Compare the results to a standard angle from Appendix V.

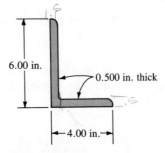

Figure 10.15   Problems 10.3 and 10.19.

10.4 The equal leg angle shown in figure 10.16 is made by bending 5-mm-thick sheet metal. What is the cross-sectional area and where is the centroid located?

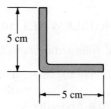

Figure 10.16   Problems 10.4 and 10.20.

10.5 The zee-shaped roof purlin, as shown in figure 10.17, is formed from 0.125-in.-thick galvanized sheet metal. What is the area of the cross section and where is the centroid located?

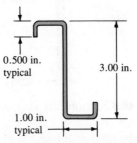

Figure 10.17   Problems 10.5 and 10.21.

10.6 An aluminum extrusion for an aircraft stringer has a cross section shape as shown in figure 10.18. What is the cross-sectional area and where is the centroid?

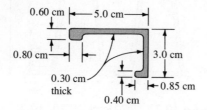

Figure 10.18   Problems 10.6 and 10.22.

10.7 A pre-stressed concrete beam is cast to the cross-sectional dimensions shown in figure 10.19. What is the area and where is the centroid?

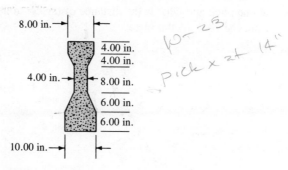

Figure 10.19   Problems 10.7 and 10.23.

10.8 An extruded neoprene seal has the cross-sectional dimensions shown in figure 10.20. What is the area and where is the centroid located?

10.9 A solid circular shaft has a flat for the attachment of a pulley, as shown in figure 10.21. What is the cross-sectional area of the shaft and what is the location of the centroid?

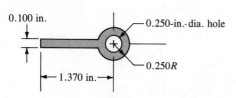

Figure 10.20   Problems 10.8 and 10.24.

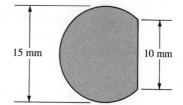

Figure 10.21   Problems 10.9, 10.26, and 10.33.

10.10 A 4-mm-thick sheet of steel is reinforced for bending resistance with a corrugated sheet of the same thickness as shown in figure 10.22. What is the area of a single repetition of the assembly? At what distance above the plate is the centroid located?

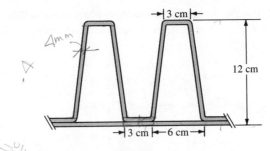

Figure 10.22 Problems 10.10, 10.25, and 10.34.

10.11 The $\frac{1}{4}$-in. steel plate (ℝ) bottom of a coal hopper is reinforced for bending by means of angles welded on 12-in. pitch as shown in figure 10.23. What is the area in one pitch and what is the distance of the centroid of the combined cross section from the back of the plate?

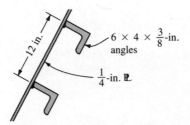

Figure 10.23 Problem 10.11.

10.12 Using calculus, find the area and location to the centroid of the region bounded by the parabola

$$y = h\left[1 - \left(\frac{x-a}{b-a}\right)^2\right]$$

and the x and y axes as shown in figure 10.24.

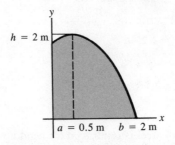

Figure 10.24 Problem 10.12.

10.13 Using calculus, find the area and location of the centroid of the region bounded by the parabola $y = h(x/b)^2$ shown in figure 10.25. Also find the x axis and the line $x = b$.

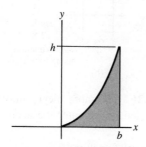

Figure 10.25   Problem 10.13.

10.9 SECOND MOMENT OF AREA

The *second moment of area* is often called the *moment of inertia of area*. Since both terms are widely used, they are used interchangeably throughout the remainder of this book.

However, moment of inertia of area is a misnomer. It comes by analogy from the concept of rotary motion in dynamics, where moment of inertia has a physical interpretation. But for areas there is no physical interpretation. Inertia is the tendency for matter to resist a change in velocity. A plane area has no thickness; therefore, it contains no mass and has no inertia.

Now, having stated what moment of inertia is not, what is it? The second moment of area (or moment of inertia of area) is a mathematical property of an area that is defined by the following relationship.

$$I_x = \sum y^2 \Delta A \quad \text{and} \quad I_y = \sum x^2 \Delta A \qquad (10.5)$$

where I_x is the second moment with the moment arm, y, defined from the x axis. I_y is the second moment with the moment arm, x, defined from the y axis and the second moment of area has length units to the fourth power, such as in.^{4}, mm^4, or cm^4.

ΔA must be interpreted as an infinitesimally small area in order for relationships 10.5 to be exact. If the ΔA's are finite in size, relationships 10.5 are then an approximation, the accuracy of which is inversely related to the size of the increments. In other words, the smaller the ΔA, or the more finely divided the area, the more accurate is the approximation of I.

10.10 RECTANGULAR AREAS

The rectangular area shown in figure 10.26a can be divided into two equal areas, ΔA, each with a height of $h/2$ and a base of b. The moment of inertia about an x axis through

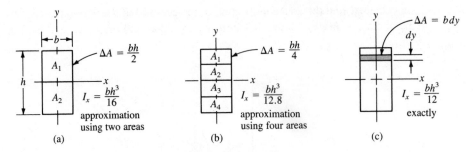

Figure 10.26 The more pieces into which an area is divided, the greater the accuracy.

the centroid of the area is from relationship 10.5.

$$I_x = \left(\frac{h}{4}\right)^2 \left(\frac{bh}{2}\right) + \left(-\frac{h}{4}\right)^2 \left(\frac{bh}{2}\right)$$

$$I_x = \frac{bh^3}{16} \text{ as a first approximation}$$

A more accurate approximation would be gained by dividing the area into four pieces, as in figure 10.26b. Each area would have a height of $h/4$ and a base of b. Again, using relationship 10.5,

$$I_x = \left(\frac{3h}{8}\right)^2 \left(\frac{bh}{4}\right) + \left(\frac{h}{8}\right)^2 \left(\frac{bh}{4}\right) + \left(-\frac{h}{8}\right)^2 \left(\frac{bh}{4}\right) + \left(-\frac{3h}{8}\right)^2 \left(\frac{bh}{8}\right)$$

$$I_x = \frac{bh^3}{12.8} \text{ for four equal areas}$$

Continuing in this fashion using eight equal areas,

$$I_x = \frac{bh^3}{12.19} \text{ for eight equal areas}$$

As increasingly finer slices of area are used, the approximate values of I_x approach $bh^3/12$ as a limit, which can be shown by taking the integral as depicted in figure 10.26c:

$$I_x = y^2 \, dA = b \int_{-h/2}^{h/2} y^2 \, dy$$

$$I_x = \frac{1}{3} by^3 \Big]_{-h/2}^{h/2} = \frac{b}{3} \left[\frac{h^3}{8} - \left(-\frac{h^3}{8}\right) \right]$$

$$\boxed{I_x = \frac{bh^3}{12}} \quad \text{exactly} \qquad \textbf{(10.6)}$$

This is the moment of inertia of area of a rectangular section of base b and height h about an x axis through the centroid. It is well worth memorizing for future work.

Note that the height is cubed. The moment of inertia of a cross section can be increased much more rapidly by adding to the height rather than to the base. Doubling the height increases I_x by a factor of eight, whereas doubling the base merely doubles the second moment.

In the other direction, the moment of inertia can be obtained by exchanging b and h.

$$I_y = \frac{hb^3}{12}$$ **(10.7)**

10.11 THE CIRCULAR CROSS SECTION

The moment of inertia of area of a circular cross section can be computed using calculus as shown in figure 10.27.

The x and y axes are drawn through the centroid of the circle. The incremental area, ΔA, for the computation of I_x can be used if the integral is doubled to include the left half of the circle. From relationship 10.5,

$$I_x = \Sigma y^2 \Delta A$$

If $\Delta A = x\,dy$, the integral can be written,

$$I_x = 2\int_{-R}^{+R} y^2 x\,dy$$

For the circle, $x^2 + y^2 = R^2$. Therefore,

$$x = \sqrt{R^2 - y^2}$$

$$I_x = 2\int_{-R}^{+R} y^2\sqrt{R^2 - y^2}\,dy$$

From a table of integrals,

$$I_x = 2\left[-\frac{y}{4}\sqrt{(R^2 - y^2)^3} + \frac{R^2}{8}\left(y\sqrt{R^2 - y^2} + R^2 \sin^{-1}\frac{y}{R}\right)\right]_{-R}^{+R}$$

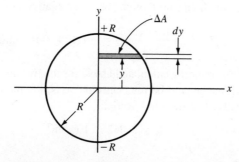

Figure 10.27 Computation of the second moment of area of a circle about an x axis through the centroid.

The quantity $(R^2 - y^2)$ is zero if $y = R$ or $-R$. Therefore, all terms containing this quantity can be omitted for these limits of integration. When substituting limits this expression becomes

$$I_x = \frac{R^4}{4}\left[\sin^{-1}(1) - \sin^{-1}(-1)\right] = \frac{R^4}{4}\left[\frac{\pi}{2} - \left(-\frac{\pi}{2}\right)\right]$$

Therefore,

$$\boxed{I_x = \frac{\pi R^4}{4}}$$ (10.8)

for a circular area of radius R about its centroid.

The I_y is identical to I_x for a circle. In fact, the second moment of area about any axis in any direction through the centroid of a circle is identical to I_x.

ILLUSTRATIVE PROBLEM

10.10 What is the moment of inertia of area of a concrete pipe with a 90-cm outside diameter and a 70-cm inside diameter about an axis through the centroid?

Solution:
Using relationship 10.8,

$$I_x = \frac{\pi R^4}{4}$$

The pipe can be thought of as two areas, a solid circle of radius 45 cm and a hollow of radius 35 cm.

For the solid circle,

$$I_x = \frac{\pi(45 \text{ cm})^4}{4} = 3.221 \times 10^6 \text{ cm}^4$$

For the hollow, which can be thought of as a negative area,

$$I_x = -\frac{\pi(35 \text{ cm})^4}{4} = -1.179 \times 10^6 \text{ cm}^4$$

Since the moments of inertia calculated above for the two areas are taken about the same axis, which happens to be the centroidal axis, the moment of inertia can be added algebraically. For the total section,

$$I_x = 2.042 \times 10^6 \text{ cm}^4$$

10.12 THE PARALLEL-AXIS RULE

A very important relationship for calculating the moment of inertia of area of irregular or compound shapes is the parallel-axis rule, which can be derived as follows. For the area

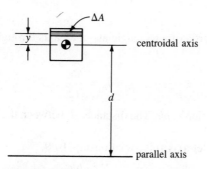

Figure 10.28   The parallel-axis rule.

shown in figure 10.28, the moment of inertia from the defining relationship 10.5 can be written,

(a)
$$I_x = \Sigma(y + d)^2 \Delta A$$

where y is the variable distance between the centroidal axis and ΔA, and d is the constant distance between the centroidal axis and the parallel axis. Expanding,

(b)
$$I_x = \Sigma(y^2 + 2yd + d^2)\Delta A$$

This equation can be broken down into three independent summations as follows:

(c)
$$I_x = \Sigma \, y^2 \Delta A + 2d\Sigma \, y\Delta A + d^2\Sigma \, \Delta A$$

The middle term contains the first moment of area defined by relationship 10.3. Since y is taken here to be the distance from the centroid, the moment of area must be zero in this case. The moment of an area about its centroid is always equal to zero. The middle term drops out.

The last term contains the total area as defined by relationship 10.1. The first term is the moment of inertia of the area about its centroid. Therefore, **(c)** can be reduced to

$$\boxed{I = I_c + d^2A} \qquad\qquad \textbf{(10.9)}$$

where I is the moment of inertia about the parallel axis, I_c is the moment of inertia about the centroidal axis, d is the distance between the centroidal axis and the parallel axis, and A is the total area of the section.

Relationship 10.9 is called the *parallel-axis rule,* or the transfer rule. It can be stated as follows: *The moment of inertia of an area about any axis is equal to the moment of inertia about a parallel axis through the centroid of the area plus the square of the distance between the axes times the area.*

A corollary to the parallel-axis rule is that the moment of inertia about the centroidal axis is smaller than the moment of inertia about any other parallel axis.

ILLUSTRATIVE PROBLEM

10.11 Calculate the second moment of area of a rectangle 12 in. high and 3 in. wide about a horizontal axis through its base.

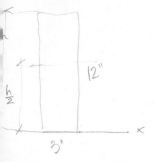

Solution:

The moment of inertia of a rectangle about an axis through its centroid is given by relationship 10.6,

(a)
$$I_c = \frac{bh^3}{12}$$

The area of the section is bh. The distance, d, between the base and an axis through the centroid is $h/2$.

Using the parallel-axis rule, relationship 10.9,

(b)
$$I = I_c + d^2A = \frac{bh^3}{12} + \left(\frac{h}{2}\right)^2 bh$$

$$I = \frac{bh^3}{12} + \frac{bh^3}{4}$$

(c)
$$I = \frac{bh^3}{3}$$

For the dimensions given in this problem

$$I = \frac{3 \text{ in.} \times (12 \text{ in.})^3}{3} = 1\,728 \text{ in.}^4$$

The moment of inertia of the rectangular area about its base is $1\,728$ in.[4]

10.13 MOMENT OF INERTIA OF COMPOUND AREAS AND STRUCTURAL SHAPES

Often complex structural shapes can be considered as combinations of other basic shapes. In these cases, the moments of inertia of the basic shapes can be calculated. Using the parallel-axis rule, the moments of inertia can be effectively transferred from the centroid of the basic shape to the centroid of the compound area. The moments of inertia of the component shapes about the centroid of the total area can then be added to give the moment of inertia of the total area. This process is shown in the next illustrative problem.

ILLUSTRATIVE PROBLEMS

10.12 Calculate the moments of inertia of area, I_x and I_y, about the centroid of the wide-flange section shown in figure 10.29. Compare the results with those given for standard structural shapes given in Appendix V.

Solution:

First the area must be broken down into basic shapes. Three rectangles can be used.

Number	Description	Dimensions (in. × in.)	Area (in.²)
1	upper flange	0.718 × 12.00	8.616
2	web	0.428 × 12.624	5.403
3	lower flange	0.718 × 12.00	8.616
		total area =	22.635

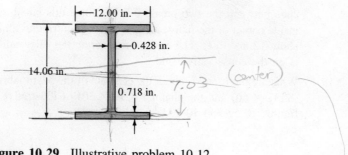

Figure 10.29 Illustrative problem 10.12.

This is a doubly symmetric area; therefore, the centroid is located at the junction of the two axes of symmetry. It is at the center of the web.

Computation of I_x:

Number	b (in.)	h (in.)	$bh^3/12$ (in.⁴)	d (in.)	$I_c + Ad^2$ (in.⁴)
1	12.00	0.718	0.37	6.671	383.80
2	0.428	12.624	71.75	0	71.75
3	12.00	0.718	0.37	6.671	383.80
				total $I_x =$	839.35

The column $bh^3/12$ is the basic I_x for each complete rectangular area about its own centroid per relationship 10.6. The column $I_c + Ad^2$ is the parallel-axis rule, relationship 10.9, moving the I_x to the x axis through the centroid of the total section. The columns can be added. Note that the largest part of the moment of inertia is actually due to the Ad^2 term of the parallel-axis rule. The flange areas times the squares of their distances from the centroid make up more than 91 percent of the moment of inertia of this area.

Computation of I_y:
In the other direction, no translation of axis is required, because the centroids of the component areas share the same y axis as the centroid of the overall area. Therefore:

Number	I_y (in.⁴)
1	103.39
2	0.08
3	103.39
total $I_y =$	206.86

Conclusion:
From Appendix V the moment of inertia of the area, I_x and I_y, are 851 and 207 in.⁴ The calculated area, 22.6 in.², compares to 22.9 in.² from Appendix V. The discrepancy results because the actual W14 × 78 structural shape has inside fillet radii that are ignored in this calculation. This additional area amounts to the differ-

ence between the two areas, 0.3 in. Since this small amount of area lies at the inside corner of the flange and the web, it contributes almost nothing to I_y but adds about 12 in.[4] to I_x. The error in neglecting the fillet radii, only 1.4 percent for I_x, is on the conservative side.

10.13 What are the I_x and I_y about the centroid of the beam fabricated of the structural tee (ST12 × 60) and the channel (C10 × 30) of illustrative problem 10.9, shown in figures 10.12 and 10.30?

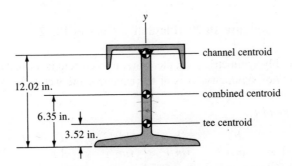

Figure 10.30 Illustrative problems 10.13 and 10.14.

Solution:
The location of the centroid was previously calculated in illustrative problem 10.9 and is shown in figure 10.30.

Computation of I_x:
Proceeding in tabular fashion and using the I_x, I_y, and A for the two component pieces from Appendix V,

Component	Area (in.²)	I_c (in.⁴)	d (in.)	$I_c + Ad^2$ (in.⁴)
ST12 × 60	17.6	245	2.83	385.9
C10 × 30	8.82	3.94	5.67	288
				I_x total = 673.9

$$Q = 12.02/2 = 6.01 - \frac{6.35}{2} = 2.83$$

Computation of I_y:
The centroids of the two components and of the total section share the same y axis. I_y for the combined section is therefore the sum of the I_y for the two components.

$$I_y = 42.1 + 103.0 = 145.1 \text{ in.}^4$$

Result:
The moments of inertia of the area made by combining the ST12 × 60 and the C10 × 30 standard structural shapes, as shown in figure 10.12, are

$$I_x = 674 \text{ in.}^4 \quad \text{and} \quad I_y = 145.1 \text{ in.}^4$$

PROBLEM SET 10.2

In the following problems, ignore all fillet radii.

10.14 Calculate the moment of inertia of area of a circular shaft 25 mm in diameter about its centroid.

10.15 A rectangular beam cross section has a height twice as great as its width. What is the ratio of I_x to I_y?

10.16 Calculate the second moment of area of a standard steel pipe with an outside diameter of 10.750 in. and an inside diameter of 10.020 in. Compare the result with the table of Appendix IV.

10.17 Calculate I_x and I_y about the centroid for the structural tee section shown in figure 10.13. Compare to the table in Appendix V.

10.18 Calculate I_x and the I_y about the centroid for the aluminum channel shown in figure 10.14.

10.19 Calculate I_x and I_y about the centroid for the structural steel angle shown in figure 10.15. Compare the results to those for a standard angle from Appendix V.

10.20 What are I_x and I_y about the centroid for the angle shown in figure 10.16 made from 5-mm-thick sheet metal?

10.21 What are I_x and I_y about the centroid of the zee-shaped roof purlin shown in figure 10.17 formed from 0.125-in.-thick galvanized sheet metal?

10.22 What are I_x and I_y about the centroid of the extruded aluminum aircraft stringer, the cross section of which is shown in figure 10.18?

10.23 What is the moment of inertia of area of the pre-stressed concrete beam section shown in figure 10.19 about a horizontal axis through the centroid?

10.24 Calculate I_x and I_y about the centroid of the extruded neoprene seal shown in figure 10.20.

10.25 Calculate the moment of inertia about a horizontal axis through the centroid of one repetition of the 4-mm-thick corrugated reinforced plate shown in figure 10.22.

10.26 Using Appendix III, compute the moments of inertia, I_x and I_y, of the 15-mm-diameter circular shaft with the 10-mm-wide flat shown in figure 10.21.

10.14 BENDING STRESSES IN NON-RECTANGULAR CROSS SECTIONS

In chapter 9, the stresses induced in a beam of rectangular cross sections by a bending moment were studied. In this chapter, we intend to use the properties of cross sections developed so far to generalize the bending theory to beams of any cross-sectional shape.

The basic assumption, which turns out to be highly accurate, is that plane sections remain plane during bending regardless of the shape of the cross section, as shown in figure 10.31.

For plane sections to remain plane, there must be a linear variation of strain from the top of the section, where in this case there is compression, to the bottom of the section, where there is tension. If the material is elastic, stress is proportional to strain. Therefore, the stress distribution must be linear, as shown in figure 10.32.

$$\sigma = \frac{\sigma_{max}}{c}y \qquad\qquad (10.10)$$

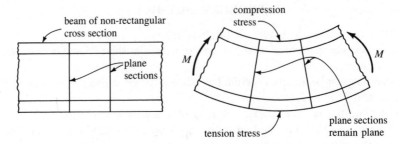

Figure 10.31 Deformation of a beam acted upon by a bending moment.

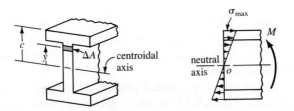

Figure 10.32 Elastic stress distribution in a non-rectangular beam.

The stress acting over an increment of area, ΔA, produces an increment of horizontal force, $\sigma \Delta A$. The sum of horizontal forces produced by all increments of area over the entire cross section must be zero in order to have horizontal equilibrium.

(a)
$$\Sigma H \mapsto : \Sigma \sigma \Delta A = 0$$

From relationship 10.10,

(b)
$$\Sigma \frac{\sigma_{max}}{c} y \Delta A = 0$$

Since σ_{max} and c are constants, they may be moved in front of the Σ sign.

(c)
$$\frac{\sigma_{max}}{c} \Sigma y \Delta A = 0$$

Relationship (c) will be true if $\Sigma y \Delta A = 0$. $\Sigma y \Delta A$ is the definition of the first moment of area, relationship 10.3. For the first moment of area of the entire section to be zero, y must be the distance from the centroid of the section. Therefore, *the neutral axis is at the centroid of the cross section.*

Now, as to the moments due to the incremental forces, $\sigma \Delta A$, equilibrium dictates that:

(e)
$$\Sigma M_o \oplus : \Sigma y \sigma \Delta A - M = 0$$

Substituting for σ,

$$\sum y \frac{\sigma_{max}}{c} y \Delta A = M$$

(f)
$$\frac{\sigma_{max}}{c} \sum y^2 \Delta A = M$$

The quantity $\sum y^2 \Delta A$ is the definition of the second moment of area about the centroid I_x, relationship 10.5. In this case, we will call the quantity I. Remember that if M is about an x axis, use I_x; if about the y axis, use I_y. Therefore, **(f)** can be simplified as follows:

$$\boxed{\sigma_{max} = \frac{Mc}{I}} \qquad \text{(10.11)}$$

Relationship 10.11 is the very important relationship for the maximum bending stress in an elastic beam. In this case, σ_{max} is compressive at the top and tensile at the bottom. Relationship 10.11 is the general case for any cross section shape. The previously derived relationship 9.2 is for the special case of rectangular beams.

10.15 SINGLY SYMMETRIC CROSS SECTIONS

A singly symmetric cross section is symmetric about one axis only. If the cross section is unsymmetrical about one axis, such as a horizontal axis as in a tee section, the top and the bottom have different maximum stresses. The distances between the neutral axes and the centroid will be different for the top, c_T, and the bottom, c_B, as shown on figure 10.33.

The maximum stress on the top of the beam will be

$$\sigma_T = -\frac{Mc_T}{I} \qquad \text{(10.12)}$$

The maximum stress at the bottom of the beam will be

$$\sigma_B = \frac{Mc_B}{I} \qquad \text{(10.13)}$$

where compression is negative and tension is positive.

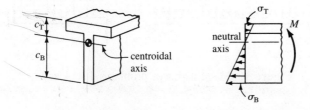

Figure 10.33 Elastic stress distribution in a nonsymmetrical beam.

10.16 SECTION MODULUS

Since both c and I are properties of the cross-sectional area, they can be combined into a single property, Z (called the section modulus), which is defined as

$$Z = \frac{I}{c} \qquad\qquad \textbf{(10.14)}$$

Relationship 10.11 can be rewritten in terms of Z as follows

$$\sigma_{max} = \frac{M}{Z} \qquad\qquad \textbf{(10.15)}$$

Section modulus means *measure of the section*. The most direct property that relates cross section to stress is the section modulus. Section modulus can be obtained directly from the tables of standard structural sections in Appendix V.

The units of section modulus are length units cubed, such as in.^{3} or cm^3. In some publications and codes, the symbol S is used to indicate section modulus.

ILLUSTRATIVE PROBLEMS

10.14 What is the maximum bending stress in a simply supported beam 240 in. long carrying a uniformly distributed load of 20 lb/in.? The beam is fabricated of a structural tee and a channel as shown in figure 10.30.

Solution:
The cross-sectional properties were found in illustrative problem 10.13.

$$I_x = 674 \text{ in.}^4$$

The maximum stresses are at the top and the bottom of the beam at mid-span. The bending moment can be obtained from the free-body diagram in figure 10.34, which includes the half of the beam to the left of mid-span.

$$\Sigma \, M_o\text{↺}: 2\,400 \times 120 - 20 \times 120 \times 60 - M = 0$$
$$M = 1.44\text{E5 lb} \cdot \text{in.}$$

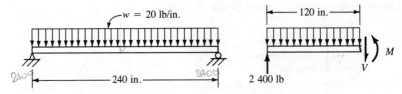

Figure 10.34 Illustrative problems 10.14 and 10.19.

Using relationships 10.12 and 10.13,

$$\sigma_T = -\frac{1.44E5 \text{ lb} \cdot \text{in.} \times 6.323 \text{ in.}}{674 \text{ in.}^4}$$

$$\sigma_T = -1\ 351 \text{ psi}$$

$$\sigma_B = \frac{1.44E5 \times 6.35}{674}$$

$$\sigma_B = 1\ 357 \text{ psi}$$

The stress at the top of the section at mid-span is 1 351 psi compression. That at the bottom is 1 357 psi tension.

10.15 Choose the lightest standard wide-flange section that can support a concentrated load of 24 000 lb at the mid-span of a simply supported beam 15 ft long. The allowable bending stress for structural steel is 22 000 psi.

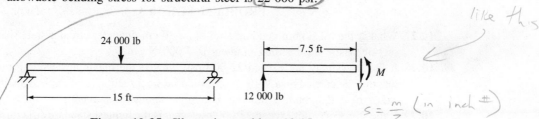

Figure 10.35 Illustrative problem 10.15.

Solution:

The bending moment can be computed using the free-body diagram from figure 10.35. The maximum bending moment is at the mid-span of the beam.

$$\Sigma \ M_o \circlearrowleft: \ 12\ 000 \text{ lb} \times 7.5 \text{ ft} \times 12 \text{ in./ft} - M = 0$$
$$M = 1.080E6 \text{ lb} \cdot \text{in.}$$

From relationship 10.15,

$$\sigma_{max} = \frac{M}{Z}$$

The required section modulus must be

$$Z = \frac{M}{\sigma_{max}} = \frac{1.080E6 \text{ lb} \cdot \text{in.}}{22\ 000 \text{ psi}}$$
$$Z = 49.09 \text{ in.}^3$$

Starting with the 8-in.-wide flange beams from the section tables of Appendix V, the lightest 8-in. beam that has a section modulus exceeding 49.09 in.³ is the W8 × 58. This means that its weight is 58 lb/ft. Preparing a table of choices for larger beams:

Size	Section Modulus $(in.^3)$
W8 × 58	52
W10 × 45	49.1
W12 × 40	51.9
W14 × 38	54.7
W16 × 36	56.5
W18 × 35	57.6
W21 × 44	81.6

As the beam sizes get larger, lighter sections can be used until the 18-in. beam is encountered. Then the weights begin to increase. The lightest section is the W18 × 35, which is 18 in. deep, weighs 35 lb/ft, and has plenty of excess section modulus over that which is required.

PROBLEM SET 10.3

10.27 What is the maximum bending stress in a 20-cm-diameter circular steel shaft at a section where the bending moment is 7 000 N · m?

10.28 If the stress may not exceed 22 000 psi, what is the maximum bending moment that may be carried by a W10 × 45 structural steel section (a) if applied about the *x* axis? (b) if the bending moment is applied about the *y* axis?

10.29 If the stress may not exceed 22 000 psi, what is the lightest standard structural steel wide-flange section that can span 15 ft as a simply supported beam and support a uniform distributed load of 500 lb/ft of length? Ignore the dead weight of the beam itself in computing the bending moment.

10.30 What is the stress at the top and at the bottom of a standard structural tee ST12 × 45 that is 18 ft long and simply supported at the ends? The stress should be computed at the point of a single concentrated load of 15 kips, 6 ft from the left end of the beam. The tee is positioned so that the flange is on top.

10.31 A hollow circular shaft has an inside diameter of 10 mm and an outside diameter of 15 mm. What is the bending stress at the bearing *B* in figure 10.36?

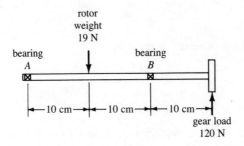

Figure 10.36 Problem 10.31.

10.32 How many feet may a 12-in. Schedule 40 steel pipe span as a simply supported beam if it is filled with water having a density of 65 lb/ft³ and if the stress may not be greater than 22 000 psi?

10.33 What is the maximum allowable bending moment on the round circular shaft with the flat shown in figure 10.21, if the allowable stress is 160 MPa? Use the results of problems 10.9 and 10.26.

10.34 If the corrugated reinforced plate shown in figure 10.22 carries a bending moment of 100 N · m per 9-cm repetition, what is the bending stress at the top of the corrugation and at the bottom of the plate? The bending moment is such that the plate is stressed in tension. Use the results of problems 10.10 and 10.25.

10.17 SHEAR STRESS IN NON-RECTANGULAR CROSS SECTIONS

Shear stress can be important in beams of non-rectangular cross sections. Since the sections are shaped to put the most material at the maximum distance from the neutral axis, the thinned webs are left to resist shear stress.

Shear stress is needed for a section of the beam, as shown in figure 10.37d, to be in horizontal equilibrium. Since the bending moment is variable from section 1 to section 2, the bending stresses on these sections will be unequal, as shown in figure 10.37c. The area over which the shear stress acts is $b\Delta x$. Therefore, the horizontal equilibrium force at any level is $\tau b\Delta x$, where b is the width of the cross section.

The stress over the cross section faces is given from relationships 10.10 and 10.11,

$$\sigma = \frac{\sigma_{max}}{c}y \quad \text{and} \quad \sigma_{max} = \frac{Mc}{I}$$

Combining these algebraically,

(a)
$$\sigma = \frac{M}{I}y$$

The force caused by the stress over face 1 in figure 10.37d can be determined by summing the stress times the ΔA, as shown in figure 10.38.

(b)
$$F_1 = \sum \sigma \Delta A$$

Substituting relationship **(a)**,

(c)
$$F_1 = \sum \frac{M}{I}y\Delta A$$

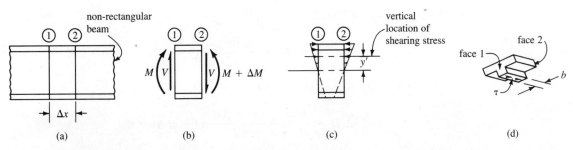

Figure 10.37 Shear stress, τ, required for horizontal equilibrium of the portion of the beam between sections 1 and 2 and above level y'.

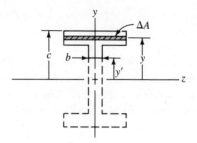

Figure 10.38 The force on face 1 of the equilibrium element is $\sigma \times \Delta A$.

Since M is constant over face 1 and I is a constant for the cross section,

(d) $$F_1 = \frac{M}{I} \sum y\Delta A$$

From relationship 10.3, $Q_x = \sum y\Delta A$, which is by definition the moment of area.

Here we must deal with an unfortunate change of notation that pervades most of structural technology. When determining cross section properties, the horizontal axis in the plane of the cross section is usually taken to be the x axis. When determining shearing forces and bending moments, the horizontal axis perpendicular to the cross section is also commonly called the x axis.

It turns out that this double, contradictory notation is rarely a problem in structural technology. Except in advanced stress analysis, the calculation of section properties and the computation of shear force and bending moments are done separately. Only in this derivation do they come together. And here, the horizontal axis in the plane of the cross section as shown in figure 10.38 is called the z axis rather than the x axis. However, rather than change the subscript for Q, the convention is to omit the subscript entirely, defining

$$Q = \sum y\Delta A$$

Then:

(e) $$F_1 = \frac{M}{I}Q$$

Similarly, the force on face 2 is

$$F_2 = \frac{(M + \Delta M)}{I}Q$$

acting to the left. For horizontal equilibrium,

$$\sum H \mapsto: \quad \frac{M}{I}Q - \frac{(M + \Delta M)}{I}Q + \tau b\Delta x = 0$$

(f) $$-\frac{\Delta M Q}{I} + \tau b\Delta x = 0$$

Solving **(f)** for τ,

(g)
$$\tau = \left(\frac{Q}{Ib}\right)\left(\frac{\Delta M}{\Delta x}\right)$$

From relationship 9.5, $V = \Delta M/\Delta x$. Therefore:

$$\tau = \frac{QV}{Ib} \qquad \textbf{(10.16)}$$

Relationship 10.16 is the very important relationship for shear stress in a beam having a cross section of *any* shape, where τ is the shearing stress at any level in the cross section y', Q is the moment of the area above y' defined as $\sum y\Delta A$, I is the second moment of area of the total cross section, and b is the width of the cross section at level y'.

ILLUSTRATIVE PROBLEM

10.16 What is the shearing stress in the web of the beam with the cross section shown in figure 10.39 (a) at the neutral axis, (b) at the junction of the web and the flange, (c) 3 in. above the neutral axis, and (d) 3 in. below the neutral axis? The shearing force is 20 000 lb.

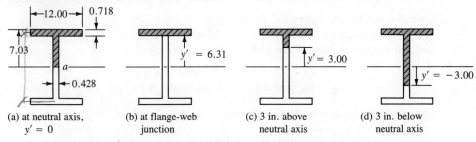

(a) at neutral axis, $y' = 0$	(b) at flange-web junction	(c) 3 in. above neutral axis	(d) 3 in. below neutral axis

$14.06 - .718 = 13.342 = \frac{}{2}$ **Figure 10.39** Illustrative problem 10.16.

6.671

Solution:

Referring to illustrative problem 10.12, the area was already calculated to be 22.6 in.^{2}, and the moment of inertia about the horizontal axis through the centroid was calculated to be 839 in.^{4} From relationship 10.16,

$$\tau = \frac{QV}{Ib}$$

where $V = 20\,000$ lb, $I = 839$ in.^{4}, and b = web thickness = 0.428 in.

All that remains to be calculated is the moment of area, Q, which depends upon the elevation, y'. Q has units of in.^{3} Therefore,

$$\tau = \frac{Q \text{ in.}^3 \times 20\,000 \text{ lb}}{839 \text{ in.}^4 \times 0.428 \text{ in.}}$$

$$\tau = 55.70Q \text{ psi}$$

Now finding Q for cases (a), (b), (c), and (d) as shown in figure 10.39, the areas for Q in each case are the shaded part of the cross section.

Case (a) at the Neutral Axis:
The area is that of the flange and of the web above the neutral axis.
The area of the flange is $12.00 \times 0.718 = 8.616$ in.^{2}
The area of the web is $6.312 \times 0.428 = 2.702$ in.^{2}
The moment arm of the flange area is 6.671 in. and that of the web area is 3.156 in.
The moment of area of the flange is $8.616 \times 6.673 = 57.49$ in.^{3}
The moment of area of the web is $2.702 \times 3.156 = 8.527$ in.^{3}
The total moment of area, Q, is 66.02 in.^{3}

$$\tau = 55.70Q = 55.70 \times 66.02 = 3\ 680 \text{ psi}$$

The shearing stress at the neutral axis is 3 680 psi.

Case (b) at the Junction of Web and Flange:
The moment of area is simply that of the flange,

$$\tau = 55.70Q = 55.70 \times 57.49 = 3\ 200 \text{ psi}$$

The shearing stress in the web at the flange is 3 200 psi.

Case (c) at 3.00 in. Above the Neutral Axis:
The moment of area is that of the flange, 57.49 in.^{3}, plus 3.31 in. of the web. The moment of that portion of the web is $3.31 \times 0.428 \times 4.655$.

$$Q = 57.49 + 6.595 = 64.085 \text{ in.}^3$$
$$\tau = 55.70Q = 55.70 \times 64.084 = 3\ 570 \text{ psi}$$

The shearing stress 3 in. above the neutral axis is 3 570 psi.

Case (d) at 3.00 in. Below the Neutral Axis:
The moment of area is that of the flange, 181.45 in.^{3}, plus 9.31 in. of the web. The moment of that portion of the web is

$$9.31 \times 0.428 \times (6.31 - 3.00)/2 = 6.595 \text{ in.}^3$$
$$Q = 57.49 + 6.595 = 64.085 \text{ in.}^3$$
$$\tau = 3\ 570 \text{ psi}$$

The shearing stress 3 in. below the neutral axis is identical to that 3 in. above the axis.

10.18 SHEAR STRESS IN SYMMETRICAL FLANGED SECTIONS

Illustrative problem 10.16 shows how to calculate shear stress using relationship 10.16. The stress distribution through the web is as shown in figure 10.40.

The maximum shear stress occurs at the neutral axis. The shear stress is symmetrical about the neutral axis. Since the web is of constant thickness, the distribution of stress is

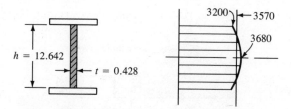

Figure 10.40 Shear stress distribution in the web of the beam of illustrative problem 10.16.

parabolic in shape just as was the shear stress in a rectangular beam. The shear stress at the flange is only about 13 percent lower than at the maximum. The shear stresses in the flanges themselves are much lower. The higher shear stresses are in the web, and although they are variable, they do not vary much over the web.

For this reason, a web shear formula that is a simple approximation is often used. The vertical shear force is assumed to be carried uniformly by the web area taken to be the total height of the web, h, times the thickness, t. The shear stress, therefore, is

$$\tau = \frac{V}{ht} \qquad \qquad (10.17)$$

This formula is the web shear stress approximation for flanged sections such as wide-flanged beams, I beams, or even channel sections when used in their symmetrical direction.

ILLUSTRATIVE PROBLEM

10.17 Calculate the approximate shear stress in the web for illustrative problem 10.16.

Solution:
As shown in figure 10.40, h is 12.642 in. and t is 0.428 in. Using the approximate relationship 10.17,

$$\tau = \frac{V}{ht} = \frac{20\ 000\ \text{lb}}{12.642\ \text{in.} \times 0.428\ \text{in.}}$$

$$\tau = 3\ 696\ \text{psi}$$

The approximation is 3 700 psi after rounding, which is very close to the "exact" solution of 3 680 psi.

10.19 SHEAR STRESS IN CIRCULAR SECTIONS

Another very important cross section shape is the circle. Maximum shear stress will occur at the neutral axis. Q for relationship 10.16 is moment of area above the neutral axis, as shown in figure 10.41.

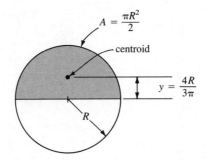

Figure 10.41 $Q = Ay$ for deriving the maximum shear stress due to bending of a circular cross section.

This is a semicircular area of $\pi R^2/2$. From Appendix III, the distance to the centroid from the neutral axis is $4R/3\pi$. Therefore,

$$Q = \frac{\pi R^2}{2} \times \frac{4R}{3\pi} = \frac{2R^3}{3}$$

$$I = \pi R^4/4$$

Now substituting into relationship 10.16,

$$\tau_{max} = \frac{4V}{3\pi R^2} \qquad (10.18)$$

Relationship 10.18 is the formula for maximum shearing due to bending in a solid cross section of radius R when the shear force is V. The maximum shear stress due to bending occurs at the neutral axis and is 33 percent higher than would be predicted by simply dividing the cross-shear force, V, by the circular area, πR^2.

10.20 LIGHTENING AND CLEARANCE HOLES

Holes are sometimes made in the web of a flanged section to lighten a structure and sometimes to provide a passage for pipes, tubes, cables, or other utilities. Substantial amounts of material are often removed without unduly weakening the beam. Such holes are best centered upon the neutral axis of a beam, as shown in figure 10.42, so that they do not weaken the beam's capacity to withstand tension and compression stresses due to bending. However, when located at the neutral axis, holes do decrease the shearing capacity of the web. This decrease in strength is totally independent of any stress concentration that the holes may cause.

The increased web shear stress due to a series of holes at the neutral axis can be estimated by dividing the shear stress calculated by relationship 10.17 by the ratio of the neutral axis length that is covered by web material.

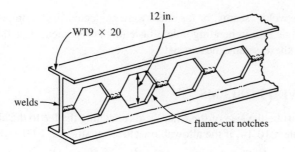

Figure 10.42 Illustrative problem 10.18.

ILLUSTRATIVE PROBLEM

10.18 A beam is fabricated by welding together two structural tees, WT9 × 20, with the stems "notched-out" as shown in figure 10.42. The beam must be able to carry a shear force due to bending of 45 000 lb. What maximum shear stress will be produced in the welds of the fabricated beam due to the shear load? (See figure 10.43.)

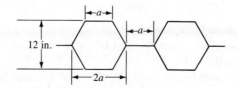

Figure 10.43 Illustrative problem 10.18.
Two-thirds of neutral axis is cut away.

Solution:

The stress in the web without the holes can be estimated by the approximate relationship 10.17. Using the information from Appendix V, $t = 0.316$ in. (Assuming the welds to be full-penetration butt welds) and $h = (8.95 - 0.524) \times 2 = 16.852$ in.

$\tau = 45\ 000$ lb 16.852 in. × 0.316 in.d = 8 450 psi (without holes)

The ratio of the neutral axis remaining after the holes is 1/3. The shear stress will be increased accordingly.

$$\tau = \frac{8\ 450 \text{ psi}}{1/3} = 25\ 400 \text{ psi}$$

10.21 FABRICATED BEAMS

Many non-standard shapes of special materials are used for beams and girders. These are called *fabricated* or, as discussed earlier, "built-up" beams. In the moment of inertia calculations, the welds, seams, or fasteners were neglected. But, components will not act together as a single unit in bending unless fastening is sufficient. It is the shear stress due to bending that these seams must resist.

ILLUSTRATIVE PROBLEMS

10.19 What must be the size of the fillet welds attaching the tee to the channel of illustrative problem 10.14, if the allowable weld shear stress is 10 000 psi (figure 10.44)?

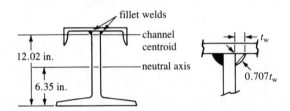

Figure 10.44   Illustrative problem 10.19.

Solution:

The maximum shear force due to the 20-lb/in. uniformly distributed load occurs just inside the supports. There V must be equal to the entire support reaction of 2 400 lb (figure 10.34).

In this case the weld seam is the boundary of the area for Q. All area on the upper side of that boundary contributes to Q. This side is the entire area of the channel, even though the flanges of the channel extend back down below the seam level. The area of the channel from Appendix V is 8.82 in.². From illustrative problem 10.9, the centroid of the entire cross section is an elevation of 6.35. Using data from Appendix V, the elevation of the centroid of the channel can be found to be 12.02 in. above the base of the cross section. Therefore,

$$Q = 8.82(12.02 - 6.35) = 50.01 \text{ in.}^3$$

I for the entire cross section was calculated in illustrative problem 10.13 to be 674 in.⁴ In the case of the weld seam, b is not the thickness of the stem of the tee but rather the thickness of the weld metal, $2t_w$, multiplied by the factor 0.707, which takes into account that the weld must shear through the throat dimension at 45°.

$$b = 2t_w \times 0.707 = 1.414t_w$$

Substituting into relationship 10.16,

$$\tau = \frac{QV}{Ib} = \frac{QV}{I \times 1.414t_w}$$

Solving for t_w,

$$t_w = \frac{QV}{1.414I\tau}$$

$$t_w = \frac{50.01 \text{ in.}^3 \times 2\,400 \text{ lb}}{1.414 \times 674 \text{ in.}^4 \times 10\,000 \text{ lb/in.}^2}$$

$$t_w = 0.012\,59 \text{ in.}$$

This result indicates that a very small weld of less than 1/16 in. is sufficient. Perhaps it would be better to use intermittent welds of larger size than continuous welds, since intermittent welds are preferable for controlling warping of long parts.

10.20 The aluminum fabricated beam shown in figure 10.45 is to be designed for a shear force due to bending of 0.75 MN. What diameter rivets must be used if they are to be spaced 10 cm on centers along the length of the beam and have an allowable stress of 100 MPa?

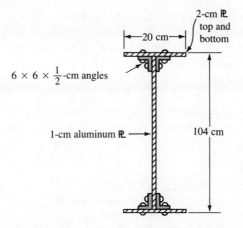

Figure 10.45 Illustrative problem 10.20.

Solution:

The shear stress in the web can be found from the approximate relationship 10.17.

$$\tau = \frac{V}{ht} = \frac{0.75 \text{ MN}}{100 \text{ cm} \times 1 \text{ cm}} \times \frac{(100 \text{ cm})^2}{1 \text{ m}^2}$$

$$\tau = 75 \text{ MPa}$$

Each 10 cm of beam length will accumulate a shear force at the joint of

$$F = \tau A = 75 \text{ MPa} \times 10 \text{ cm} \times 1 \text{ cm} \times \frac{1 \text{ m}^2}{(100 \text{ cm})^2}$$

$$F = 0.075 \text{ MN}$$

In each 10 cm of beam length, there will be two rivet areas to transfer the shear force from the web to the angles. The rivet area is $2 \times \pi d^2/4 = \pi d^2/2$, where d is the rivet diameter.

$$\tau = F/A = 2F/\pi d^2$$

Solving for d,

$$d = \sqrt{2F/\pi\tau}$$

If the rivet stress is set equal to the allowable stress,

$$d = \sqrt{2 \times 0.075 \text{ MN}/\pi \times 100 \text{ MPa}}$$
$$d = 0.021\ 85 \text{ m}$$

Use 22-mm-diameter rivets.

PROBLEM SET 10.4

10.35 What is the maximum shear stress due to bending on a 20-mm-diameter circular steel shaft at a section where the shearing force is 50 kN?

10.36 What is the shear stress in the web of a W8 × 31 beam spanning 17 ft, simply supported, with a concentrated load of 20 kN³ at its mid-span?

10.37 If the allowable shear stress is 10 000 psi, what is the maximum shear force due to bending that can be carried by an ST6 × 25 standard steel shape?

10.38 A fabricated beam is made of a $\frac{1}{4}$-in.-thick plate connecting a pair of 4-in.-diameter standard steel pipes as shown in figure 10.46. If the shearing force due to bending is 5 000 lb, what is the shear stress (a) at the weld and (b) at the neutral axis? Use the exact relationship.

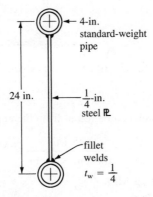

Figure 10.46 Problem 10.38.

10.39 A beam is fabricated of plexiglass to the cross-sectional dimensions shown in figure 10.47. Cement with a shearing stress capacity of 10 MPa is used to bond the web to the flanges. What is the maximum shear force due to bending that the beam can carry if the cement strength governs? Use approximate relationships.

10.40 A fabricated beam is made of aluminum angles riveted to a sheet aluminum web as shown in figure 10.48. If the rivets are 6 mm in diameter, are spaced 2 cm apart,

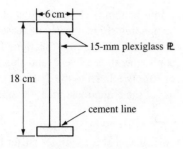

Figure 10.47 Problem 10.39.

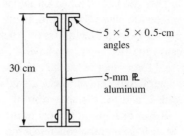

Figure 10.48 Problem 10.40.

and have an allowable stress of 25 MPa, what is the allowable cross shear on the beam due to bending if the rivets govern the strength? Use the approximate formula.

10.41 Circular lightening holes 8 in. in diameter are cut into the web of an S12 × 42.9 standard I beam. If the web shear due to bending may not exceed 7 000 psi, what is the smallest spacing of the lightening holes on the neutral axis? The shear force is expected to be 10 kips. Use the approximate formula.

10.42 Make a graph of the shear stress distribution in the stem of a WT8 × 18 in terms of V.

10.43 What is the maximum shear stress in a 12-in. standard-weight pipe forming a simply supported span of 8 ft and carrying a uniformly distributed load of 1 000 lb/ft? Describe the location of maximum stress.

10.22 BEAMS OF TWO MATERIALS: EQUIVALENT CROSS SECTIONS

Often beams are made of more than one material. A common example is steel-reinforced concrete. Since concrete has good compression strength but poor tension strength, the tension side of the beam is reinforced with a material having good tension properties, such as steel. The effect of the steel on the concrete can be envisioned using the concept of the *equivalent cross section* shown in figure 10.49.

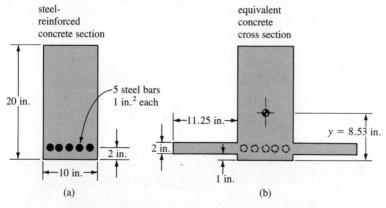

Figure 10.49 The concept of equivalent cross section.

This concept is based on the assumption that, while bending, plane sections of the beam remain plane. Therefore, in the elastic range, the bending strain linearly increases as one goes from the neutral axis toward the outer extremities of the cross section. But, because the two materials have different elastic moduli, the stress levels (and thus the internal forces that resist the bending moment) are not simply related to the distance from the neutral axis. These forces are related to the product of the component area, the elastic modulus, and the distance from the neutral axis.

If, for example, the steel in the cross section of figure 10.49 has an elastic modulus of 30.0E6 psi and the concrete a modulus of 3.00E6 psi, the steel is ten times stiffer than the concrete. Therefore, for a given strain level such as that occurring at the level of the center of the five reinforcing bars, the stress level in the steel is ten times that of the concrete. The steel, therefore, has the same effect as ten times that area of concrete. So, in the equivalent cross section, the beam is considered to be composed entirely of concrete with an extra 50 in.^{2} at the level of the reinforcing bars. Five in.^{2} are considered to fill the voids of the bars, leaving 45 in.^{2} to be placed as fins, in this case 2 in. thick to either side of the beam. This is a cross section of a single material, concrete, that has an elastic effect equivalent to the two materials for bending about the x axis.

ILLUSTRATIVE PROBLEM

10.21 What is the stress in the concrete of the beam shown in figure 10.49a when subjected to 500 kip · in. of bending moment? Find the stress in the concrete at the top and at the bottom of the beam. Also, find the stress in the steel reinforcement. The elastic modulus for the steel is 30E6 psi and for the concrete is 3E6 psi. What would be the stress in an unreinforced concrete beam?

Solution:
Due to the ten-to-one ratio of the elastic modulus of steel to that of concrete, the equivalent cross section shown in figure 10.49b applies. The fins are taken to be 2 in. thick. Fifty in.^{2} of equivalent concrete area are required less the 5 in. needed to fill the voids left by the steel. The fins need to extend 11.25 in. to either side.

The location of the centroid from the bottom of the beam can be computed for the equivalent cross section from relationships 10.1, 10.3, and 10.4.

$$A = 20 \times 10 + 2 \times 22.5 = 245 \text{ in.}^2$$

$$Q_x = 200 \times 10 + 45 \times 2 = 2\,090 \text{ in.}^3$$

$$\bar{y} = \frac{2\,090}{245} = 8.530$$

The moment of inertia about the centroid can be computed as follows:

$$I \text{ (for the base cross section)} = \frac{10 \times 20^3}{12} + 200(10 - 8.530)^2$$

$$= 6\,667 + 432 = 7\,099 \text{ in.}^4$$

$$I \text{ (for the two fins)} = \frac{22.5 \times 2^3}{12} + 45 \times (8.530 - 2)^2$$

$$= 15 + 1\,920 = 1\,935 \text{ in.}^4$$

$$I \text{ (total equivalent section)} = 9\,034 \text{ in.}^4$$

The stress in the concrete can be determined from relationships 10.12 and 10.13. At the top of the beam:

$$\sigma_T = -\frac{Mc_T}{I} = -\frac{500 \text{ kip} \cdot \text{in.} \times 11.47 \text{ in.}}{9\,034 \text{ in.}^4}$$

$$\sigma_T = 634 \text{ psi compression}$$

At the bottom of the beam:

$$\sigma_B = \frac{Mc_B}{I} = \frac{500 \text{ kip} \cdot \text{in.} \times 8.530 \text{ in.}}{9\,034 \text{ in.}^4}$$

$$\sigma_B = 472 \text{ psi tension}$$

In the steel reinforcement bars, due to the ratio of elastic moduli, the steel stress is ten times what the concrete stress would be at the location of the bars. From relationship 10.10,

$$\sigma = 10\frac{\sigma_{max}y}{c} = \frac{10 \times 472 \times 6.530}{8.530}$$

$$\sigma = 3\,613 \text{ psi tension}$$

In the unreinforced beam the moment of inertia would be

$$I = 6\,667 \text{ in.}^4$$

And, from relationship 10.11, the stress would be

$$\sigma = \frac{Mc}{I} = \frac{500 \text{ kip} \cdot \text{in.} \times 10 \text{ in.}}{6\,667 \text{ in.}^4}$$

$$\sigma = 750 \text{ psi}$$

compression on top and tension on the bottom.

Summary:
In the case of the reinforced beam, the compression stress in the concrete and the tension stress in the steel are very low compared to the strength of the materials involved. The tension stress of 472 is actually large for concrete. However, should the concrete develop cracks in the tension area of the cross section, there is great reserve tension strength in the reinforcing bars. The cracks would be microscopic and would be restrained from enlarging by the steel. If there would be no reinforcement, the maximum tension stress, as well as the compression stress, would be

750 psi. A tension stress of that level would surely crack the beam and it would fully collapse because there is no reserve strength.

SUMMARY

□ The *properties of areas* relevant to beam cross sections in bending are the centroid, first moment of area, and second moment of area. Complex areas can be divided into segments to determine these properties.

$$(10.1) \quad A = \Sigma \, \Delta A \quad \text{or} \quad A = \int y \, dx = \int x \, dy = 2\pi \int_0^R r \, dr$$

□ The *first moments of area* are *defined* by:
$$(10.3) \quad Q_x = \Sigma \, y\Delta A \quad \text{and} \quad Q_y = \Sigma \, x\Delta A$$

□ The coordinates of the *centroid* can be determined from relationship 10.4:
$$(10.4) \quad \bar{x} = Q_y/A \quad \text{and} \quad \bar{y} = Q_x/A$$

□ Centroids will fall on any axis of symmetry that may exist.

□ The *second moments of area,* or moments of inertia, are *defined* by:
$$(10.5) \quad I_x = \Sigma \, y^2\Delta A \quad \text{and} \quad I_y = \Sigma \, x^2\Delta A$$

□ For rectangular and circular cross sections, relationships **10.6** and **10.8** give:

$$I_x = \frac{bh^3}{12} \quad \text{and} \quad I_x = \frac{\pi R^4}{4}, \text{ respectively}$$

□ The *parallel-axis rule* transfers moments of inertia of areas.
$$(10.9) \quad I = I_c + d^2A$$

□ The maximum *tension* or *compression stress* in a beam cross section is:

$$(10.11) \quad \sigma_{max} = \frac{Mc}{I} \quad \text{or} \quad \sigma_{max} = \frac{M}{Z} \quad \text{where} \quad Z = I/c$$

□ The *section modulus* is defined as I/c.

□ The *shear stress* in a beam cross section due to bending is:

$$(10.16) \quad \tau = \frac{QV}{Ib}$$

□ For a flanged section, the maximum shear stress can be *approximated* by:

$$(10.17) \quad \tau = \frac{V}{ht}$$

□ For a solid circular cross section, the maximum shear is given by:

$$(10.18) \quad \tau_{max} = \frac{4V}{3\pi R^2}$$

REFERENCES

1. Timoshenko, S. and MacCullough, G. H. *Elements of Strength of Materials*. Princeton, NJ: D. Van Nostrand, 1958.

2. Wynne, G. B. *Reinforced Concrete*. Reston, VA: Reston, 1981.

CHAPTER ELEVEN

Shear Force and Bending Moment Diagrams

Shear force and bending moments, previously treated from the standpoint of equilibrium of a portion of a beam as a free body, can be summarized over the entire length of a beam. These summaries are called shear force and moment diagrams, and they are studied three ways in this chapter.

First, the portion of the beam taken as a free body is defined in terms of a variable length x. The equilibrium equations are then written in terms of x, yielding equations for shear force and bending moment. Second, the load is defined in terms of x, and calculus is used to derive equations describing diagrams, without making a free-body diagram of portions of a beam. Third, the slope-area method is developed as a means of directly drawing the shear force and bending moment diagrams, without making either a free-body diagram or writing equations.

In the course of the illustrations in this chapter, several basic beam cases typically found in engineering handbooks are developed. Illustrative and exercise problems from both civil and mechanical engineering technology are included.

11.1 ARRANGEMENT AND SIGN CONVENTION

In chapters 9 and 10, shear force and bending moment were determined at single points on beams. This determination was made by cutting through the beam at one point and making a free-body diagram. Ideally, the point described by the diagram was the point along the beam at which either the shear force or bending moment was maximum. For simply supported beams with symmetrical downward loads, the maximum bending moment is at mid-span, and the maximum shear force occurs just inside the supports. For cantilever beams, both the maximum shear force and the maximum bending moment for any combination of downward loads occur at the point where the beam enters the supporting wall.

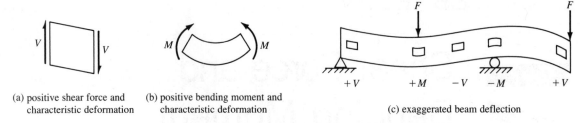

(a) positive shear force and characteristic deformation

(b) positive bending moment and characteristic deformation

(c) exaggerated beam deflection

Figure 11.1 Sign conventions for shear force and bending moments.

But in other cases, the location of the maximums is not known at the outset, making it necessary to search the entire length of the beam for them. It is good practice to construct a complete diagram of the shear and moment in such cases. Also, there are times when a complete understanding of the internal force distribution over the entire beam is desirable.

There are three ways to construct complete diagrams: algebraically (from a free-body diagram), by using calculus, and by using the slope-area method. These methods will be discussed in turn. In all cases, diagrams are constructed with the beam axis as the x axis. Normally, the diagrams are placed one exactly over the other, in sets of three, as in figures 11.3, 11.5, 11.7, and 11.9. The top diagram usually shows the loads, concentrated and distributed, acting upon the beam. The load diagram may be a free-body diagram of the entire beam, including its support reactions. The second diagram is the shear force. The bottom diagram, directly below the other two, is the bending-moment diagram. Positive loads are taken to be upward loads. Positive shear force is shown in figure 11.1a. Positive bending moment is shown in figure 11.1b. This convention regarding signs is used throughout the remainder of this book.

11.2 FOUR IMPORTANT CASES SOLVED ALGEBRAICALLY FROM FREE-BODY DIAGRAMS

Both the method and the results are very important in the following four cases. The method, writing algebraic expressions describing equilibrium conditions on a free-body diagram, is extremely useful for many problems. But additionally, the cases chosen here for illustration are important because they constitute four basic cases that exemplify the majority of beams one encounters. Therefore, the shapes and maximum values of the diagrams are worth committing to memory. The four cases are:

□ The cantilever beam with a single concentrated load at its free end.

□ The cantilever beam with a uniformly distributed load over its entire length.

□ The simply supported beam with a uniformly distributed load.

□ The simply supported beam with a concentrated load at mid-span.

The Cantilever Beam with a Single Concentrated Load at Its Free End

A portion of the beam, cut at an arbitrary distance, x, from the built-in end is shown as a free-body diagram in figure 11.2.

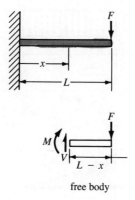

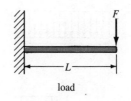

load

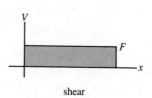

shear

Figure 11.2 Free-body diagram for algebraic construction of the shear force and bending moment diagrams for a cantilever beam of length L with a concentrated load F applied to the free end.

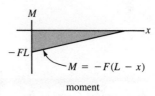

moment

Figure 11.3 The set of shear force and bending moment diagrams for a cantilever beam of length L with a concentrated load F at the free end.

For the vertical equilibrium of the forces,

$$V = F \qquad (11.1)$$

For moment equilibrium,

$$\sum M_x \circlearrowright: \quad M + F(L - x) = 0$$
$$M = -F(L - x) \qquad (11.2)$$

By graphing relationships 11.1 and 11.2, one can see that the shearing force is a constant positive value of F. The bending moment has a maximum value of $-FL$ at the support that reduces linearly to zero at the free end. The diagrams are shown in figure 11.3.

The Cantilever Beam with a Uniformly Distributed Load over Its Entire Length

Again, a portion of the beam cut at an arbitrary distance from the built-in end, x, is drawn as a free body in figure 11.4.

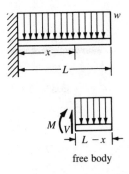

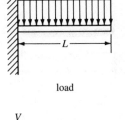

load

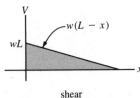

shear

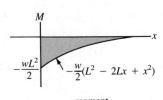

moment

Figure 11.4 Free-body diagram for algebraic construction of the shear force and bending moment diagrams for a cantilever beam of length L with a distributed load w over its length.

Figure 11.5 Shear force and bending moment diagrams for a cantilever beam with a uniformly distributed load.

For equilibrium of the vertical forces,

$$\Sigma V \updownarrow: \quad V - w(L - x) = 0$$
$$V = w(L - x) \tag{11.3}$$

For moment equilibrium,

$$\Sigma M_x \circlearrowleft: \quad M + \frac{w(L - x)(L - x)}{2} = 0$$

$$M = -\frac{w}{2}(L^2 - 2Lx + x^2) \tag{11.4}$$

The maximum shear force and bending moment occur at the built-in end when $x = 0$ and have values of wL and $-wL^2/2$, respectively. The diagrams that can be constructed from relationships 11.3 and 11.4 are shown in figure 11.5.

The Simply Supported Beam with a Uniformly Distributed Load

A portion of the beam, cut at an arbitrary distance, x, from the left end of the beam is shown as a free-body diagram in figure 11.6.

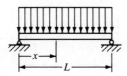

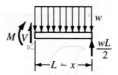

Figure 11.6 Free-body diagram for construction of the shear force and bending moment diagram for a simply supported beam under a uniformly distributed load, w.

By symmetry, the support reactions must share the total applied load and, therefore, have values of $wL/2$.

For equilibrium of the vertical forces on the free-body diagram,

$$\Sigma V \uparrow: \quad V + \frac{wL}{2} - w(L - x) = 0$$

$$V = w\left(\frac{L}{2} - x\right) \tag{11.5}$$

For equilibrium of the moments,

$$\Sigma M_x \circlearrowright: \quad M + \frac{w(L - x)(L - x)}{2} - \frac{wL(L - x)}{2} = 0$$

$$M + \frac{wL^2}{2} - wLx + \frac{wx^2}{2} - \frac{wL^2}{2} + \frac{wLx}{2} = 0$$

Collecting terms,

$$M - \frac{wLx}{2} + \frac{wx^2}{2} = 0$$

$$M = \frac{w}{2}(Lx - x^2) \tag{11.6}$$

The maximum shear force occurs at either end of the beam, when x equals 0 or L. The maximum values are $wL/2$ or $-wL/2$, respectively.

The maximum bending moment occurs at the center of the beam when $x = L/2$ and has the value of $wL^2/8$. This relationship should be memorized.

The diagrams constructed from relationships 11.5 and 11.6 are shown in figure 11.7.

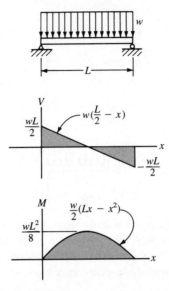

Figure 11.7 Shear force and bending moment diagrams for a simply supported beam under a uniformly distributed load.

The Simply Supported Beam with a Concentrated Load at Mid-Span

This problem requires two free-body diagrams, one with the beam cut to the left of the concentrated load, F, and one with the beam cut to the right. They are shown in figure 11.8.

For the left side of the beam, when x is less than $L/2$, free-body diagram 11.8a applies.

$$\Sigma\,V\uparrow:\ V - F + \frac{F}{2} = 0$$

$$V = \frac{F}{2} \tag{11.7}$$

$$\Sigma\,M_x\circlearrowleft:\ M + F\left(\frac{L}{2} - x\right) - \frac{F}{2}(L - x) = 0$$

$$M = \frac{Fx}{2} \tag{11.8}$$

For the right side of the beam, when x is greater than $L/2$, free-body diagram 11.8b applies.

$$\Sigma\,V\uparrow:\ V + \frac{F}{2} = 0$$

$$V = -\frac{F}{2} \tag{11.9}$$

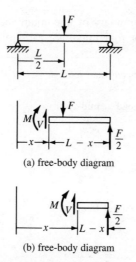

(a) free-body diagram

(b) free-body diagram

Figure 11.8 Free-body diagram for construction of the shear force and bending moment diagrams for a simply supported beam with a concentrated load at mid-span.

$$\sum M_x \circlearrowright : \quad M - \frac{F}{2}(L - x) = 0$$

$$M = \frac{F(L - x)}{2} \tag{11.10}$$

The diagrams constructed from relationships 11.7 through 11.10 are shown in figure 11.9. The maximum bending moment occurs at mid-span, $x = L/2$. Using either relationship 11.8 or 11.10, the maximum bending moment can be calculated to be equal to $FL/4$.

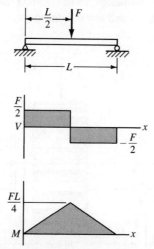

Figure 11.9 Shear force and bending moment diagrams for a simply supported beam under a concentrated load at mid-span.

Table 11.1 Summary of maximum shear and moment relationships for four important cases of beam loading

		Concentrated load, F	UNIFORMLY Distributed load, w
Simply supported		Load at mid-span $V_{max} = \pm \dfrac{F}{2}$ $M_{max} = \dfrac{FL}{4}$	$V_{max} = \pm \dfrac{wL}{2}$ $M_{max} = \dfrac{wL^2}{8}$
Cantilevered		Load at free end $V_{max} = F$ $M_{max} = -FL$	$V_{max} = wL$ $M_{max} = -\dfrac{wL^2}{2}$

The maximum values of shear and moment for these four important cases are summarized in table 11.1.

ILLUSTRATIVE PROBLEM

11.1 The reverse idler gear slides to the position on the shaft shown in figure 11.10 when shifting a power transmission into reverse. The maximum component of force acting in the direction that tends to bend the shaft is calculated to be 2.48 kN. The ball bearings are self-aligning. The allowable stress for the solid steel shaft is 85 MPa. What must be the shaft diameters, to the nearest mm, at the slide and at the bearings? (Neglect shear stress and stress concentration.)

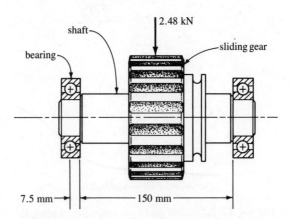

Figure 11.10 Illustrative problem 11.1.

Solution:

The self-aligning bearings behave as simple supports. Since the maximum load occurs while the gear is at mid-span, the bending moment diagram as shown in figure 11.9 applies to the shaft. Relationship 11.8 describes the bending moment diagram from the center of the left bearing support to the gear at mid-span,

$$M = \frac{Fx}{2}$$

At Mid-span:

$$x = 82.5 \text{ mm}$$

(a)
$$M = 2.48 \text{ kN} \times \frac{82.5 \text{ mm}}{2} = 102.3 \text{ N} \cdot \text{m}$$

From relationship 10.8,

$$I_x = \frac{\pi R^4}{4}$$

For the circular of the shaft, $c = R$ and $R = D/2$.

Substituting for I and c in relationship 10.11,

(b)
$$\sigma_{\text{max}} = \frac{Mc}{I} = \frac{32 \, M}{\pi D^3}$$

Solving for D,

(c)
$$D = \sqrt[3]{\frac{32M}{\pi \sigma_{\text{max}}}}$$

Letting $\sigma_{\text{max}} = 85$ MPa,

$$D = \sqrt[3]{\frac{32 \times 102.3 \text{ N} \cdot \text{m}}{\pi \times 85\text{E}6 \text{ N/m}^2}}$$

$$D = 2.31\text{E}{-2} \text{ m} = 23.1 \text{ mm}$$

Use 24 mm for the shaft diameter at the gear slide.

At the Bearing:

$$x = 7.5 \text{ mm}$$

(d)
$$M = 2.48 \text{ kN} \times \frac{7.5 \text{ mm}}{2} = 9.30 \text{ N} \cdot \text{m}$$

$$D = \sqrt[3]{\frac{32 \times 9.30 \text{ N} \cdot \text{m}}{\pi \times 85\text{E}6 \text{ N/m}^2}}$$

$$D = 1.036\text{E}{-2} \text{ m} = 10.36 \text{ mm}$$

Use 11 mm, or perhaps 12 mm, for the shaft diameter at the bearings, left and right, since the shaft loading is symmetrical.

PROBLEM SET 11.1

11.1 Calculate the bending moment and shear force 3 ft from the left end and at the mid-span of a simply supported beam 7 ft long with a uniformly distributed load of 10 kips/ft. (Use the derived relationships.)

11.2 What shear force and bending moment are caused at the support of a diving board extending 3 m out over the water by a diver who weighs 100 kg? What shear force and bending moment occur 2 m from the end of the diving board? (Use the derived relationships.)

11.3 A flag pole in a hurricane is acted upon by a wind load of 60 lb/ft. The pole is 100 ft high. Sketch a set of shear force and bending moment diagrams giving the critical values.

11.4 A 90-m pre-stressed concrete beam is simply supported. What bending moment and shear force are produced at 10-m intervals along the span due to a concentrated load of 100 MN at mid-span?

11.5 Derive an expression for the variable section height, h, of a rectangular cantilever beam with a concentrated force F at the end if the cross section is to be rectangular and of constant width b. The allowable bending stress is S. Disregard shear stress.

11.6 What is the load capacity of a W8 × 31 beam supported as a cantilever beam with a uniformly distributed load if the length is 10 ft and the allowable stress in bending is 22 000 psi?

11.7 Using a free-body diagram and algebra, construct the shear force and bending moment diagrams for a cantilever beam supporting a distributed load that varies linearly from 0 at the support to W at the free end.

11.3 SHEAR FORCE AND BENDING MOMENT DIAGRAMS USING CALCULUS

Shear force and bending moment diagrams can also be constructed in some cases by using calculus.

Figure 11.11 shows a free-body diagram of an element of length Δx cut from a beam under a distributed load w.

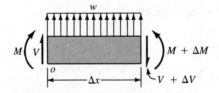

Figure 11.11 Element of a beam of length Δx shown as a free body. The bending moment is increasing by an amount ΔM from the left to the right faces of the element. The shear force increases by ΔV.

The element must be in vertical and moment equilibrium.

$$\Sigma\, V\updownarrow:\quad V - (V + \Delta V) + w\Delta x = 0$$
$$-\Delta V + w\Delta x = 0$$
$$w = \frac{\Delta V}{\Delta x} \qquad\qquad (11.11)$$

Expressed as a derivative,

$$w = \frac{dV}{dx} \qquad\qquad (11.12)$$

The derivative of the shear force equals the distributed load.
   Conversely,

$$V = \int w\, dx \qquad\qquad (11.13)$$

The shear force is equal to the integral of the distributed load.
   For moment equilibrium,

$$M_o\circlearrowleft:\quad M - (M + \Delta M) + (V + \Delta V)\Delta x + \frac{w\Delta x\Delta x}{2} = 0$$

The products of two Δ quantities such as $\Delta V\Delta x$ or Δx^2 are *second-order effects*. This term means that if Δx is small, Δx^2 is an order of magnitude again smaller. The terms containing second-order effects are neglected and dropped out of the equation. Therefore, using algebra on the remaining terms,

$$V = \frac{\Delta M}{\Delta x} \qquad\qquad (11.14)$$

Relationship 11.14 is identical to relationship 9.5. Written as a derivative,

$$V = \frac{dM}{dx} \qquad\qquad (11.15)$$

The shear force is equal to the derivative of the bending moment.

$$M = \int V\, dx \qquad\qquad (11.16)$$

Conversely, the integral of the shear force is equal to the bending moment. Since the maximum value of a function occurs at a location where its derivative equals zero, relationship 11.15 leads to a very important corollary:

> The maximum bending moment will occur at some location along the length of the beam where the shear force is equal to zero.

This corollary is important because it enables one to determine the maximum bending moment by searching out the likely location or locations by constructing a shear force diagram and looking for the places where the shear force is zero. Check figures 11.7 and 11.9 as examples.

ILLUSTRATIVE PROBLEM

11.2 Find the location of the maximum bending moment and construct (using calculus) the bending moment diagram for the simply supported beam shown in figure 11.12.

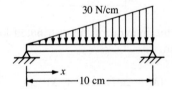

Figure 11.12   Illustrative problem 11.2.

Solution:
The load, w, increases uniformly from 0 at the left end of the beam to 30 N/cm at the right end of the beam. This effect can be expressed algebraically by the relationship,

$$w = \frac{-30 \text{ N/cm}}{10 \text{ cm}} x$$

(a) $$w = -3.0x \text{ N/cm}^2$$

The shear force can be determined from relationship 11.13,

$$V = \int w \, dx = -3.0 \int x \, dx$$

Integrating,

(b) $$V = \frac{-3x^2}{2} + C_1$$

where C_1 is a constant of integration. The constant of integration cannot yet be evaluated, because at no point on the beam is the shear force known at this time in the solution.

Going on to relationship 11.16 for the bending moment,

$$M = \int V \, dx = \int \left(-\frac{3x^2}{2} + C_1 \right) dx$$

Integrating,

(c)
$$M = -\frac{x^3}{2} + C_1 x + C_2$$

where C_2 is the second constant of integration.

C_1 and C_2 can be found because M must be zero at both ends of the beam. M is zero because, at simply supported ends, there is nothing to resist bending moments.

Letting M in relationship (c) equal zero when $x = 0$,

$$0 = -\frac{0}{2} + C_1(0) + C_2$$

Therefore, C_2 must equal 0. Relationship (c) becomes

(d)
$$M = -\frac{x^3}{2} + C_1 x$$

Letting M equal zero when $x = 10$ cm,

$$0 = -\frac{1000}{2} + C_1(10)$$

$$C_1 = 50$$

Therefore, relationship (d) becomes

(e)
$$M = -\frac{x^3}{2} + 50x$$

And relationship (b) for the shear force becomes

(f)
$$V = -\frac{3x^2}{2} + 50$$

The maximum bending moment is where the shear force equals zero.

$$0 = -\frac{3x^2}{2} + 50$$

$$x = \sqrt{33.32} = 5.77 \text{ cm}$$

The maximum bending moment is 5.77 cm from the left end of the beam. Its value can be determined by substituting 5.77 for x into relationship (e).

$$M_{max} = -\frac{5.77^3}{2} + 50(5.77)$$

$$M_{max} = 192.6 \text{ N} \cdot \text{cm}$$

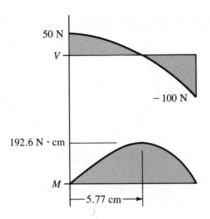

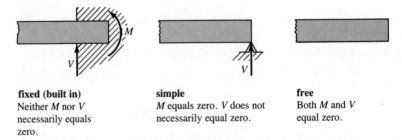

Figure 11.13 Shear and moment diagrams for illustrative problem 11.2.

The graphs of relationships **(f)** and **(e)** give the shear and moment diagrams, respectively, shown in figure 11.13.

11.4 BOUNDARY (END) CONDITIONS

The constants of integration are usually solved using the conditions at the boundary between the beam and the rest of the structure. These conditions depend on the kinds of supports. Thus far we have dealt with three kinds of supports or end conditions: fixed (or built-in), simple, and free, as shown in figure 11.14.

Fixed, or built-in, beam supports are capable of providing both a moment and a vertical force reaction to the beam.

Simple supports can provide only a vertical force. They resist no moments. Therefore, $M = 0$.

Free ends of beams can provide neither a vertical reaction nor a moment. Therefore, $V = 0$ and $M = 0$.

fixed (built in)
Neither M nor V necessarily equals zero.

simple
M equals zero. V does not necessarily equal zero.

free
Both M and V equal zero.

Figure 11.14 Boundary (end) conditions for various types of beam supports or ends.

ILLUSTRATIVE PROBLEM

11.3 An airplane wing has the lift forces distributed over its length in a fashion that can be approximated by the relationship $f = f_o(1 - x^2/L^2)$ shown in figure 11.15, where f_o

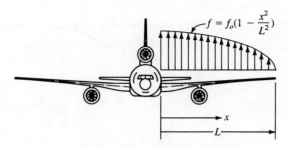

Figure 11.15 Illustrative problem 11.3.

is the maximum lift force of 53 lb/in. and L is the length of the wing of 840 in. Construct the shear and moment diagrams. What is the maximum bending moment at the root of the wing?

Solution:
From relationship 11.13,

$$V = \int w \, dx$$

Where $w = f$,

(a)
$$V = \int f_o \left(1 - \frac{x^2}{L^2}\right) dx$$

Integrating,

(b)
$$V = f_o \left(x - \frac{x^3}{3L^2} + C_1\right)$$

Because there is a free end at $x = L$, $V = 0$ at that point. Substituting this boundary condition into relationship **(b)**,

$$0 = f_o \left(L - \frac{L}{3} + C_1\right)$$

Therefore,

$$C_1 = -\frac{2L}{3}$$

and

(c)
$$V = f_o \left(x - \frac{x^3}{3L^2} - \frac{2L}{3}\right)$$

From relationship 11.16,

$$M = \int V \, dx$$

(d)
$$M = \int f_o \left(x - \frac{x^3}{3L^2} - \frac{2L}{3}\right) dx$$

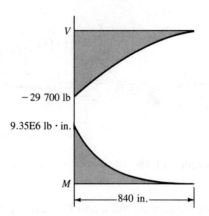

Figure 11.16   Shear force and bending moment diagrams for illustrative problem 11.3.

Integrating,

(e)
$$M = f_o\left(\frac{x^2}{2} - \frac{x^4}{12L^2} - \frac{2Lx}{3} + C_2\right)$$

At $x = L$, M is also equal to 0. Substituting this boundary condition into (e),

$$0 = f_o\left(\frac{L^2}{2} - \frac{L^2}{12} - \frac{2L^2}{3} + C_2\right)$$

$$C_2 = \frac{L^2}{4}$$

(f)
$$M = f_o\left(x^2 - \frac{x^4}{12L^2} - \frac{2Lx}{3} + \frac{L^2}{4}\right)$$

Relationships (c) and (f) give the shear and moment diagrams. The diagrams are plotted in figure 11.16.

The maximum values of both V and M will be at $x = 0$. Substituting this into relationships (c) and (f),

$$V_{max} = \frac{-2f_oL}{3} \quad \text{and} \quad M_{max} = \frac{+f_oL^2}{4}$$

When $f_o = 53$ lb/in.   and $L = 840$ in.,

$$V_{max} = -29\,700 \text{ lb} \quad \text{and} \quad M_{max} = 9.35\text{E6 lb} \cdot \text{in.}$$

PROBLEM SET 11.2

Using calculus, plot the shear force and bending moment diagrams for the following problems. Compute the maximum values.

11.8 A cantilever beam is 15 ft long with a load uniformly increasing from 0 at the built-in end to 1 kip/ft at the free end as shown in figure 11.17.

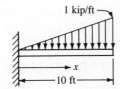

Figure 11.17 Problem 11.8.

11.9 A simply supported beam is 10 m long with a uniformly distributed load of 100 kN/m as shown in figure 11.18.

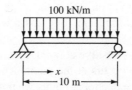

Figure 11.18 Problem 11.9.

11.10 A parabolically distributed load is 1 N/cm, maximum, as shown in figure 11.19.

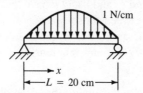

Figure 11.19 Problem 11.10.

11.11 The hull of a ship is, among other things, a beam that must resist the bending moments caused by waves, as shown on the supertanker of figure 11.20. The variations in water level along the hull from the mean water level cause variations in buoyant forces, idealized as a sine wave in this problem. One of the worst cases occurs when a wave has a pitch exactly the length of the hull. The distributed load

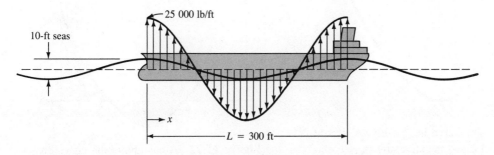

Figure 11.20 Problem 11.11.

is approximated as $w = 25\ 000 \cos (2\pi x/L)$ lb/ft. In this problem, the hull is to be idealized as a beam with both ends free.

11.5 SHEAR FORCE AND BENDING MOMENT DIAGRAMS USING THE SLOPE-AREA METHOD

Of the two methods previously discussed for constructing shear diagrams, the algebraic method stems most directly from the theory of static equilibrium. The calculus method is more elegant mathematically. Both methods, however, become complicated when discontinuities in the loading upon the beam are present. Discontinuities are abrupt changes in load distribution. Concentrated loads are themselves discontinuities. For this reason, the method discussed from this point on, called the slope-area method, is most commonly used when concentrated loads are present. But first a word about positive and negative slopes and areas is necessary.

A curve has slope defined as positive at points where the tangents to the curve go up and to the right. Negative slope occurs at the points where the tangents to the curve go down and to the right. Zero slope occurs at points where the tangent is horizontal. These cases are shown in figure 11.21. The derivative of the function has the same sign as the slope of its curve.

A curve encloses area defined as positive when the area is above the x axis and negative when the area is below the x axis (figure 11.22). The integrals of the function between the limits a and b, b and c, and c and d have the same sign as the areas enclosed by the curve of the function between those points.

The four rules of the slope-area method are as follows:

Relationship 11.12, $w = dV/dx$, and relationship 11.15, $V = dM/dx$, are the bases for the slope rules.

1. The slope of the shear diagram is equal to the value of the load diagram at any point.
2. The slope of the moment diagram is equal to the value of the shear diagram at any point.

Relationship 11.13, $V = \int w\ dx$, and relationship 11.16, $M = \int V\ dx$, are the bases for the area rules.

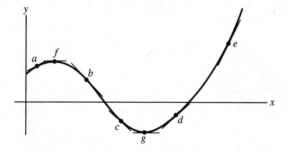

Figure 11.21 The curve has positive slopes at points a, d, and e and negative slopes at b and c. Zero slopes occur at points f and g.

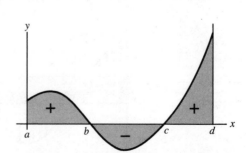

Figure 11.22 Areas above the curve are positive and below the curve, negative.

3. The area under the load diagram between two points on the x axis is equal to the value of the change in the shear diagram between those two points.

4. The area under the shear diagram between two points on the x axis is equal to the value of the change in the moment diagram between those two points.

The area rules can be thought of using the analogy of a squeegee, shown in figure 11.23.

If the squeegee is passed from left to right over a wet area shaped as shown, some area would be wiped dry at any position of the squeegee. The moisture from the area would be collected in the container. The graph below the squeegee indicates the level of water collected in the container as the squeegee passes from left to right. The lower graph is the integral of the initial ''wet'' region above. The constant of integration is the starting level of water in the container.

If the container is empty at the start of the pass, the level of water in the container is zero at a. In moving the squeegee from a to b, moisture is collected at a uniform rate. Therefore, the lower graph increases from zero, in a straight line, to a positive value proportional to the area of the rectangle under the upper diagram between a and b.

In moving from b to c, the misted area tapers off to zero at c. Therefore, the curve representing water collected continues to increase at the same rate at b but begins to level off. At c, the lower curve is horizontal, indicating that as the squeegee moves incrementally through c, there is essentially no misted area being swept and no increase of moisture collected.

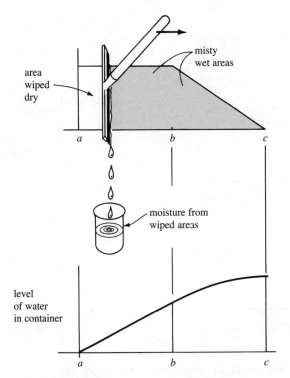

Figure 11.23 The squeegee analogy.

The squeegee analogy breaks down when negative areas are encountered, because the physical process is irreversible. But ''negative'' areas are encountered in using the slope-area method to construct shear and moment diagrams. One just has to think of the analogous process as working in reverse when wiping negative areas (those below the x axis).

PROBLEM SET 11.3

11.12 The diagrams on the right side of figure 11.24 are integrals of those on the left. They all start at zero when $x = 0$. Using the area principle, match the diagrams on the left with their integrals on the right.

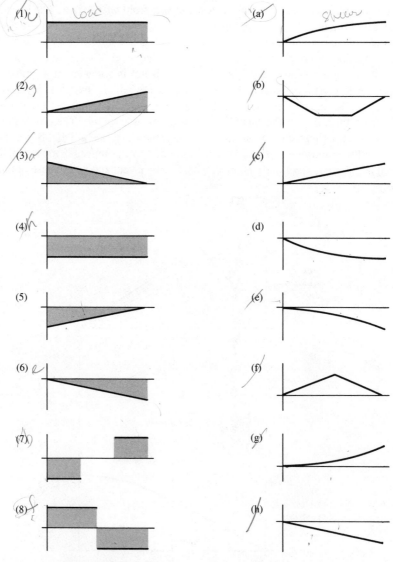

Figure 11.24 Problem 11.12. Match the diagrams on the left with their integrals on the right.

11.13 Diagrams (a) through (g) on the right side of figure 11.25 are derivatives of diagrams 1 through 11. Using the slope principle, match the diagrams on the left with their derivatives. More than one diagram may have the same derivative.

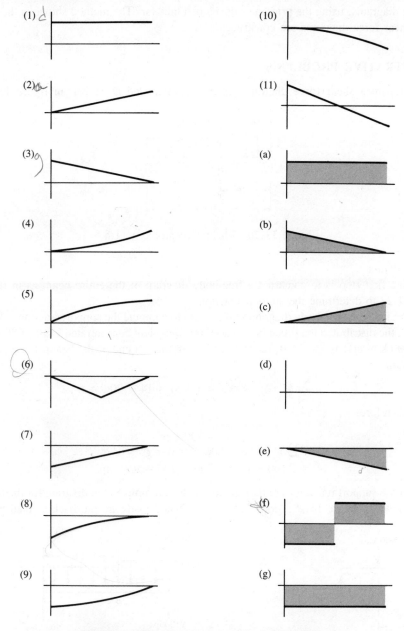

Figure 11.25 Problem 11.13. Using slopes, match the diagrams with their derivatives, diagrams (a) through (g). Several diagrams may have the same derivative.

11.6 CONSTRUCTING THE SHEAR DIAGRAM USING THE AREA RULES

The two examples that follow show the direct construction of shear force diagrams from loading diagrams, using the principles developed thus far. The method shown is the way most shear force diagrams are constructed.

ILLUSTRATIVE PROBLEMS

11.4 Prepare a shear force diagram for the beam loaded as shown in figure 11.26.

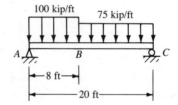

Figure 11.26   Illustrative problem 11.4.

Solution:
The first step is to construct a free-body diagram of the entire beam as in figure 11.27 to determine the support reactions.

Since this free-body diagram will be used to compute the support reactions, A and C, the distributed loads can be replaced by equivalent concentrated loads of 800 and 900 kips at 4 ft and 14 ft, respectively, from the left end of the beam. For equilibrium,

$$\Sigma \, M_A \circlearrowleft: \quad 800 \times 4 + 900 \times 14 - 20C = 0$$

Therefore,

$$C = 790 \text{ kips}$$
$$\Sigma \, V \uparrow: \quad A - 800 - 900 + C = 0$$
$$A = 1\,700 - C = 1\,700 - 790 = 910 \text{ kips}$$

These support reactions are then entered onto a complete load diagram for the beam shown in figure 11.28. Note that the distributed loads are put back as distributed

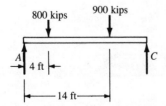

Figure 11.27   Free-body diagram of the beam of illustrative problem 11.4 with distributed loads reduced to equivalent concentrated loads.

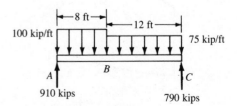

Figure 11.28   Complete loading diagram.

loads rather than equivalent concentrated loads. Distributed loads can be replaced with concentrated loads only for the purposes of computing *external* reactions. Shear force is an *internal* force; therefore, the distributed loads may not be replaced.

Now, using the area rules to construct the shear force diagram, one moves from left to right on this load diagram accumulating all the load as though using a squeegee.

The first load encountered is the 910-kip concentrated reaction at A. It produces a positive shear on the end of the beam. Therefore, the shear diagram begins with a positive 910 kips at end A.

As you go from end A, think of the squeegee accumulating negative load at a rate linear with the position x. When point B is reached,

$$100 \text{ kips/ft} \times 8 \text{ ft} = 800 \text{ kips}$$

of distributed load have been accumulated. This new accumulation must be taken as negative according to relationship 11.13, leaving 110 kips of positive shear at B. This result is shown by figure 11.29.

Beyond point B, more negative load is being accumulated, but at a slightly lesser rate than before B. So, there is a change in slope at point B. By the time C is reached, an additional

$$75 \text{ kips/ft} \times 12 \text{ ft} = 900 \text{ kips}$$

of negative shear has been accumulated in a linear fashion. This amount subtracts from the shear of 110 kips at B, leaving -790 kips of shear at C as shown in figure 11.30.

On approaching C, the support reaction of 790 kips upward is then encountered, returning the diagram abruptly to zero at the end of the beam. This return to zero at the end of the beam is a check on the calculations of the support reactions and load accumulation.

Before leaving the shear diagram, using the slope rules, also check to see if the load diagram is in fact the slope of the resulting shear diagram at every point. This check is usually a good way to find errors if the shear diagram does not zero out.

Finally, before leaving the diagram, note that it goes through zero at point D. Point D is 1.467 ft to the right of B, or 9.467 ft from the left end of the beam. This point will be the location of the maximum bending moment in the beam. The point of maximum bending moment is usually the most important conclusion to be drawn from a shear force diagram.

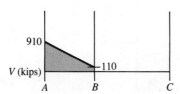

Figure 11.29 Construction of shear force diagram through point B.

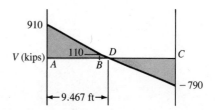

Figure 11.30 Completion of the shear force diagram to C.

Conclusion:

The shear diagram is shown in figure 11.30. The maximum shear force is 910 kips, occurring just inside the support at the left end of the beam. The maximum bending moment will be at $x = 9.467$ ft.

11.5 Draw directly the shear diagram for the beam loaded as shown in figure 11.31.

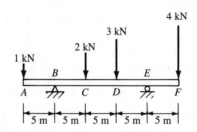

Figure 11.31 Illustrative problem 11.5.

Solution:

This beam has two overhangs and several concentrated loads. First, computing the reactions at B and at E,

$$\Sigma\, M_B\circlearrowright:\ -1 \times 5 + 2 \times 5 + 3 \times 10 - E \times 15 + 4 \times 20 = 0$$
$$E = 7.667 \text{ kN}$$
$$\Sigma\, V\uparrow:\ -1 + B - 2 - 3 + E - 4 = 0$$
$$B = 10 - E = 2.333 \text{ kN}$$

These reactions are put on the load diagram as shown in figure 11.32.

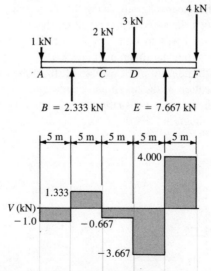

Figure 11.32 Load and shear diagrams for illustrative problems 11.5 and 11.7.

The shear diagram is constructed moving from the left end of the beam to the right. The first load encountered is the 1-kN downward concentrated force. In accordance with the sign convention, this load is a negative shear force. Therefore, the diagram starts with a negative 1 kN.

No other loads are encountered until point B on the beam. No distributed loads exist in this case. Therefore, the diagram remains at -1 from A to B. At B, an upward concentrated force of 2.333 kN is provided by the support. Coming from -1, the diagram is increased by 2.333 to a positive 1.333 kN, where it is constant until the next load downward at C. This process is continued through points D and E until the end of the beam, where the 4-kN downward load brings the shear diagram back to zero as it should.

Conclusion:
The shear diagram is given in figure 11.32. The maximum shear force is 4 kN, occurring from points E to F on the beam. The diagram goes through zero at three points, B, C, and E. The maximum bending moment will occur at one of these three points.

11.7 CONSTRUCTING THE BENDING MOMENT DIAGRAM DIRECTLY FROM THE SHEAR DIAGRAM USING THE SLOPE-AREA METHOD

The slope-area method can be used to construct the bending moment diagram from a shear diagram. This is done by moving from left to right and accumulating all the area under the shear diagram. Illustrative problems 11.4 and 11.5 are continued below in illustrative problems 11.6 and 11.7, respectively, to demonstrate the extension of the process for the construction of bending moment diagrams.

ILLUSTRATIVE PROBLEMS

11.6 Using the previously constructed shear diagram, make the bending moment diagram for the beam of illustrative problem 11.4 by the slope-area method. What is the maximum bending moment?

Solution:
We begin the bending moment diagram on the left end at zero, because the bending moment must be zero at a simply supported end as in figure 11.33.

Area under the shear diagram is accumulated, using the squeegee analogy, when moving to the right. However, since the shear diagram is declining, the area is accumulating but at an ever-decreasing rate; hence, the parabolic shape drawn between A and B. At B, positive shear area is still being accumulated. Therefore, the slope of the moment curve is still positive. The height at point B is determined by calculating the area under the shear diagram accumulated from A to B.

$$M_B = \frac{(910 + 110)}{2} \times 8 = 4\ 080 \text{ kip} \cdot \text{ft}$$

A parabola from zero at A to 4 080 at B is sketched on the moment diagram.

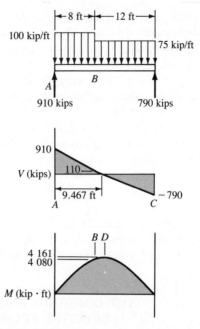

Figure 11.33 Illustrative problem 11.6.

The moment diagram continues up to a maximum at D, where the shear is zero. It continues to be parabolic in shape, although the parabola is of different shape from B to D. However, the breaking point between the two parabolas A to B and B to D may be difficult to discern. The additional rise is due to the area under the shear diagram between B and D, which can be calculated to be

$$M_{BD} = 110 \times \frac{(9.467 - 8)}{2} = 80.7 \text{ kip} \cdot \text{ft}$$

This additional accumulation of area must be added to the previous accumulation at B, giving a bending moment of 4 161 kip · ft at D.

$$M_D = M_B + M_{BD} = 4\ 080 + 80.7 = 4\ 161 \text{ kip} \cdot \text{ft}$$

After D, negative area under the shear diagram is accumulated at an ever-increasing rate. This accumulation causes the moment curve to increasingly slope downward until the moment returns to zero. The bending moment must be zero at C, because this end, too, is simply supported. The parabola from B to D is sketched to continue on to zero at C. It must be checked to see that it reaches zero.

$$M_{DC} = -\frac{790}{2}(20 - 9.467) = -4\ 161$$

$$M_C = M_D + M_{DC} = 4\ 161 - 4\ 161 = 0$$

Conclusion:
The bending moment diagram is sketched below the shear diagram in figure 11.33. It has a maximum value of 4 161 kip · ft.

11.7 Using the shear diagram previously constructed for the beam loaded as in illustrative problem 11.5, prepare a bending moment diagram. What is the maximum bending moment and what is its location?

Solution:
The bending moment must begin at zero as shown in figure 11.32, because the left end is free.

Moving from A to B, negative area accumulates linearly under the shear diagram. Upon reaching B, -5.000 kN $\cdot$ m of area have been swept. This amount is the bending moment at B.

After passing B, positive area accumulates at a constant rate. Therefore, the slope of the bending moment diagram abruptly changes from negative to positive. On reaching C, the net accumulated area is $+1.667$ kN $\cdot$ m.

At C, the slope of the moment diagram must go negative again. Area will accumulate at the rate of -0.667 kN $\cdot$ m/m. After 5 m, upon reaching D, the bending moment diagram is once again negative and has a value of -1.667 kN $\cdot$ m. (Note, more than three decimal places are being carried in intermediate calculations so successive round-off errors are not accumulated.)

After D, the slope of the moment diagram is strikingly more negative. It goes down to -20 at point E. After E, positive area under the shear diagram is accumulated linearly. This accumulation is sufficient to bring the moment back to zero at F. This result makes sense, because F is a free end and can have no moment.

Conclusion:
The bending moment diagram is as shown in figure 11.34. The maximum moment is negative. It has a value of -20 kN $\cdot$ m and occurs at point E, just at the support on the right.

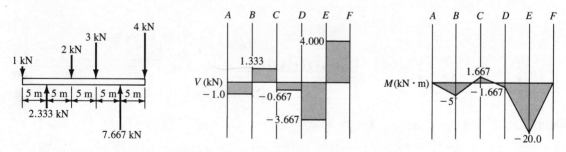

Figure 11.34 Illustrative problem 11.7.

11.8 A FEW OBSERVATIONS REGARDING DISCONTINUITIES

Integration is a smoothing process. Shear diagrams must, therefore, be smoother than load diagrams, and moment diagrams must be smoother than shear diagrams. It helps to remember this continuum when constructing shear and moment diagrams. The ultimate discontinuity is a spike, or a concentrated load. This case is called a *singularity*. If there is a singularity on the load diagram, the shear diagram will have a *step discontinuity,* which is at least somewhat smoother than the spike. And, the moment diagram will have a *slope discontinuity,* which is a bit smoother than the step. Figure 11.34 is an example of this. The concentrated loads cause steps on the shear diagram and abrupt changes of slope on the moment diagram.

Where there is a step discontinuity on the load diagram as in the earlier problem of figure 11.33, the shear diagram will show the smoother slope discontinuity. The moment diagram may appear to have no discontinuity whatsoever. Actually, there is an abrupt change of curvature, but this may not be discernible to the eye.

Physically, discontinuities do not exist in nature. Concentrated loads are, in fact, distributed over a finite though small area. Real structures spread out the spikes and round all the corners.

Whereas integration is ultimately a smoothing process, its reverse, differentiation, is a roughening process. Philosophically, there may be a lesson in there somewhere.

PROBLEM SET 11.4

11.14–11.25

In each of the beams loaded as in figures 11.35 through 11.46, draw the load, shear, and moment diagrams.

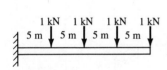

Figure 11.35 Problem 11.14.

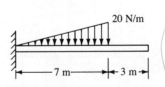

Figure 11.36 Problem 11.15.

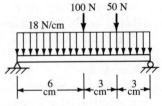

Figure 11.37 Problem 11.16.

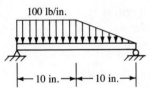

Figure 11.38 Problem 11.17.

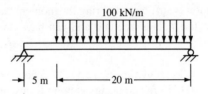

Figure 11.39 Problem 11.18.

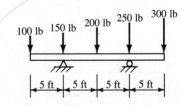

Figure 11.40 Problem 11.19.

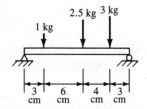

Figure 11.41 Problem 11.20.

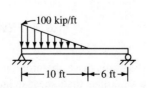

Figure 11.42 Problem 11.21.

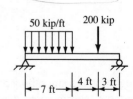

Figure 11.43 Problem 11.22.

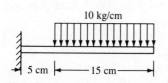

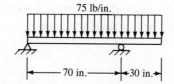

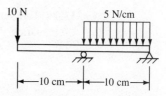

Figure 11.44 Problem 11.23.   **Figure 11.45** Problem 11.24.   **Figure 11.46** Problem 11.25.

11.26 Draw the shear force and bending moment diagrams for the lever shown in figure 11.47.

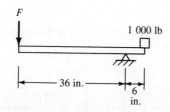

Figure 11.47 Problem 11.26.

11.27 Draw the shear force and bending moment diagrams for part *A* of the pliers shown in figure 11.48.

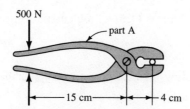

Figure 11.48 Problem 11.27.

11.28 The rail car axle in figure 11.49 has a diameter of 4 in. Draw the shear and moment diagrams, and compute the maximum bending stress.

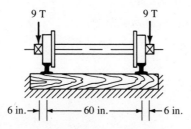

Figure 11.49 Problem 11.28.

11.29 The semi-trailer truck shown in figure 11.50 carries a 10 000-lb load package. Draw the shear and moment diagrams for the trailer.

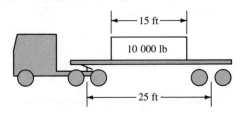

Figure 11.50 Problem 11.29.

11.30 Draw the shear and bending moment diagram for the airplane wing at rest upon its landing gear shown in figure 11.51.

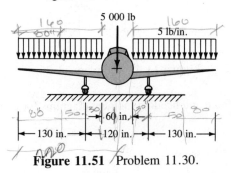

Figure 11.51 Problem 11.30.

SUMMARY

□ The shear force and bending moment diagrams for four basic cases are as follows:

Cantilever beam loaded at its free end:
(11.1) $V = F$
(11.2) $M = -F(L - x)$

Cantilever beam with uniform load:
(11.3) $V = w(L - x)$
(11.4) $M = -\dfrac{w}{2}(L^2 - 2Lx + x^2)$

Simply supported beam with uniform load:
(11.5) $V = w\left(\dfrac{L}{2} - x\right)$
(11.6) $M = \dfrac{w}{2}(Lx - x^2)$

Simply supported beam with a mid-span load:

(11.7) $V = \dfrac{F}{2}$ (left of load)   or $-\dfrac{F}{2}$ (to the right)

(11.8) $M = \dfrac{Fx}{2}$ (left of load)   or $\dfrac{F(L - x)}{2}$ (to the right)

☐ Shear force and bending moment diagrams can be derived from load diagrams using calculus with the following:

(11.12) $w = \dfrac{dV}{dx}$,   conversely, $V = w \displaystyle\int dx$

(11.15) $V = \dfrac{dM}{dx}$,   conversely, $M = V \displaystyle\int dx$

☐ The *boundary conditions,* or end conditions, yield the constants of integration:

    free end: $M = 0$ and $V = 0$

    simple support: $M = 0$

☐ The slope of the shear diagram is equal to the value of the load diagram.

☐ The slope of the moment diagram is equal to the value of the shear diagram.

☐ The area under the load diagram equals the change of the shear diagram.

☐ The area under the shear diagram equals the change of the moment diagram.

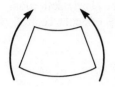 # Deformation of Beams

Do #'s 1, 3, 5, 7, 9

Our study of deformation of beams begins with the relationship between bending moment and curvature. The differential and integral relationships among curvature, slope, and deflection are developed. They are used with calculus to derive the slope and deflection equations of the four most common, and therefore most useful, beam cases. The bending tests of both steel and reinforced concrete beams are described in order to explain the behavior of these materials in bending. The moment-area method of obtaining slope and deflection diagrams is discussed and applied in an example problem. The use of the trapezoidal method of approximate integration is demonstrated for variable beam stiffness, or *EI*. The need for and use of computers for variable *EI* is demonstrated with an illustrative problem. Illustrative problems and exercise problems are included from both civil technology and mechanical technology.

12.1 THE RELATIONSHIP BETWEEN BENDING MOMENT AND CURVATURE

Curvature in beams is caused by bending moment. This curvature over the length of a beam produces significant elastic deflections. Shear forces also cause deformation in beams, but unless the beam is extremely short and stocky, the deformation due to shear is negligible compared to that caused by the bending moment. Therefore, only deformation due to bending moment is considered in this book.

Beams are actually very flexible. In fact, they are frequently used as springs. Leaf springs are an example of beam springs. Because beams are flexible, their deformation can sometimes be the controlling factor in structural design. For example, plaster ceilings may crack if their support beams are too flexible, even though the beams may be satisfactory from the standpoint of stress.

Flat roofs supported on beams that are too flexible may deform excessively, allowing rain puddles to grow into pools. Floors that are too flexible may seem too springy to the occupants. Long, heavy beams are frequently *cambered;* that is, made in a slight arched bend, so that when they deform under their own load they will become straight rather than sag. These cases all demonstrate the necessity of information about the elastic deflection of beams, in addition to stresses. Rigidity, the opposite of flexibility, is often the controlling characteristic of a beam in a building or machine structure.

Elastic Deformation Caused by Bending Moment

An element, or short length, of a straight beam having a length of Δx, will deform to a circular shape as shown in figure 12.1, when acted upon by a bending moment, M.

Figure 12.1 shows an exaggerated amount of deformation so that the geometry becomes obvious. The neutral axis of the beam will have a radius of curvature, ρ. The faces of the element will tilt so they point to the center of curvature at o. The extreme "fiber" of the beam will lengthen by an amount $\delta/2$ on each side of the element.

The fiber stress is given by relationship 10.11.

(a)
$$\sigma = \frac{Mc}{I}$$

From the definition of elastic modulus, the strain of the extreme fiber can be calculated as follows:

(b)
$$E = \sigma/\epsilon$$

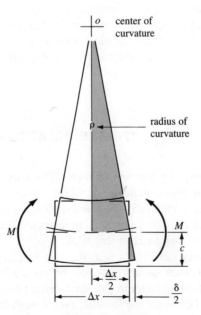

Figure 12.1 An element of a beam deformed to a circular shape by bending moment.

Substituting **(a)** into **(b)** and solving for ϵ,

(c)
$$\epsilon = \frac{Mc}{EI}$$

The strain, ϵ, is the lengthening of the fiber divided by its original length, $\delta/\Delta x$. Therefore,

(d)
$$\frac{\delta}{\Delta x} = \frac{Mc}{EI}$$

Solving for δ and dividing by 2, the lengthening of the fiber on one side of the element is related to the bending moment by the following:

(e)
$$\frac{\delta}{2} = \frac{Mc\Delta x}{2EI}$$

Applying the law of similar triangles to the two shaded areas,

(f)
$$\frac{\Delta x/2}{\rho} = \frac{\delta/2}{c}$$

Substituting **(e)** into **(f)**,

(g)
$$\frac{\Delta x}{2\rho} = \frac{Mc\Delta x}{2EIc}$$

Cancelling c and dividing through by $\Delta x/2$ yields the very important relationship between bending moment and curvature for a beam in the elastic range,

$$\boxed{\frac{1}{\rho} = \frac{M}{EI}}$$
(12.1)

where ρ is the radius of curvature of the beam. The product, EI, is often referred to as the *stiffness* of the beam.

ILLUSTRATIVE PROBLEM

12.1 A wire rope is braided of 0.010-in.-diameter high-strength steel wires having a yield strength of 120 000 psi. What may be the minimum sheave or pulley diameter to avoid alternating plasticity as this cable runs back and forth over the sheave?

Solution:
As the cable is run over the sheave, a bending moment is produced to form it to the radius of the sheave. Taking each wire in the bundle as a beam, the bending moment is given by relationship 12.1,

(a)
$$\frac{1}{\rho} = \frac{M}{EI}$$

where ρ is the radius of the sheave. The wire closest to the sheave has the most extreme bending curvature.

The stress due to bending is given by relationship 10.11,

(b)
$$\sigma = \frac{Mc}{I}$$

If σ remains less than the yield stress, the wire will remain elastic as the cable goes over the pulley. The wire will have the tendency to spring back straight after being bent around the pulley. If it were to have a plastic set, the wire would then have "permanent" curvature and would have to yield again in the opposite direction as the cable is pulled straight. This condition is known as alternating plasticity. Bending a wire back and forth could result in a fatigue failure in few cycles. Therefore, σ must be no greater than the yield strength of 120 000 psi.

Solving **(b)** for M and substituting into **(a)** gives

(c)
$$\frac{1}{\rho} = \frac{\sigma}{Ec}$$

The moment of inertia cancels out. Inverting **(c)**,

(d)
$$\rho = Ec/\sigma.$$

In other words, the radius of the pulley is equal to the radius of the wire times the ratio of the elastic modulus over the stress. In terms of diameters, letting D equal the pulley diameter and d the wire diameter,

(e)
$$D = Ed/\sigma$$
$$D = 30\text{E}6 \text{ psi} \times 0.010 \text{ in.}/120\text{E}3 \text{ psi}$$
$$D = 2.50 \text{ in.}$$

Conclusion:
The pulley diameter must be 2.50 in. or greater if the maximum wire stress is to be less than 120E3 psi. Cables with heavier wire diameter require proportionally larger pulleys. Cable wires are wound into bundles and spun so that the wires on the inside will not be crimped as the cable is bent. Since substantial sliding of one wire over another occurs during bending, cables must be well lubricated for long wear.

PROBLEM SET 12.1

12.1 A W24 $\times$ 94 wide-flange steel beam is simply supported and spans 25 ft. It carries a total load of 141 kips distributed uniformly over its length. What is the radius of curvature of the beam at the mid-span?

12.2 What bending moment is required to produce a radius of curvature of 10 200 in. on a 2-in.-square steel bar?

12.3 A steel band saw blade is 1 mm thick and must not be stressed in bending more than 400 MPa in order to conform to warranty and safety specifications. What minimum pulley diameter may be used for the blade drive?

12.2 DEFLECTION OF BEAMS

Although relationship 11.1 is the fundamental theory of deformation of beams, it is not by itself directly useful. Normally, the radius of curvature of a beam is not of direct interest. However, one frequently wants to know how much a beam will deflect or sag under load. To find this value, it is necessary to "gather" the effects of the curvature over the entire length of the beam. This gathering is done by the process of integration from calculus. Later on, a diagram method will be presented that can be used for discontinuous problems. First, the more sophisticated method of calculus must be understood.

The deflection, y, depends on the location, x, along the length of the beam, as shown in figure 12.2.

Consider that the beam in the figure has some load causing it to bend upward. The slope of the beam, θ, is a function of position, x. From calculus, θ is the derivative of y.

$$\theta = \frac{dy}{dx} \tag{12.2}$$

The curvature of the beam is the reciprocal of the radius of curvature, $1/\rho$, and is approximately equal to the second derivative of y. This assumption is extremely accurate for the small deformations normally encountered in beams of common structural materials.

$$\frac{1}{\rho} = \frac{d^2y}{dx^2} \tag{12.3}$$

The positive curvature is as shown in figure 12.1.

From the fundamental relationship 12.1,

$$\frac{d^2y}{dx^2} = \frac{d\theta}{dx} = \frac{M}{EI} \tag{12.4}$$

Figure 12.2 The sign convention for deflection and slope of the deformed beam. In this book positive y is taken to be upward to conform to most mathematics sign conventions.

The bending moment, M, is a function of x. Integrating $d\theta/dx$ from relationship 12.4 yields

$$\theta = \int \frac{M}{EI}\, dx \tag{12.5}$$

If EI, the stiffness of the beam, is constant along its length,

$$\theta = \frac{dy}{dx} = \frac{1}{EI}\int M\, dx \tag{12.6}$$

Relationship 12.6 can in turn be integrated,

$$y = \int \theta\, dx \tag{12.7}$$

Relationships 12.6 and 12.7 allow the construction of slope and deflection diagrams from bending moment diagrams.

12.3 BOUNDARY CONDITIONS

When relationships 12.6 and 12.7 are integrated in any specific case, the constants of integration must be evaluated. This evaluation is done using the conditions at the supports of the beam. Again, as in chapter 11, the three common types of supports are the fixed (or built-in), simple, and free, as shown in figure 12.3.

Fixed-beam supports permit neither vertical displacement nor rotational displacement. Therefore, both y and θ must equal zero at such supports.

Simple supports permit only rotary displacement. Therefore, y must equal zero at such supports.

Free supports allow any combination of rotational and vertical displacement.

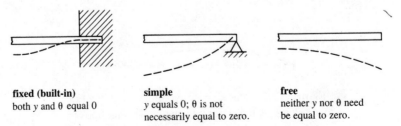

fixed (built-in)
both y and θ equal 0

simple
y equals 0; θ is not necessarily equal to zero.

free
neither y nor θ need be equal to zero.

Figure 12.3 Boundary conditions for deflection and slope for various types of beam supports or ends.

12.4 FOUR MOST COMMON CASES OF BEAM LOADING SOLVED

The four most common cases encountered in structural design are the cantilever beam with a uniformly distributed load over its entire length, the cantilever beam with a concentrated load at its free end, the simply supported beam under a uniformly distributed load, and the simply supported beam with a concentrated load at mid-span. They are solved here both to illustrate the application of relationships 12.2 through 12.7 and to provide some results for later use.

12.5 THE CANTILEVER BEAM WITH A UNIFORMLY DISTRIBUTED LOAD OVER ITS ENTIRE LENGTH

The bending moment from relationship 11.4 is

$$M = -\frac{w}{2}(L^2 - 2Lx + x^2)$$

Integrating, according to relationship 12.6, to find the slope,

(a) $$\theta = -\frac{w}{2EI}\int(L^2 - 2Lx + x^2)\,dx$$

(b) $$\theta = -\frac{w}{2EI}\left(L^2x - Lx^2 + \frac{x^3}{3} + C_1\right)$$

Due to the boundary condition at the fixed end, θ equals zero when x equals zero. Therefore, C_1 equals zero and

$$\theta = \frac{dy}{dx} = -\frac{w}{2EI}\left(L^2x - Lx^2 + \frac{x^3}{3}\right) \tag{12.8}$$

Integrating again according to relationship 12.7,

(c) $$y = -\frac{w}{2EI}\int\left(L^2x - Lx^2 + \frac{x^3}{3}\right)dx$$

(d) $$y = -\frac{w}{2EI}\left(\frac{L^2x^2}{2} - \frac{Lx^3}{3} + \frac{x^4}{12} + C_2\right)$$

Due to the boundary condition at the fixed end, y equals zero when x equals zero. Therefore, C_2 equals zero and

$$y = -\frac{w}{2EI}\left(\frac{L^2x^2}{2} - \frac{Lx^3}{3} + \frac{x^4}{12}\right) \tag{12.9}$$

Relationships 12.8 and 12.9 are plotted in figure 12.4.
The maximum deflection occurs at the end of the beam where $x = L$.

$$\boxed{y_{max} = -\frac{wL^4}{8EI}} \tag{12.10}$$

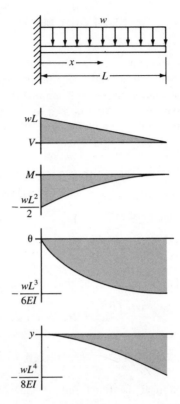

Figure 12.4 A complete set of beam diagrams for a cantilever beam of length L, with a uniformly distributed load w.

12.6 SOME COMMENTS ON THE PREVIOUS EXAMPLE

The example of the cantilever beam under a uniformly distributed load clearly demonstrates that the load, w, the shear force, V, the bending moment, M, the slope, θ, and the deflection, y, are all related by differential-integral operations. In chapter 11, and so far in chapter 12, this interrelationship was developed primarily by integration. Usually, one knows the load and must work toward the bending moment and then continue further toward the deflection, a sequence of four integrations. However, the process in reverse is equally valid. If a curve of deflection as a function of x is known, it is possible to differentiate four times to get the load distribution. A person who loves calculus can find much amusement in beams.

The derivative relationship can be summarized as follows:

$$\text{deflection} = y \qquad\qquad \text{moment} = M = \frac{EI}{\rho} = EI\frac{d^2y}{dx^2}$$

$$\text{slope} = \theta = \frac{dy}{dx} \qquad\qquad \text{shear} = V = \frac{dM}{dx} = EI\frac{d^3y}{dx^3} \qquad \textbf{(12.11)}$$

$$\text{curvature} = \frac{1}{\rho} = \frac{d\theta}{dx} = \frac{d^2y}{dx^2} \qquad\qquad \text{load} = w = \frac{dV}{dx} = EI\frac{d^4y}{dx^4}$$

Unfortunately, multiple sign conventions are in common use. Each has its advantages and disadvantages.

In civil engineering practice, the load, w, is usually defined as positive in the downward direction. In static structures, most loads are caused by gravity, which produces downward forces. The deflection, y, is also defined as positive downward. This convention supports the rationality that positive loads should produce positive deflections. In order to have positive deflection in the same direction as positive forces, the fundamental relationship 12.1 must be written with a negative sign, as follows:

$$\frac{1}{\rho} = -\frac{M}{EI}$$

In aircraft work, it would be highly pessimistic to define y positive downward. Also, aerodynamic loads should have a net upward force, at least most of the time during flight. In aircraft, therefore, upward loads are defined as positive. Hence, the aircraft industry tends to use the inverse of the civil engineering practice.

In this book we use a combination of the civil and the aeronautical sign conventions. In chapter 11, downward concentrated and distributed loads upon beams, F and w, were introduced as positive, which is the civil engineering sign convention. But in this chapter, upward deflection, y, is defined positive, as an airframe designer would do it. Though this system could be regarded as inconsistent, it favors students encountering the beam for the first time as a structural element. Loads are positive downward, as students are apt to "feel" them. The values of y are positive upward corresponding with their mathematics books. Slopes and curvatures are thus signed exactly as in math.

A final advantage of this sign convention is that, in going from loads to deflections, one need not stumble over a contrived negative sign in relationship 12.1. The price of these advantages is that we have to accept that positive F and w produce negative deflections. Also, we must accept that most standard sign conventions are different from those used here.

12.8 THE CANTILEVER BEAM WITH A CONCENTRATED LOAD AT ITS FREE END

The bending moment from relationship 11.2 is

$$M = -F(L - x)$$

Dividing by EI and integrating according to relationship 12.6,

(a)
$$\theta = -\frac{F}{EI}\left(Lx - \frac{x^2}{2} + C_1\right)$$

Due to the condition at the built-in end, θ must equal zero when x equals zero. When substituting this into **(a)**, the constant of integration, C_1, is found to be zero. Therefore,

$$\theta = -\frac{F}{EI}\left(Lx - \frac{x^2}{2}\right) \tag{12.12}$$

Integrating again according to 12.7, to get the deflection y,

(b)
$$y = -\frac{F}{EI}\left(\frac{Lx^2}{2} - \frac{x^3}{6} + C_2\right)$$

At the built-in end (at $x = 0$) y must equal zero. Therefore, C_2 is also equal to zero. And,

$$y = -\frac{F}{EI}\left(\frac{Lx^2}{2} - \frac{x^3}{6}\right) \qquad\qquad \textbf{(12.13)}$$

Relationships 12.12 and 12.13 are plotted in figure 12.5.

The maximum deflection occurs at the end of the beam where $x = L$.

$$y_{max} = -\frac{FL^3}{3EI} \qquad\qquad \textbf{(12.14)}$$

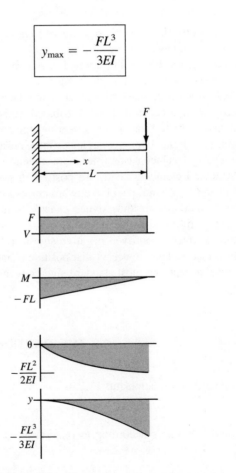

Figure 12.5 A complete set of beam diagrams for a cantilever beam of length L with a concentrated load at the free end.

12.9 THE SIMPLY SUPPORTED BEAM UNDER A UNIFORMLY DISTRIBUTED LOAD

Bending moment relationship 11.6 applies to the important case shown in figure 12.6.

$$M = \frac{w}{2}(Lx - x^2)$$

Integrating to find θ,

(a)
$$\theta = \frac{w}{2EI} \int (Lx - x^2)\, dx$$

(b)
$$\theta = \frac{w}{2EI}\left(\frac{Lx^2}{2} - \frac{x^3}{3} + C_1\right)$$

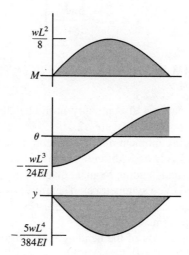

Figure 12.6 A complete set of beam diagrams for a simply supported beam of length L under a uniformly distributed load.

In this case, no boundary condition involving slope exists. The two boundary conditions are due to the simple supports at the ends of the beams. At these supports, where $x = 0$ and $x = L$, y must be zero. Integrating again,

(c)
$$y = \frac{w}{2EI} \int \left(\frac{Lx^2}{2} - \frac{x^3}{3} + C_1 \right)$$

(d)
$$y = \frac{w}{2EI} \left(\frac{Lx^3}{6} - \frac{x^4}{12} + C_1x + C_2 \right)$$

At $x = 0$, $y = 0$. Therefore C_2 must equal zero.

From the second boundary condition, $y = 0$ at $x = L$,

$$0 = \frac{w}{2EI} \left(\frac{L^4}{6} - \frac{L^4}{12} + C_1L \right)$$

Solving,

$$C_1 = -L^3/12$$

Therefore, relationships **(c)** and **(d)** become

$$\theta = \frac{w}{24EI} (6Lx^2 - 4x^3 - L^3) \tag{12.15}$$

$$y = \frac{w}{24EI} (2Lx^3 - x^4 - L^3x) \tag{12.16}$$

The maximum deflection is at mid-span when $x = L/2$,

$$\boxed{y_{max} = -\frac{5wL^4}{384EI}} \tag{12.17}$$

12.10 THE SIMPLY SUPPORTED BEAM WITH A CONCENTRATED LOAD AT MID-SPAN

Concentrated loads always cause a discontinuity in the bending moment diagram. This case has a discontinuity at point b in figure 12.7.

Integration of the bending moment diagram is more complicated because the equation on the left side of the load, F, is different from that on the right side of the load.

Since the load diagram and bending moment diagram are symmetrical, the deformation, y, can also be expected to be symmetrical. Therefore, only the left side of the beam will be considered. The right side will have a deformation curve of the same shape but described by a different equation.

From chapter 11, the equation of the bending moment to the left of the load, F, at mid-span is:

$$M = Fx/2$$

Dividing by EI and integrating gives the slope:

(a)
$$\theta = \frac{F}{2EI} \left(\frac{x^2}{2} + C_1 \right)$$

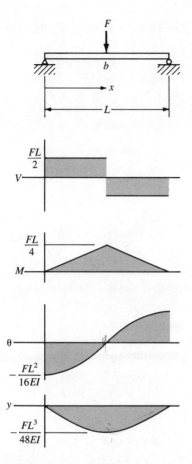

Figure 12.7 Beam diagrams for a simply supported beam acted upon by a concentrated load at mid-span.

The slope, θ, must be zero at mid-span in order for the deformation to be symmetrical. This fact can be used as a boundary condition to solve for C_1. When solved, a value for C_1 equal to $-L^2/8$ is obtained. Therefore,

$$\theta = \frac{F}{4EI}\left(x^2 - \frac{L^2}{4}\right) \qquad \textbf{(12.18)}$$

Relationship 12.18 describes the left half of the beam. This is the relationship for slope at the load and to the left. The slope on the right will have a different equation, but it will be *antimetrical* rather than symmetrical as shown in figure 12.7.

Now, integrating the slope to find the deflection of the beam,

(b)
$$y = \frac{F}{4EI}\left(\frac{x^3}{3} - \frac{L^2 x}{4} + C_2\right)$$

At $x = 0$, the left end of the beam, $y = 0$. Therefore, for the left half of the beam, C_2 must equal zero.

$$y = \frac{F}{4EI}\left(\frac{x^3}{3} - \frac{L^2x}{4}\right)$$ (12.19)

The maximum deflection will occur at mid-span where $x = L/2$. Substituting this into 12.19 gives the very important relationship,

$$y_{max} = -\frac{FL^3}{48EI}$$ (12.20)

The preceding four cases are summarized in Table 12.1.

Table 12.1 Summary of maximum deflection relationships for four important cases of beam loading

	Concentrated load, F	Distributed load, w
Simply supported	F at mid-span $$y_{max} = -\frac{FL^3}{48EI}$$	$$y_{max} = -\frac{5wL^4}{384EI}$$
Cantilevered	F at free end $$y_{max} = -\frac{FL^3}{3EI}$$	$$y_{max} = -\frac{wL^4}{8EI}$$

ILLUSTRATIVE PROBLEMS

12.2 A high-strength steel simply supported W24 × 76 beam carries a uniformly distributed load over its entire length of 25 ft. (a) If the allowable stress is 33 ksi, what may be the value of the load, in addition to the beam's own weight? (b) What is the maximum deflection of the beam under that load?

Solution:
Part A: Using the data in Appendix VI,

$$Z = 176 \text{ in.}^3 \quad \text{and} \quad I = 2\,100 \text{ in.}^4$$

For a simply supported beam under uniform load, the maximum bending moment occurs at mid-span and has the value:

$$M = \frac{wL^2}{8}$$

Since

$$\sigma = \frac{M}{Z}$$

(a)
$$w = \frac{8\sigma Z}{L^2}$$

$$w = \frac{8 \times 33 \text{ kips/in.}^2 \times 176 \text{ in.}^3}{(25 \text{ ft})^2} \times \frac{1 \text{ ft}}{12 \text{ in.}}$$

$$w = 6.20 \text{ kips/ft}$$

Since the beam itself weighs 76 lb/ft, the additional allowable load it may carry is 6.12 kips/ft.

Part B: From relationship 12.17,

$$y_{max} = \frac{-5wL^4}{384EI}$$

$$y_{max} = \frac{-5 \times 6.20 \text{ kips/ft} \times (25 \text{ ft})^4}{384 \times 30\text{E}6 \text{ psi} \times 2\ 100 \text{ in.}^4} \times \frac{1\ 728 \text{ in.}^3}{1 \text{ ft}^3}$$

$$y_{max} = -0.865 \text{ in.}$$

Conclusion:
The beam may carry 6.12 kips/ft in addition to its own weight. Under this load, it will sag elastically 0.865 in. at mid-span.

12.3 A rectangular aluminum bar 4 cm high by 2 cm wide is cantilevered 80 cm horizontally. A 100-kg load is suspended from the free end. What are the slopes in degrees 20, 40, 60, and 80 cm from the built-in end? What is the deflection of the free end?

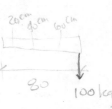

Solution:

$$I = \frac{bh^3}{12} = \frac{2 \text{ cm} \times (4 \text{ cm})^3}{12} = 10.67 \text{ cm}^4$$

The modulus of elasticity for aluminum is 69E9 Pa.
    From relationship 12.12,

$$\theta = -\frac{F}{EI}(Lx - x^2/2)$$

Entering everything but x and L in cm,

$$\theta = \frac{-100 \text{ kg}(Lx - x^2/2)\text{cm}^2}{69\text{E}9 \text{ Pa} \times 10.67 \text{ cm}^4} \times \frac{9.8 \text{ N}}{1 \text{ kg}} \times \frac{1 \text{ Pa}}{1 \text{ N/m}^2} \times \frac{(100 \text{ cm})^2}{1 \text{ m}^2} \times \frac{57.3°}{1 \text{ radian}}$$

$$\theta = -7.62\text{E}{-4}(Lx - x^2/2) \text{ degrees}$$

For $L = 80$ cm and at $x = 20, 40, 60,$ and 80 cm,

$$\theta = -1.067, -1.830, -2.29, \text{ and } -2.44 \text{ degrees, respectively}$$

The deflection at the end of the beam is given by relationship 12.14,

$$y_{max} = -\frac{FL^3}{3EI}$$

$$y_{max} = -\frac{100 \text{ kg} \times (80 \text{ cm})^3}{3 \times 69\text{E}9 \text{ Pa} \times 10.67 \text{ cm}^4} \times \frac{1 \text{ Pa}}{1 \text{ N/m}^2} \times \frac{(100 \text{ cm})^2}{1 \text{ m}^2} \times \frac{9.8 \text{ N}}{1 \text{ kg}}$$

$$y_{max} = -2.27 \text{ cm}$$

Conclusion:
The slope of the beam will be increasingly negative until it reaches -2.44 degrees at the free end. The free end of the beam will deflect downward 2.27 cm.

PROBLEM SET 12.2

12.4 Calculate the vertical deflection of a southern pine 2×4, 8 ft long and simply supported at each end, due to the weight of a 150-lb man standing at mid-span. Calculate the deflection for the 2×4 oriented both the ''hard way'' and the ''soft way'' to the load. See figure 12.8.

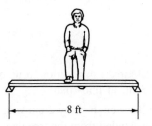

Figure 12.8 Problem 12.4.

12.5 A 12-in. standard-weight steel pipe is to span a roadway in a chemical plant. The pipe will carry a slurry weighing 70 lb/ft^3. If the pipe length is effectively simply supported for the 15-ft span, what amount of upward camber should the pipe have so that it does not appear to sag under its own weight and that of the slurry? See figure 12.9.

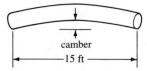

Figure 12.9 Problem 12.5.

12.6 A uniformly loaded cantilever beam is made of 6061-T6 aluminum with a square cross section 120 mm on a side. If the free end sags 10 mm under the uniform load of 1 kN/m, what is the length of the beam? See figure 12.10.

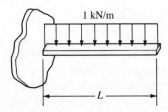

Figure 12.10   Problem 12.6.

12.7 A floor supporting a dining room of a restaurant is to be designed for 100 lb/ft². It is to be supported by 2 × 12's of douglas fir on 12-in. centers spanning the 12.5 ft between girders. What is the deflection of the 2 × 12's at mid-span? See figure 12.11.

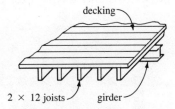

Figure 12.11   Problem 12.7.

12.8 A new composite material is tested as a simply supported beam. The test rig has a span of 18 in. The sample is rectangular in cross section, $\frac{3}{4}$ in. high by $\frac{3}{8}$ in. wide. A mid-span deflection of 0.200 in. is measured under a concentrated load of 450 lb. What is the elastic modulus of this material?

12.9 A cantilever beam is to be used as a pre-loading device in an instrumentation linkage. It is to be made of 155A titanium alloy. The beam will be 10.45 cm in free length, will be stamped from 4.00-mm-thick sheet metal, and will have a width of 0.80 cm. How far may the end of the cantilever be deflected without yielding the titanium? See figure 12.12.

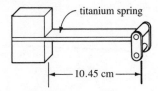

Figure 12.12   Problem 12.9.

12.10 Choose the lightest weight wide-flange structural steel beam that has sufficient rigidity to span 25 ft on simple supports while deflecting no more than 1 percent of the span distance under a load of 200 kips/ft.

12.11 TESTING A STEEL BEAM TO FAILURE:
THE PLASTIC SHAPE FACTOR

It is highly informative, as well as being a lot of fun, to conduct a bending test of a hot-rolled steel beam to failure. A large beam and special flexural test fixtures are not required. Very accurate tests can be conducted on small beams using student-lab-quality testing machines such as a Technovate 9014 materials testing system, as shown in figure 12.13.

A 3/4-by-3/8-in. bar cut to 24 in. is a good specimen. The specimen may overhang the simple support points spaced at a maximum distance of 18 in. by the 9014. The load is applied by a hydraulic cylinder. Deflections can be obtained with good accuracy by using a linear micrometer to measure the displacement of the load shackle. Accuracy can be improved by using a dial gage. Whether the test apparatus is of student quality or of more accurate professional testing lab quality, all beam tests using structural steel will produce results similar to figure 12.14, which shows a typical force-deflection curve.

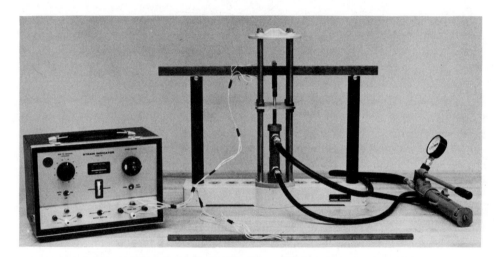

Figure 12.13   A Technovate model 9014 materials testing machine arranged for a beam test. (Source: Technovate, Inc.).

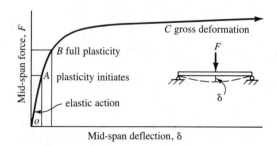

Figure 12.14   Test results from a bending test of a structural steel beam.

The curve will be truly linear through to point A. This is the elastic range of the test. After a small amount of load takes the slack out of the system, load and deflection data will correlate closely to a straight line.

From point o through point A, the relationship between the force, F, and the deflection, δ, is predictable using relationship 12.20.

$$\delta = \frac{FL^3}{48EI}$$

Such a test is an excellent way to obtain the elastic modulus, E, of an unknown material using the above relationship. If the load is relieved at any point on the elastic line, o to A, the beam will return to zero deflection.

At point A the beam begins to yield. Relationships 10.11 and 11.10 can be combined to give the limiting point, A.

$$\sigma = \frac{Mc}{I} \quad \text{and} \quad M = \frac{FL}{4}$$

Therefore, where σ equals σ_y

$$F = \frac{4I\sigma_y}{Lc} \tag{12.21}$$

The force-deflection data will go right on through yield initiation, point A, with indiscernible deviation from the straight elastic line. In bending, a hot-rolled steel bar has substantial reserve strength after yield is initiated due to the core of material that remains elastic. This situation is shown in figure 12.15. Initial yielding does not constitute a catastrophic failure of a structural steel beam. It is almost impossible to detect initial yielding from a load-deflection curve.

From points A to B on the load-deflection curve, the plastification continues inward. The affected area on a hot-rolled beam is easily visible due to scaling and the Lüder's line forming at the surface.

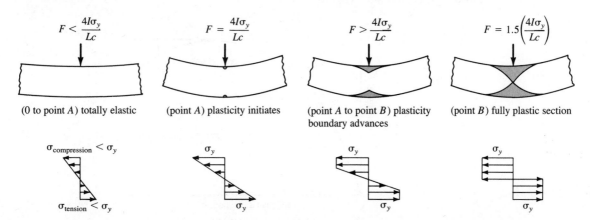

Figure 12.15 Local action at the point of load application in a force-deflection test of a structural steel beam of rectangular cross section.

Strain hardening plays a very small role in the reserve plastic strength of a beam. Strain hardening of a hot-rolled structural steel is insignificant until strains of 20 000 micros or more are reached. The levels of local strain involved between points A and B are less than 10 000 micros.

The force-deformation curve begins to deviate from a strain line as point B is reached. Here, the plastification has worked inward to the neutral axis. Beyond point B, gross deformations are observed. This fully plastic condition constitutes true structural collapse and failure.

For a rectangular section, the load due to the "stress blocks" in the cross section for point B can be shown to be exactly 50 percent greater than the load required to initiate yield. The "plastic shape factor" has a value of 1.5 for a rectangular cross section.

All cross sections do not behave identically. Tee sections have greater plastic shape factors, ranging from 1.6 to 1.8. Wide-flange shapes and I beams have shape factors of only 1.05 to 1.20 percent. In other words, gross deformation of a wide-flange shape is likely to occur at loads only slightly above the first yield load; whereas tee sections have substantially more reserve plastic strength after the onset of yield.

After gross deformation, only the small amount of material adjacent to the plastic hinge will have been affected by plasticity. In fact, a straight beam of constant cross section makes rather inefficient use of its material.

12.12 TESTING A CONCRETE BEAM TO FAILURE

Concrete has very little tension strength. Therefore, it is not used alone in bending situations. However, when reinforced with steel, concrete can be a very effective material for a beam.

A bending test of a steel-reinforced concrete beam must be conducted on heavy professional-quality equipment, because any practical test specimen will have considerable load-carrying capacity. The force-deflection test will produce a curve of similar shape to that for a steel beam even though the characteristics of concrete differ greatly from those of structural steel (figure 12.16).

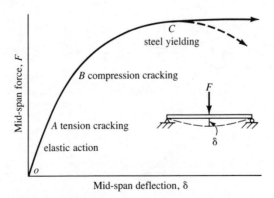

Figure 12.16 Test results from a bending test of a reinforced concrete beam.

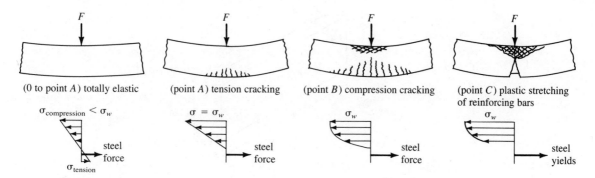

Figure 12.17 Local action at the point of load application during a force-deflection test of a reinforced concrete beam.

Like steel, concrete has a linear portion up to point A. The deflection can be predicted using relationship 12.20. However, the I and C values must be computed using an equivalent cross section for beams of two materials, as discussed in chapter 10.

After this linear portion is exceeded, the force-deflection curve deviates to the right and is shaped as though it were made of a ductile rather than a brittle material. A closer look at the region of the beam beneath the point of load application reveals that the concrete actually degrades progressively in a brittle fashion, as shown in figure 12.17.

First, tension cracks form below the neutral axis. Such cracks can be minute and do not necessarily constitute even a cosmetic failure of the beam. However, the cracked section of concrete cannot carry tension stress and, therefore, transfers all of the tension load to the steel bars below the neutral axis.

As the load is increased to point B, the compression strength of the concrete may be exceeded in the region above the neutral axis. Compression cracking would then initiate. The compression "stress block" takes the shape shown in figure 12.17. The beam may continue to have strength in reserve, but its appearance is degraded.

Gross deformation and collapse occur when the load is increased to the point C, where the steel yields in tension. This load level is accompanied by some crushing of the concrete in compression.

PROBLEM SET 12.3

12.11 A materials-testing machine is used to conduct a force-deflection test of a structural steel beam. The beam is a bar of rectangular cross section measuring 0.997 in. high by 0.498 in. wide. The span is 18.00 in. and the load is applied at mid-span. At load levels of 0, 250, 500, 750, 1 000, 1 250, 1 500, 1 750, and 2 000 lb, deflections of 0.0, 0.021, 0.044, 0.070, 0.094, 0.120, 0.140, 0.180, and 0.753 in. are measured, respectively. Plot the force-deformation curve for this test. Calculate the modulus of elasticity and estimate the yield stress for the material tested.

12.12 A reinforced concrete beam is tested using a flexure fixture in a large hydraulic testing machine. The beam has the cross section shown in figure 12.18. It is reinforced top and bottom with four No. 3 reinforcing steel bars, having $\frac{1}{2}$ in. of

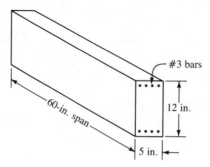

Figure 12.18 Problem 12.12.

cover. The modulus of elasticity for the bars was tested in tension and found to be 29.9E6 psi. The load is applied in the middle of the 60-in. span. Deflections of 0.0, 0.022, 0.041, 0.062, 0.079, 0.098, 0.203, 0.398, and 0.750 are measured for loads of 0, 1 000, 2 000, 3 000, 4 000, 5 000, 6 000, 7 000, and 8 000 psi. Plot the load-deflection curve. Using equivalent cross sections, estimate the modulus of elasticity of the concrete.

12.13 THE MOMENT-AREA METHOD FOR BEAM DEFLECTION (OPTIONAL)

This subject is often reserved for advanced courses in structural design or machine design. "Problems" are better described as projects, because they are lengthy even for what may appear to be obvious application. However, with the availability of personal computers, this subject may reasonably be added to introductory courses in statics and strength of materials. It offers a set of problems having sufficient complexity to justify the use of a computer, yet it is based on fundamental principles already presented. The method, called the moment-area method, is an extension of the slope-area method used to find shear and moment diagrams in chapter 11.

First, we have to know a useful fact about areas under a parabolic curve. As shown in figure 12.19, the area under a parabola is two-thirds times the base times the height.

So far, only beams having a constant cross-sectional shape throughout their lengths have been considered from the standpoint of deformation. Sometimes, especially in machine structures, bending members have areas which vary along their length. This variance can be included by treating the moment of inertia, I, as a variable. The fundamental relationship 12.1,

$$\frac{1}{\rho} = \frac{M}{EI}$$

is equally valid for the variable elastic modulus, E, and moment of inertia, I. However, relationship 12.6 is affected in that E or I must remain inside the integral if either is a variable of x. Relationship 12.6 then becomes

$$\theta = \frac{dy}{dx} = \int \frac{M}{EI} \, dx \qquad (12.22)$$

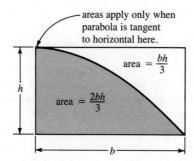

Figure 12.19 Area under parabolic curves.

where both M and I are functions of x.

Relationship 12.7,

$$y = \int \theta \, dx$$

is equally valid for variable E and I.

In practice, many cases have areas which are not only variable but also discontinuous. Hence, they are difficult to integrate. The moment-area method can be used as in the following example.

ILLUSTRATIVE PROBLEM

12.4 A high-speed turbine shaft has a step to facilitate precise location of the turbine wheel. Calculate the deflection at the turbine wheel under the loads shown in figure 12.20. E equals 207E9 Pa. The bearings are spherically seated and behave as simple supports.

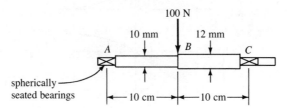

Figure 12.20 High-speed turbine shaft of illustrative problem 12.4.

Solution:

The moment of inertia from A to B is

$$I = \frac{\pi d^4}{64}$$

$$I = \frac{\pi \times (10 \text{ mm})^4}{64}$$

$$I = 0.491\text{E}3 \text{ mm}^4$$

And from B to C,

$$I = 1.018E3 \text{ mm}^4$$

The M/EI Diagram:

The reduced diameter of the shaft at the bearings is neglected as having little effect on the total deformation. The maximum bending moment will be at point B and has a value of 500 N · cm.

The *M/EI* diagram is as shown in figure 12.21. At point B, *M/EI* is calculated to be $4.92E{-4}$ cm^{-1} on the 10-mm-diameter side and $2.37E{-4}$ cm^{-1} on the 12-mm-diameter side. This discontinuity is caused by the step at B.

The Slope Diagram:

Although a negative slope would be expected at point A, the actual value of the slope at that point is unknown and will be represented by the unknown quantity θ_A.

Accumulating the triangular area under the *M/EI* diagram going from A to B,

(a)
$$\theta_B = \theta_A + (4.92E{-4} \text{ cm}^{-1} \times 10 \text{ cm})/2$$
$$\theta_B = \theta_A + 2.46E{-3}$$

Then continuing to point C,

(b)
$$\theta_C = \theta_B + (2.37E{-4} \text{ cm}^{-1} \times 10 \text{ cm})/2$$
$$\theta_C = \theta_B + 1.185E{-3}$$

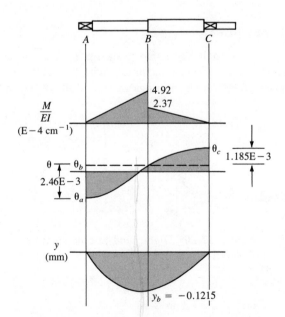

Figure 12.21 The *M/EI* slope and deflection curves for the turbine shaft of illustrative problem 12.4.

Substituting for θ_B from relationship **(a)**,

(c)
$$\theta_C = \theta_A + 3.645\text{E}{-}3$$

The Deflection Diagram:

The deflection is the area under the slope diagram. Since the slope θ_A is at this time unknown, it is necessary to find the areas in terms of θ_A.

At point A, the deflection, y, must equal zero because of the bearing support.

The deflection at B is obtained by accumulating the area under the θ curve between A and B. This area consists of a negative parabolic area of 10-mm base and a height of 2.46E$-$3 plus a rectangular area of 10-mm base and a height of θ_B.

$$y_B = -\frac{2}{3} \times 2.46\text{E}{-}3 \times 10 \text{ cm} + \theta_B \times 10 \text{ cm}$$

Substituting for θ_B from relationship **(a)**

$$y_B = -1.640\text{E}{-}2 + 10\theta_A + 2.46\text{E}{-}2$$
(d)
$$y_B = 10\theta_A + 0.820\text{E}{-}2 \text{ cm}$$

The deflection at C is obtained by accumulating the area under the θ curve from B to C and adding it to the deflection at B. This additional area consists of the parabolic area of height 1.185E$-$3 and a base of 10 mm plus the rectangular area of $10\theta_B$ cm.

$$y_C = \frac{2}{3} \times 1.185\text{E}{-}3 \times 10 \text{ cm} + 10\theta_B + y_B$$

Substituting for θ_B from relationship **(a)** and y_B from relationship **(d)**,

(e)
$$y_C = 20\theta_A + 4.069\text{E}{-}2$$

Because there is a bearing supporting the shaft at C, $y_C = 0$. Therefore, by setting relationship **(e)** equal to zero, we can solve for the unknown θ_A.

$$0 = 20\theta_A + 4.069\text{E}{-}2$$
$$\theta_A = -2.0345\text{E}{-}3$$

Substituting the now known value of θ_A into relationship **(e)**,

$$y_B = -10 \times 2.0345\text{E}{-}3 + 0.820\text{E}{-}2$$
$$y_B = -0.1\ 215 \text{ mm}$$

The deflection of the turbine wheel is 0.1 215 mm.

Note: The *critical speed* of shaft rotation is inversely related to its static deflection. The critical speed is one at which the shaft cannot operate without excessive, or even destructive, vibration. For this reason shaft deflections are often specified to be within certain limits for certain kinds of machines. Hence there are many practical applications for calculating shaft deflections, even though the deflections may seem quite small.

12.14 THE TRAPEZOIDAL METHOD (OPTIONAL)

In illustrative problem 12.4 the moment-area method employed exact methods of finding area. Yet, when the bending moment or the stiffness, EI, is variable in a complex way, or if many discontinuities exist in loading or in shape, approximate methods may be necessary.

Several methods of approximation are available. The simplest is called the *trapezoidal method*, in which the area is approximated by trapezoidal slices. This method can quickly be programmed on a personal computer. If greater accuracy is desired, an increased number of narrower slices can be used. Since repetition is no chore for a computer, there is little justification for going to more complex approximate methods to improve accuracy. As long as we are guessing, why make the guess complicated?

In the trapezoidal method, the area under the curve is approximated by a series of trapezoids as shown in figure 12.22.

The increment of area represented by each trapezoid is the base, b, times the average height of the trapezoid.

$$\Delta A_i = b(h_i + h_{i-1})/2 \tag{12.23}$$

In each trapezoid, the approximation excludes a small bit of area if the curve above is convex; conversely, the approximation overestimates the area if the curve is concave. The unaccounted area represents the error in the method. Figure 12.22 shows that the method becomes increasingly accurate as the vertical slices become thinner, although more slices are required. If the slices were to become infinitely thin, the limit that represents the definition of an integral would be reached. Hence, this technique is a method for approximate integration.

The trapezoidal method is demonstrated by the next illustrative problem. It may look complicated, but it is very easily solved using a personal computer. In fact, it is fun to make the computer actually graph the diagrams.

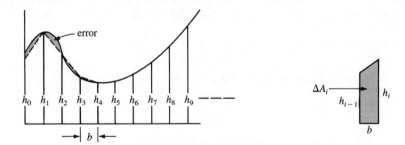

Figure 12.22 The trapezoidal method of approximating areas under a curve.

ILLUSTRATIVE PROBLEM

12.5 A student has designed a connecting rod for a small high-speed engine intended to run at 7 000 rpm and have a stroke of 2.91 in. The connecting rod is shown in figure 12.23.

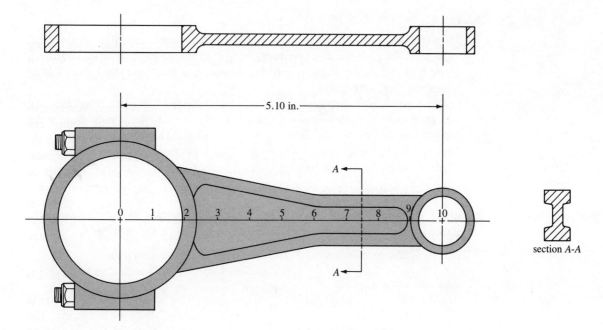

Figure 12.23 Connecting rod for illustrative problem 12.5.

Connecting rods are the important part of a reciprocating machine that attach the reciprocating pistons to the rotating crankshaft. Connecting rods experience enormous bending moments at high speeds due to the dynamic lateral forces. These rods can bend excessively or rupture at high speeds. In competition racing, this damage is called "throwing a rod" or "blowing an engine."

At "red line" velocity, the *M/EI* calculated, or estimated, at 10 equally spaced intervals between the center of the rod bearing on the crank shaft and the wrist pin bearing on the piston are given in the following table. Calculate the deflections of the connecting rod at these stations, assuming the material remains elastic.

Station number	Distance from rod bearing (in.)	M/EI $(10^{-4}$ in.$^{-1})$
0	0	0
1	0.51	0
2	1.02	0
3	1.53	1.667
4	2.04	1.939
5	2.55	2.867
6	3.06	4.583
7	3.57	3.583
8	4.08	2.500
9	4.59	1.410
10	5.10	0

Solution:

The trapezoidal method will be used. The *M/EI* data given is plotted in figure 12.24.

To solve the problem on a computer, *M/EI* is given the variable name MOEI(K), where K is the station number, 0 to 11. A routine is written to enter the 11 *M/EI* values into the computer.

The slope, θ, is the accumulation of area under the *M/EI* diagram. θ will have the variable name TH(K), where K is again the station number. Because the slope at the boundary is unknown, let TH(0) equal 0.0 initially. This value will need to be corrected later. Then, using relationship 12.22 repeatedly, letting K increment in whole numbers from 1 to 10,

(a)   TH(K)=TH(K−1)+B*[MOEI(K)+MOEI(K−1)]/2

where B = 0.51. This routine creates a table of θ values for the remaining 10 stations. They are plotted in figure 12.24b and given in the next table.

Station number	θ (Uncorrected) (10^{-4} radians)	y (Uncorrected) (10^{-4} in.)
0	0	0
1	0	0
2	0	0
3	0.425	0.108
4	1.344	0.559
5	2.569	1.557
6	4.469	3.353
7	6.551	6.133
8	8.103	9.902
9	9.100	14.287
10	9.450	19.019

Next, the deflection values can be computed using the trapezoidal method to accumulate the area under the θ diagram. First, let y (0) = 0.0. This value is the correct boundary condition for y at station zero. Then, using relationship 12.22, letting K go from 1 to 10,

(b)   Y(K)=Y(K−1)+B*[TH(K)+TH(K−1)]/2

This routine creates a table of y values for the remaining 10 stations. They are plotted in figure 2.24c and given in the table.

To correct for the boundary condition, it is necessary to recognize that θ at station 0 is not equal to zero but has some value θ_0. If the value of θ_0 were known, it would need to be added to each of the θ values. Also, y at station 10 must be equal to zero and cannot be 19.019E−4 in. as the accumulation of area indicates.

If θ_0 were added as an unknown to each of the slope values, the summation of the area under the total θ diagram would be increased by 0.51 × θ_0 × 10. This, plus the summation obtained before, would give the deformation, y, at station 10, which should be set equal to zero. Therefore,

(c) $5.1\theta_0 + 19.019E−4 = 0$

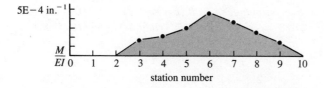

Figure 12.24 Slope and deflection diagrams for the connecting rod of illustrative problem 12.5, uncorrected for boundary conditions.

Solving for θ_0,

$$\theta_0 = -3.729\text{E}-4$$

This result is the corrector for the θ diagram and must be added to each previously computed value of θ. Programming:

$$\text{TH(K)} = \text{TH(K)} + \theta_0$$

Again, $y(0)$ must be zero. And, the process (b) must be repeated giving a corrected set of values for θ and y plotted in figure 12.25 and given next.

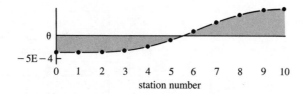

Figure 12.25 Corrected slope and deflection curve for the connecting rod of illustrative problem 12.5.

Station number	θ (Corrected) (10^{-4} radians)	y (Corrected) (10^{-4} in.)
0	−3.729	0
1	−3.729	−1.901
2	−3.729	−3.804
3	−3.304	−5.597
4	−2.384	−7.048
5	−1.159	−7.952
6	0.740	−8.059
7	2.822	−7.150
8	4.373	−5.315
9	5.370	−2.830
10	5.730	0

Figure 12.26 Flowchart for computer program for calculating slopes and deflection using the trapezoidal method.

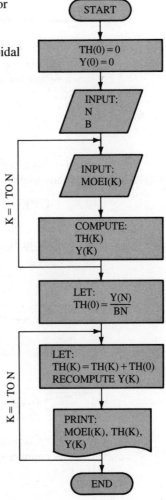

Note that the summation of area under the corrected θ curve produces the desired boundary condition at station 10. The process used in illustrative problem 12.5 can be put into a flowchart as appears in figure 12.26.

This flowchart can be used to write a program for a personal computer that applies the trapezoidal method in computing the deformation of a simply supported beam having a variable and/or discontinuous bending moment diagram and a variable and/or discontinuous cross-sectional moment of inertia.

PROBLEM SET 12.4

12.13 A structural steel beam is to be tested to failure. It is a W8 × 31, 8 ft long, is simply supported, and has a concentrated load at its center. The plastic shape factor is estimated to be 1.10. Estimate (a) the load required for gross deformation and (b) the deflection of the mid-span at the end of the elastic range.

12.14 Using the moment-area method, calculate the deflection under the load of the beam in figure 12.27. It is rectangular in cross section, 5 cm wide and 28 cm high, and is made of a material having a modulus of elasticity of 150E9 Pa. Graphically determine the x location of maximum deflection from the slope diagram.

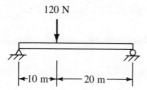

120 N

←10 m→|←— 20 m —→|

Figure 12.27   Problem 12.14.

12.15 Using the moment-area method, calculate the deflection at mid-span for the round steel shaft shown in figure 12.28. It has a diameter of 1.5 in. and a modulus of elasticity of 29E6 psi.

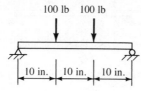

100 lb   100 lb

| 10 in. | 10 in. | 10 in. |

Figure 12.28   Problem 12.15.

12.16 Using the moment-area method, calculate the deflection at the end of the beam shown in figure 12.29. It is a redwood beam having a width of 10 in. and a height of 2 in.

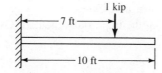

1 kip

—— 7 ft ——

—— 10 ft ——

Figure 12.29   Problem 12.16.

12.17 Using beam tables from a library reference, calculate the maximum deflection for a simply supported beam under a load uniformly increasing to 100 N/cm as shown in figure 12.30. The beam is a 4-cm-square rod made of 7075-T6 aluminum. Give as a reference the source of your beam tables.

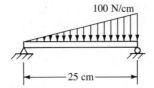

100 N/cm

25 cm

Figure 12.30 Problem 12.17.

12.18 Using beam tables from a library reference, calculate the maximum deflection for a douglas fir 2 × 10 loaded as shown in figure 12.31.

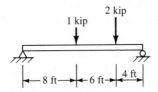

2 kip

1 kip

8 ft — 6 ft — 4 ft

Figure 12.31 Problem 12.18.

12.19 Using the trapezoidal method, find the maximum deflection of a beam under several concentrated loads as shown in figure 12.32. The beam is a WT15 × 105. Use four trapezoids.

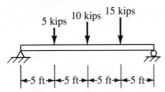

5 kips   10 kips   15 kips

5 ft — 5 ft — 5 ft — 5 ft

Figure 12.32 Problem 12.19.

12.20 Using the trapezoidal method, find the maximum deflection of the beam shown in figure 12.33. The beam is a hollow 6061-T6 aluminum tube 10 cm in diameter with a 6-mm wall thickness. Use five trapezoids and the information from the illustrative problem.

12.21 A rectangular cantilever beam $\frac{1}{2}$ in. wide has a concentrated load at its end of 1 750 lb. It is stepped down in thickness as shown in figure 12.34. Calculate the deflection at the end using the trapezoidal method. Use four trapezoids. $E = 9E6$ psi.

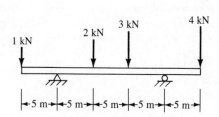

Figure 12.33 Problem 12.20.

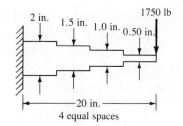

Figure 12.34 Problem 12.21.

12.22 A simply supported beam has a uniformly distributed load of 100 kN/m. It is made of low-carbon steel and has a rectangular cross section that varies linearly from 10 cm deep at its ends to 20 cm at the center. Calculate the deflection at mid-span using the trapezoidal method. The beam is shown in figure 12.35. It has a constant width of 10 cm.

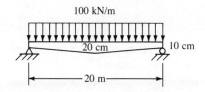

Figure 12.35 Problem 12.22.

12.23 Write a program for a personal computer that will calculate the deflection of a simply supported beam given N equally spaced values of M/EI along its length. N will be no greater than 30.

12.24 Modify the program for a personal computer from the previous problem so that it will calculate the deflection of either a cantilever or a simply supported beam.

SUMMARY

☐ The *elastic curvature* of a beam at a point is related to the bending moment divided by the flexural *stiffness, EI*, of a beam.

(12.1) $\dfrac{1}{\rho} = \dfrac{M}{EI}$

☐ The *slope* of the deflection curve can be determined from the curvature:

(12.5) $\theta = \displaystyle\int \dfrac{M}{EI}\, dx$

☐ The *deflection* can be determined from the slope.

(12.7) $y = \displaystyle\int \theta\, dx$

☐ The boundary, or end, conditions yield the constants of integration:

> fixed end: $y = 0$ and $\theta = 0$
> simple support: $y = 0$

☐ The deflection of a cantilever beam under a uniform load is:

(12.10) $y_{max} = -\dfrac{wL^4}{8EI}$

☐ The deflection of a cantilever beam with a load at its free end is:

(12.14) $y_{max} = -\dfrac{FL^3}{3EI}$

☐ The deflection of a simply supported beam with a uniform load is:

(12.17) $y_{max} = -\dfrac{5wL^4}{384EI}$

☐ The deflection of a simply supported beam with a mid-span load is:

(12.20) $y_{max} = -\dfrac{FL^3}{48EI}$

☐ The *moment-area method* is a diagram method of finding slopes and deflections usable for beams of variable as well as constant *EI*.

☐ For complex curves of *M/EI*, the *trapezoidal method* is an approximate method for performing integrations. It is particularly suitable for a computer, and can be very accurate when the slices are small.

(12.23) $\Delta A_i = b(h_i + h_{i-1})/2$

REFERENCES

1. Timoshenko, S. and MacCullough, G. H. *Elements of Strength of Materials,* 3rd ed. Princeton, NJ: D. Van Nostrand Co., 1958.

2. Neathery, R. F. *Applied Strength of Materials,* New York: John Wiley & Sons, 1982.

3. Mott, R. L. *Applied Strength of Materials,* Englewood Cliffs, NJ: Prentice-Hall, 1978.

CHAPTER THIRTEEN

Statically Indeterminate Beams (Optional)

The concept of statically indeterminate beams is introduced from the standpoint of support redundancy. The principle of superposition is described and demonstrated. Reaction forces, bending moments, and deflection curves are derived for two common, and therefore useful, cases of statically indeterminate beams: the beam on three supports with a uniform load, and the beam built-in at both ends with a uniform load. Applications of these cases are illustrated. The chapter introduces the very broad subject of statically indeterminate structures.

13.1 REDUNDANT SUPPORTS

Some structures, or members of structures, are supported in a way that offers a greater number of load paths than are necessary to satisfy equilibrium. Such structures are said to have *redundancies*. They are called *statically indeterminate* because their support reactions cannot be determined by statics alone.

Table 13.1 is a summary of the conditions of static equilibrium for various force systems. The linear and the three planar force systems were introduced in earlier chapters. The two non-planar systems at the bottom of the table were not covered.

For each type of force system, one equation can be written for each equilibrium condition. If the supports result in more load paths in the form of forces or moment reactions than there are equilibrium conditions, then the structure will be statically indeterminate, because there will be more unknown support reactions than there are equations. If there are fewer load paths than there are equilibrium conditions, the structure will be unstable. To be stable and statically determinate, a structure or member must have support load paths equal to the number of equilibrium conditions as shown in the examples of figure 13.1.

Generally speaking, redundancy strengthens a structure by providing greater reserve strength above the elastic range of the structure. However, such statically indeterminate

Table 13.1 Summary of static equilibrium conditions for various force systems

Linear	$\Sigma F = 0$
Concurrent Planar	$\Sigma H = 0, \Sigma V = 0$ or $\Sigma X = 0, \Sigma Y = 0$
Parallel Planar	$\Sigma M = 0, \Sigma V = 0$ or $\Sigma M = 0, \Sigma Y = 0$
General Planar	$\Sigma M = 0, \Sigma V = 0, \Sigma H = 0$ or $\Sigma M = 0, \Sigma Y = 0, \Sigma X = 0$
Parallel	$\Sigma M_x = 0, \Sigma M_y = 0$ and $\Sigma Z = 0$
General	$\Sigma M_x = 0, \Sigma M_y = 0, \Sigma M_z = 0$ and $\Sigma X = 0, \Sigma Y = 0, \Sigma Z = 0$

structures are more likely to have residual stresses from construction or manufacture. Also, they are more susceptible to thermal stresses caused by temperature changes. But, beyond that, statically indeterminate structures are more difficult to analyze. Some can represent formidable analytical tasks. Whole books are devoted to the subject, and powerful computer programs have been devised for such analysis. An analytically trained structural engineer must be consulted for analysis of most statically indeterminate structures.

In chapter 3, statically indeterminate structures were introduced in the form of members of two materials; for example, a reinforced concrete column. In a reinforced concrete

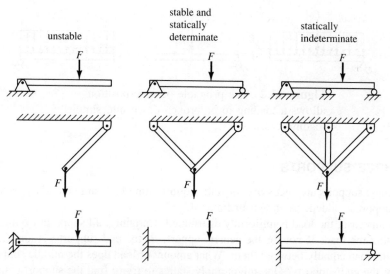

Figure 13.1 Examples of instability, stability, and static indeterminance caused by insufficient, sufficient, and redundant load paths, respectively.

column it is not possible to tell, from the statics alone, how much of the column load is carried by the steel and how much by the concrete. An additional equation describing the relative stiffness of the two materials was used in addition to equilibrium. This chapter is intended to further acquaint you with the principles required to understand a statically indeterminate structure and to make you familiar with two frequently encountered and important cases of statically indeterminate beams.

This chapter merely introduces statically indeterminant beams. Many teachers will choose to omit this chapter because it is quite ambitious to include in an introductory course in statics and strength of materials. However, for those students and teachers who do choose to tackle this chapter, the purposes are (a) to become acquainted with statically indeterminate structures so that they may be recognized, (b) to experience the frustration involved in trying to use statics to analyze a statically indeterminate beam, (c) to show how a couple of common cases of statically indeterminate beams are solved using superposition, and (d) to point the way to beam tables in the references that give the results for many other cases of statically indeterminate beams.

13.2 THE PRINCIPLE OF SUPERPOSITION

In the elastic range, the beam diagrams of one part of the loading can be added to those due to other parts of the loading to get the beam diagrams for the total loading. For example, in figure 13.2, the shear, moment, and deflection diagrams for the concentrated load can be added to those for the distributed load to get the diagrams of the combined load.

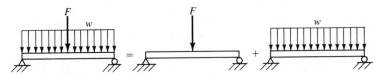

Figure 13.2 The principle of superposition allows a loading to be broken down into simpler loading cases.

13.3 A BEAM ON THREE SUPPORTS

Two simple supports are necessary for static equilibrium of a beam. What happens when a third support is added, as shown in figure 13.3?

In this case, the load is uniformly distributed. Certainly, all supports would share in supporting the load. If not for the center support, B, the end supports, A and C, would divide the load equally between them. What amount of load does the middle support pick up from the end supports? One might apply statics to try to find the support reactions at A, B, and C to find the answer.

A free-body diagram for the beam is drawn in figure 13.4. One can sum the vertical forces for an equilibrium equation,

(a) $$\sum V \uparrow : A + B + C - wL = 0$$

One can sum moments about end A for another equation,

(b) $$\sum M_A \circlearrowleft : -BL/2 - CL + wL^2/2 = 0$$

Presuming that the load, w, and the length, L, are known, equations **(a)** and **(b)** have three unknowns: A, B, and C. To solve for the unknowns, a third equation is needed. However, the two equilibrium conditions, $\sum V$ and $\sum M$, that apply to this problem have already been used.

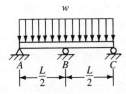

Figure 13.3 A beam on three supports is an example of a statically indeterminate structure.

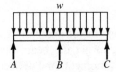

Figure 13.4 A free-body diagram for the beam on three supports shown in figure 13.3.

What if someone had the clever idea to use the moment-equilibrium condition a second time, but this time about another point on the beam, to get a third equation? Let us try it by taking moments about C.

(c)
$$\Sigma \, M_C \circlearrowleft: AL + BL/2 - wL^2/2 = 0$$

Writing the three equations so that the knowns are on the right of the equal signs and "cleaning them up" a bit, the set of equations appears as follows:

(d)
$$\begin{aligned} A + B + \quad C &= wL \\ 0A + B + 2C &= wL \\ 2A + B + 0C &= wL \end{aligned}$$

Here we have three equations in three unknowns.

One way to solve the set of equation **(a)** is with determinants using Cramer's rule[1]:

$$D = \begin{vmatrix} 1 & 1 & 1 \\ 0 & 1 & 2 \\ 2 & 1 & 0 \end{vmatrix}$$

and

$$D_A = \begin{vmatrix} wL & 1 & 1 \\ wL & 1 & 2 \\ wL & 1 & 0 \end{vmatrix}$$

$$D_B = \begin{vmatrix} 1 & wL & 2 \\ 0 & wL & 2 \\ 2 & wL & 0 \end{vmatrix}$$

and

$$D_C = \begin{vmatrix} 1 & 1 & wL \\ 0 & 1 & wL \\ 2 & 1 & wL \end{vmatrix}$$

Then, A, B, and C would be given by

(e)
$$A = \frac{D_A}{D} \qquad B = \frac{D_B}{D} \qquad C = \frac{D_C}{D}$$

Evaluating the determinate, D,

$$D = 1 \begin{vmatrix} 1 & 2 \\ 1 & 0 \end{vmatrix} - 1 \begin{vmatrix} 0 & 2 \\ 2 & 0 \end{vmatrix} + 1 \begin{vmatrix} 0 & 1 \\ 2 & 1 \end{vmatrix}$$
$$D = 1(-2) - 1(-4) + 1(-2) = 0$$

The determinate, D, is equal to zero!

To solve equations **(e)** for A, B, and C, one would need to divide by zero in each case. There is no solution to this set of equations.

The problem is that the equations in **(d)** are not *linearly independent*. One of the equations is essentially a duplicate of another. The culprit is the one that comes from applying the same equilibrium condition, the summation of moments, a second time.

Although equation (c) looks much different, it represents the same static condition as (b); it is a duplicate in disguise. The way to check for linear independence of equations is to evaluate the determinate of a matrix and see that it does not equal zero as it did here. The way to ensure linear independence is never to use the same equilibrium condition twice.

So, where does one get a third equation to handle the third unknown? The answer is ''from the beam deflections.'' The principle of superposition is useful in this solution.

13.4 THE PRINCIPLE OF SUPERPOSITION APPLIED

Suppose the beam on the supports in figure 13.3 is broken down into two familiar cases as shown in figure 13.5: a simply supported beam with a uniformly distributed load (case 1) and a simply supported beam with a concentrated load (case 2). The concentrated load of case 2 would be at the location of support B and would have the proper direction and magnitude such that the deflection upward at point B for case 2 would be exactly equal to the deflection downward for the same point in case 1. Then, when the two cases are added together, using the principle of superposition, case 3 would result. Proceeding as follows, the deflection at point B for case 1 is given by relationship 12.17,

(f)
$$y_1 = -\frac{5wL^4}{384EI}$$

The deflection at point B for case 2, using relationship 12.20 and letting $P = -B$, is given by:

(g)
$$y_2 = \frac{BL^3}{48EI}$$

By the principle of superposition, the deflection at B for case 1 plus the deflection at B for case 2 must equal that of B for case 3, which is zero.

(h)
$$y_1 + y_2 = 0$$

Substituting the values from (f) and (g) into (h),

(i)
$$-\frac{5wL^4}{384EI} + \frac{BL^3}{48EI} = 0$$

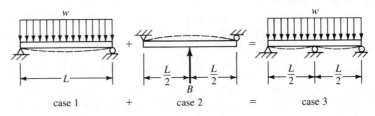

Figure 13.5 Using the principle of superposition on the beam on three supports.

Equation (i) is the needed third equation for solving for the three unknown reaction loads A, B, and C. Solving (i) for the reaction B:

(j)
$$B = \frac{5wL}{8}$$

Since the total load on the beam is wL, relationship (j) tells us that the middle support, B, carries five-eighths of the load on the beam. Knowing the value of reaction B, one can then go back to the two equilibrium equations (a) and (b) to solve for the reactions at supports A and C. In this instance it is not necessary to go back, because of the symmetry of this case. A must equal C, and together they must carry the remainder of the load. Therefore, the reactions for the beam on the supports are as follows:

$$A = C = \frac{3wL}{16} \quad \text{and} \quad B = \frac{5wL}{8} \tag{13.1}$$

The bending moment diagram is the sum of the moment diagrams of case 1 and case 2. It can be obtained by adding relationships 11.6 and 11.8, where $F = -B = -5wL/8$. For the beam to the left of the support B,

$$M = \frac{w}{2}(Lx - x^2) - \frac{5}{16}wLx$$

$$M = \frac{w}{2}\left(\frac{3}{8}Lx - x^2\right) \tag{13.2}$$

The maximum bending moment occurs either at the support B, where $x = L/2$ or where $dM/dx = 0$. The value of dM/dx is zero at $x = 3L/16$. At $x = L/2$, $M = -wL^2/32$. At $x = 3L/16$, $M = 9wL^2/512$. The former value is larger. Therefore,

$$M_{max} = -\frac{wL^2}{32} \tag{13.3}$$

Similarly, by superposition one can also add the slope and the deflection relationships for cases 1 and 2 to obtain the respective relationships for case 3, as shown in figure 13.6.

ILLUSTRATIVE PROBLEM

13.1 What is the lightest wide-flange shape that can be used for a continuous beam 75 ft long that is simply supported at three equally spaced points with no overhang? The beam must carry a uniformly distributed load of 1.080 kips/ft without exceeding an allowable stress of 22 000 psi.

Solution:
From relationship 13.3,

$$M_{max} = -\frac{wL^2}{32}$$

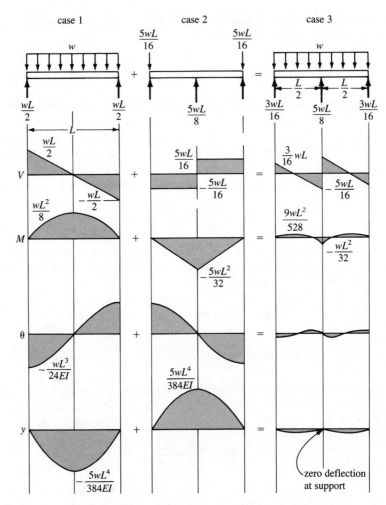

Figure 13.6 The superposition method for developing all beam diagrams for the uniformly loaded beam on three equally spaced supports.

which is over the support point, B. Since the allowable stress may be either positive or negative, only absolute values are important here,

(a)
$$\sigma = \frac{M}{Z} = \frac{wL^2}{32Z}$$

$$Z = \frac{wL^2}{32\sigma}$$

$$Z = \frac{1.080 \text{ kips/ft} \times 75^2 \text{ ft}^2}{32 \times 22 \text{ kips/in.}^2} \times \frac{12 \text{ in.}}{1 \text{ ft}}$$

$$Z = 103.6 \text{ in.}^3$$

A scan of the beam tables for the section modulus of wide-flange shapes shows that either the W21 × 55 or the W24 × 55 will be the lightest-available wide-flange shape that can provide the required Z. Both have weights of 55 lb/ft. But the 21-in. beam has a Z of 110 in.^{3}, whereas the Z of the 24-in. beam is greater at 114 in.^{3}

Trying a W21 × 55 beam, Z = 110 in.^{3} Returning to relationship (a), where $w = 1.080 + 0.055 = 1.135$ kips/ft, the required Z is 108.8, which does not exceed the section modulus of this beam. A W21 × 55 beam is sufficient. It provides enough excess section modulus to cover the weight of the beam in addition to the design load of 1.080 kips/ft. There is no need to use the larger 24-in. beam.

Conclusion:
A W21 × 55 beam is the lightest wide-flange shape that will do the job.

PROBLEM SET 13.1

13.1 A 4 × 6 timber beam 20 ft long is supported on three uniformly spaced simple supports. The allowable stress in bending for this material is 1 700 psi. What is the uniformly distributed load capacity for this beam?

13.2 A solid 20-mm-diameter steel shaft is supported on three equally spaced bearings and is 120 cm long. Under dynamic machine loading, the shaft experiences the equivalent of a 35 N/cm uniform load. What is the maximum stress due to bending in the shaft?

13.3 A W8 × 31 beam is supported on three equally spaced supports. The 35-ft total length of the beam carries a uniform load of 100 lb/ft. Use superposition to determine the location and the value of the maximum deflection of the beam.

13.4 Using the principle of superposition as demonstrated in the preceding section, calculate the maximum bending moment in a cantilever beam 30 m long with the end propped as shown in figure 13.7. The beam has a distributed load of 15 kN/m over its entire length. Hint: Superimpose a uniformly loaded cantilever upon a cantilever loaded with the proper concentrated load at A to bring the free end to zero deflection.

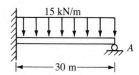

Figure 13.7 Problem 13.4.

13.5 Using the principle of superposition as demonstrated in the preceding section, calculate the maximum bending moment in a propped cantilever 15 ft long with a simple concentrated load of 105 kips at mid-span (figure 13.8). Hint: Use beam diagrams to find the deflection of the free end of a cantilever.

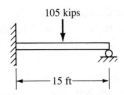

105 kips

15 ft

Figure 13.8 Problem 13.5.

13.5 BEAMS BUILT-IN AT BOTH ENDS

Frequently, beams are built-in at both ends as shown in figure 13.9. They are also called *fixed* or *clamped* beams. These are statically indeterminate structures, because each end support provides both a moment and a vertical force as shown in the free-body diagram. The most frequently encountered load conditions are symmetrical. In those cases, the support loads and moments on the two ends are equal.

It is useful to first consider the case of a simply supported beam with equal applied moments on both ends as shown in figure 13.10a. To get a feel for this loading condition, think of applying two moments with your hands to the ends of some flexible piece of material as shown in figure 13.10b. This example roughly approximates two concentrated edge moments. If the two moments are equal and opposite, no vertical forces whatsoever are needed at the supports for equilibrium.

Since there are no vertical forces, *there are no shear forces* in the beam. This is a case of "pure" bending. The bending moment diagram is a constant value over the entire length as shown in figure 13.11. Using the moment-area method, the slope and deflection diagrams shown in figure 13.11 can be determined.

The maximum deflection, y_{max}, occurs at mid-span and has the value,

$$y_{max} = -\frac{M_o L^2}{8EI} \qquad (13.4)$$

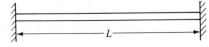

L

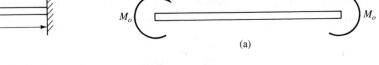

M_o M_o

(a)

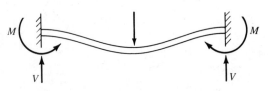

M   M

V   V

Figure 13.9 A beam built-in at both ends.

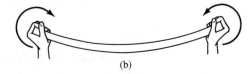

(b)

Figure 13.10 A beam with edge moments applied.

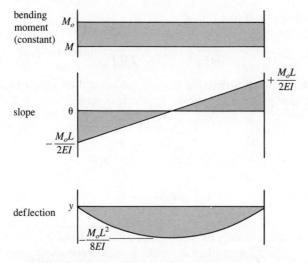

Figure 13.11 The moment, slope, and deflection diagrams for the beam with edge moments applied.

13.6 THE BUILT-IN BEAM WITH A UNIFORMLY DISTRIBUTED LOAD

The beam diagrams for a built-in beam with a distributed load can be obtained by superimposing two statically determinate beam cases as shown in figure 13.12: the uniformly loaded, simply supported beam (case 1), and the equal end moment (case 2).

The slope at the ends of the beam must be zero to meet the boundary conditions for the built-in ends. The slope for case 1 is given by relationship 12.15,

(a)
$$\theta = \frac{w}{24EI}(6Lx^2 - 4x^3 - L^3)$$

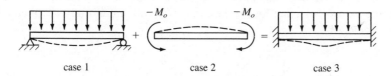

Figure 13.12 Achieving the built-in end condition by superimposing upon case 1 the loading case 2 with concentrated moments at the ends such that the slope is zero at the ends of the beam.

At the ends, when $x = 0$ or L,

$$\theta = -\frac{wL^3}{24EI} \quad \text{and} \quad +\frac{wL^3}{24EI} \quad \text{respectively}$$

By the principle of superposition, the slope at the left end for case 1 plus the slope at the left for case 2 must equal that for case 3, which is zero.

(b)

$$\theta_1 + \theta_2 = 0$$

$$-\frac{wL^3}{24EI} - \frac{M_o L}{2EI} = 0$$

(c)

$$M_o = -\frac{wL^2}{12}$$

Due to symmetry, the boundary moment must be identical on the other end. The boundary moments, M_o, can be thought of as the edge moment necessary to bend the ends of the beam back to a zero slope after the uniform load, w, is applied.

The bending moment diagram for case 3 is the sum of that for cases 1 and 2, and can be obtained by adding relationships 11.6 and a constant M_o, where M_o equals $-wL^2/12$.

$$M = \frac{w}{2}(Lx - x^2) - \frac{wL^2}{12}$$

$$M = -\frac{w}{12}(L^2 - 6Lx + 6x^2) \tag{13.5}$$

The maximum bending moment is at the supports, where $x = 0$ or L.

$$M_{\text{max}} = -\frac{wL^2}{12} \tag{13.6}$$

Building in the ends of the beam reduces the mid-span bending moment from $wL^2/8$ for a simply supported beam to $wL^2/24$. However, the bending moment at the ends goes from zero to $-wL^2/12$. The bending moment at the ends now governs. However, the net effect of the fixed ends is to redistribute the bending moment so that the beam becomes stronger. The bending moment diagram for the built-in beams with a uniformly distributed load is shown in figure 13.13.

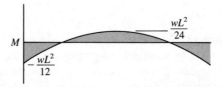

Figure 13.13 The bending moment diagram for a uniformly distributed beam with built-in ends.

The maximum deflection occurs at mid-span and can be determined by adding relationships 12.17 and 13.4, where

$$M_o = -\frac{wL^2}{12}$$

$$y_{max} = -\frac{5wL^4}{384EI} + \frac{wL^4}{96EI}$$

$$y_{max} = -\frac{wL^4}{384EI} \qquad \textbf{(13.7)}$$

ILLUSTRATIVE PROBLEM

13.2 A 32-in.-diameter steel petroleum pipeline is shown in figure 13.14 with a $\frac{1}{4}$-in. wall thickness of continuous welded construction. It spans tundra and has a roller support every 100 ft. Estimating the weight of the pipe, insulation, and contents to be the equivalent of a cylinder of water, calculate the maximum stress in the pipe due to bending under the weight of the pipe and its contents. Also calculate the maximum deflection.

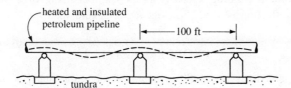

Figure 13.14 Illustrative problem 13.2.

Solution:
A continuous beam on equally spaced simple supports, such as the pipeline in this problem, is structurally identical to a built-in beam. The slope must be zero at the supports, just as in the case of a built-in beam.

Calculating the moment of inertia and the section modulus for the pipe, using the formula for a thin-wall cylinder from Appendix III,

$$I = \pi R^3 t$$
$$I = \pi \times 16^3 \times 0.25 = 3\,217 \text{ in.}^4$$
$$Z = \frac{3\,217}{16} = 201 \text{ in.}^3$$

Calculating the value of w, which is equal to the density of water times the area of the pipe,

$$w = \frac{\pi}{4} \times 32^2 \text{ in.}^2 \times \frac{62.4 \text{ lb}}{\text{ft}^3} \times \frac{1 \text{ ft}^2}{144 \text{ in.}^2}$$
$$w = 349 \text{ lb/ft}$$

The maximum bending moment is over the supports. From relationship 13.6,

$$M_{max} = -\frac{wL^2}{12} = \frac{-349 \text{ lb/ft} \times 100^2 \text{ ft}^2}{12} = -291 \text{ kip} \cdot \text{ft}$$

The maximum stress is given by

$$\sigma = \frac{M}{Z} = \frac{291 \text{ kip} \cdot \text{ft}}{201 \text{ in.}^3} \times \frac{12 \text{ in.}}{1 \text{ ft}} = 17.37 \text{ ksi}$$

The maximum deflection is between supports and is given by relationship 13.7,

$$y_{max} = -\frac{wL^4}{384EI}$$

$$y_{max} = -\frac{349 \text{ lb/ft} \times 100^4 \text{ ft}^4}{384 \times 30\text{E}6 \text{ lb/in.}^2 \times 3\,217 \text{ in.}^4} \times \frac{1\,728 \text{ in.}^3}{\text{ft}^3}$$

$$y_{max} = -1.627 \text{ in.}$$

Conclusion:

The maximum bending stress is 17.37 ksi, occurring over the supports. The maximum deflection is 1.627 in., occurring at mid-span.

PROBLEM SET 13.2

13.6 By what percentage is a built-in beam of the same material and cross section stronger than a simply supported beam under a uniformly distributed load?

13.7 What is the load capacity of a wooden plank, 4 cm high by 20 cm wide, for a uniformly distributed load, if the allowable stress is 18 MPa? The plank is 4 m long and is built-in at both ends.

13.8 What is the deflection of a built-in W8 × 31, if it has a length of 20 ft and is loaded by a uniform load to the beam material's allowable stress of 22 ksi?

13.9 A grid of concrete girders supported on 12-in.-square columns supports a garage floor (figure 13.15). The distance between columns is 25 ft. The concrete girders are 24 in. high by 12 in. wide. They are reinforced by four No. 4 steel bars top and

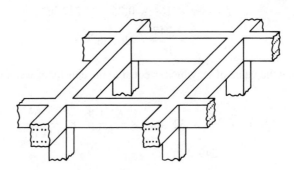

Figure 13.15 Problem 13.9.

bottom with a 1-in. cover. If the concrete may be stressed to 1 000 psi, what is the allowable floor load including the weight of the structure? Hint: Use the equivalent cross section.

13.10 Using the techniques of the preceding section, derive a relationship for the maximum bending moment in a beam of length L carrying a single concentrated load, F, at mid-span and having built-in ends as shown in figure 13.16.

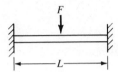

Figure 13.16 Problems 13.10 and 13.11.

13.11 Using the techniques of the preceding sections, derive a relationship for the mid-span deflection of a built-in beam as shown in figure 13.16.

13.7 OTHER STATICALLY INDETERMINATE STRUCTURES

Structural engineers take entire courses, both graduate and undergraduate, in statically indeterminate structures. Such structures can be very complex. Only a few common cases have been presented herein. Statically indeterminate structures are beyond the scope of this book.

However, you should know that statically indeterminate structures are solved using a combination of equilibrium equations and deflection equations. Powerful computer programs exist for the stress analysis of complex statically indeterminate structures. These programs and methods have reduced the work involved in design. They have also made statically indeterminate structures more common; consequently, it is more likely that field personnel will encounter statically indeterminate structures and be challenged by their peculiarities.

For more beam diagrams for other cases of statically indeterminate beams, consult references 2 and 3.

SUMMARY

☐ Beams with *redundant supports* have load paths in excess of the number of equilibrium equations that can be written.

☐ Structures with redundant supports are *statically indeterminate*.

☐ The *principle of superposition* can be used in the elastic range to introduce a deformation equation in the case of a redundancy so that the number of equations can be made equal to the number of unknowns. The reactions on the beam can then be determined.

☐ Two common cases of statically indeterminate beams are:
    A *beam on three supports under a uniform load:*

(13.2) $M = \dfrac{w}{2}\left(\dfrac{3}{8}Lx - x^2\right)$

(13.3) $M_{max} = -\dfrac{wL^2}{32}$

The maximum bending moment occurs over the middle support.

   A beam built-in at both ends under a uniform load:

(13.5) $M = -\dfrac{w}{12}(L^2 - 6Lx + 6x^2)$

(13.6) $M_{max} = -\dfrac{wL^2}{12}$

The maximum bending moment occurs at the built-in ends.

REFERENCES

1. Goodson, C. E. and Miertschin, S. L. *Technical Mathematics with Calculus*. New York: John Wiley & Sons, 1985.

2. Griffel, W. *Handbook of Formulas for Stress and Strain*. New York: F. Ungar Publishing Co., 1966.

3. American Institute of Timber Construction. *Timber Construction Manual*. New York: John Wiley & Sons, 1966.

CHAPTER FOURTEEN

Columns

The understanding of bending deflection developed in the preceding chapters allows compression members to be revisited for the case of columns. The differential equation is formulated and the sine wave is shown to be the solution. The critical load relationship is derived and discussed from the standpoint of experimental results. The concept of effective length is developed from the sine wave. Critical stress is derived in terms of the slenderness ratio. The effect of plasticity is discussed. Two codes are introduced as examples of practical design of long and intermediate design, one for wood and the other for structural steel.

14.1 BUCKLING

As discussed in chapter 3, one of the ways a compression member may fail is by buckling. Buckling need not be a material failure; it may be purely elastic. *Elastic failure* means that no yielding or fracture accompanies buckling and the column can spring back to its original straight shape after the load is removed. In most practical column configurations, however, some yielding or fracture will occur with buckling. Yet, the phenomenon of buckling is best understood in terms of a long, slender column in which purely elastic buckling can occur.

Buckling is actually bending. For this reason, the discussion of elastic buckling was not presented until bending had been thoroughly treated. The column is shown horizontally in figure 14.1.

In the horizontal position, figure 14.1 looks like a beam, because we are accustomed to seeing beams horizontally and columns vertically. Actually, columns may be oriented in any direction and may be made of the same materials or structural shapes as beams. The two are distinguished in that a column is loaded axially in compression, whereas a beam is loaded laterally. In figure 14.1, the load is axial; therefore, the member is a column. The

419

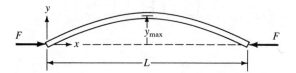

Figure 14.1 A column (shown horizontally as a beam) buckling under an axial load, F.

column was drawn horizontally so that the bending relationships, derived earlier, can be used without alteration. The column is shown in a buckled state with y being the deflection at any horizontal point, x, along the length, L. The maximum deflection is indicated by y_{max}.

A free-body diagram of a portion of the buckled column cut at an arbitrary point, o, on the horizontal axis is shown in figure 14.2.

For moment equilibrium about point o:

$$\sum M_o \circlearrowleft: \; -Fy - M = 0$$
$$M = -Fy \tag{14.1}$$

From relationship 12.4,

$$\frac{d^2y}{dx^2} = \frac{M}{EI}$$

Solving for M and substituting into relationship 14.1,

$$EI\frac{d^2y}{dx^2} = -Fy$$

Dividing through by EI,

$$\frac{d^2y}{dx^2} = -\frac{F}{EI}y \tag{14.2}$$

This kind of equation is called a *differential equation*. It is the fundamental relationship for buckling. The solution of a differential equation is itself an equation. Several techniques are used to find the solutions to a differential equation, but they are not of primary interest here. A particular solution to this differential equation (14.2) is known to be

$$y = y_{max} \sin \frac{\pi x}{L} \tag{14.3}$$

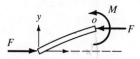

Figure 14.2 A free-body diagram of a portion of a buckled column.

This equation states that the deformation curve is a sine wave with a maximum deflection of y_{max}. By differentiation, this relation can be shown to be the solution to the differential equation as follows: The slope of the column is the derivative of relationship 14.3,

$$\frac{dy}{dx} = \frac{\pi}{L} y_{max} \cos \frac{\pi x}{L}$$

The curvature of the column is the second derivative of 14.3,

$$\frac{d^2 y}{dx^2} = -\left(\frac{\pi}{L}\right)^2 y_{max} \sin \frac{\pi x}{L} \qquad (14.4)$$

Substituting the expression for y from relationship 14.3 and that for $d^2 y / dx^2$ from 14.4 into the differential equation 14.2:

$$-\left(\frac{\pi}{L}\right)^2 y_{max} \sin \frac{\pi x}{L} = -\frac{F}{EI} y_{max} \sin \frac{\pi x}{L}$$

Dividing through by

$$-y_{max} \sin \frac{\pi x}{L}$$

yields

$$\left(\frac{\pi}{L}\right)^2 = \frac{F}{EI}$$

Solving for F, which we will now call F_c, because it is a constant independent of both x and y,

$$\boxed{F_c = \frac{\pi^2 EI}{L^2}} \qquad (14.5)$$

Relationship 14.3 is a solution to the differential equation provided that $F_c = \pi^2 EI/L^2$. This relationship is called the _critical load_ for elastic buckling. It is also called the _Euler buckling load_ in honor of Leonhard Euler, the mathematician who first conceived the derivation.[1]

Critical load, $\pi^2 EI/L^2$, according to the assumptions used in the free-body diagram, is the precise load required to hold the column in equilibrium in a bent state. If the load is slightly less than the critical load, the column will spring back to its straight shape. If the load is slightly more than the critical load, the column will not be stable. It will tend to increase its maximum deformation, y_{max}, which increases the bending moment at all points. Increased bending moment, in turn, causes more deformation; and so on, until either the load is relieved or the column collapses. Structural designers respect buckling because it can occur with no warning in a catastrophic fashion.

Note that the amplitude of the deflection curve, y_{max}, cancels out and does not appear in relationship 14.5 for critical load. Critical load is independent of y_{max}. This result can be interpreted as follows:

If $F \leq F_c$, y_{max} goes to zero.
If $F \geq F_c$, y_{max} becomes progressively larger until full collapse occurs.
If $F = F_c$, the column is stable at any value of y_{max}.

An astute person might ask, "What causes the initial bend, or deformation, in the beam shown in figure 14.1? If there is no initial deformation, there will be no bending moment, and buckling cannot occur."

The answer is that the initial deformation could be infinitesimally small. Such small deformation could be caused either by a minor geometric imperfection or by a slight vibration, both of which are always present. In the case of instability, one invokes Murphy's Law, "If something can go wrong, it will."

14.2 EXPERIMENTAL RESULTS

Laboratory experiments are extremely useful in order to become familiar with the phenomenon of buckling. As shown in figure 14.3, testing of small, inexpensive tubular aluminum columns is convenient for instructional purposes. Professional-quality universal testing machines can also be used for buckling tests. Column specimens for such machines need not be expensive. Many schools use pieces of commercially available lumber for column specimens with the larger testing machines.

In a buckling test, the axial load is increased slowly until buckling occurs. The measurement of either axial or lateral deflection is of little value in a buckling test of a

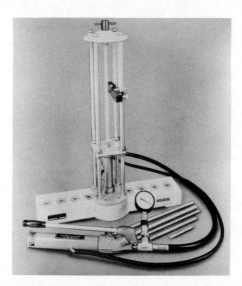

Figure 14.3 A testing machine rigged for a buckling test. (Source: Technovate, Inc.)

concentrically loaded straight column. Buckling usually occurs with no warning. The column under load suddenly bends off to one side or the other, immediately relieving the load. Therefore, it is essential that close attention be paid and the load level be increased in very small increments until buckling occurs. A plot of load versus time obtained with electronic instrumentation is extremely useful in a buckling test.

14.3 EFFECTIVE LENGTH

The buckling shape, also called the *buckling mode,* of a uniform column is always a sine wave or a cosine wave. But the column can be restrained so that the effective column length, L, may be different from the true length, L_t, of the column. Figure 14.4 shows the effect of intermediate restraints upon a column.

The governing mode shape is the sine wave that can be passed among the lateral supports in any case.

In figure 14.4, case (a) is the fundamental case for which relationship 14.5 was derived. In this case, the effective length equals the true length, L, of the beam. Case (b) has an intermediate restraint causing the effective length to be equal to $L/2$. It can be shown that the critical load for this case becomes $4\pi^2 EI/L^2$ by substituting $L/2$ for L in relationship 14.5. In other words, the column with the single intermediate restraint has a critical load four times greater than the unrestrained column. Similarly, the column with two restraints, case (c), has a critical load nine times greater, and that with three restraints, case (d), sixteen times greater. There is a great return for intermediate lateral restraint in a column.

Intermediate restraint can be achieved in several ways. In multi-storied structures, the floors may provide intermediate restraint for continuous columns. Often, intermediate restraint is achieved by cross-diagonal bracing such as in the structure shown in figure 14.5.

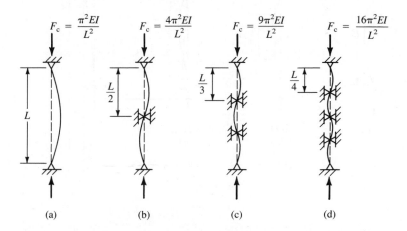

$$F_c = \frac{\pi^2 EI}{L^2} \qquad F_c = \frac{4\pi^2 EI}{L^2} \qquad F_c = \frac{9\pi^2 EI}{L^2} \qquad F_c = \frac{16\pi^2 EI}{L^2}$$

(a)   (b)   (c)   (d)

Figure 14.4 Intermediate lateral restraints reduce the effective length of a column.

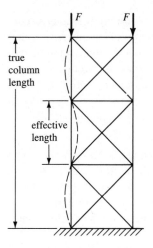

Figure 14.5 Crossed-diagonal bracing used to shorten the effective length of a column.

14.4 BOUNDARY CONDITIONS

Effective length may also be influenced by the boundary conditions. The derivation of the elastic buckling critical load, relationship 14.5, was based upon the assumption of pinned or simple supports at the ends of the columns as shown in figure 14.6a. A pinned support

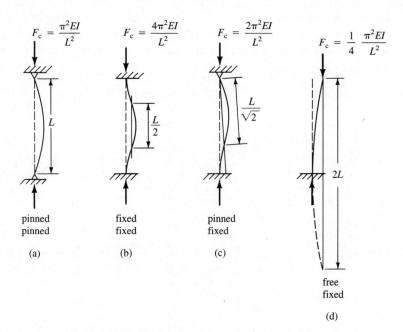

Figure 14.6 Influence of supports on the effective length of a column.

permits the column to rotate freely during buckling at the support point. The opposite boundary condition is the fixed or built-in support that allows no rotation and causes the column to assume the buckling mode shown in figure 14.6b. The fixed-fixed column has length equal to one-half its true length. Theoretically, built-in boundary conditions have the same effect in strengthening a column as an intermediate support. They increase the critical buckling load by a factor of four.

If only one end is fixed and the other is pinned, a sine wave develops about an axis that is at an angle to the original beam axis. But, the wave is still sinusoidal in shape with a length of one lobe of the wave equal to one-half the true length of the column. This case results in a critical buckling load twice that of the pinned-pinned column, case (a).

The deformation of a column with one free end, as shown in figure 14.6d, is one-fourth of a sine wave. It has an effective length twice that of the true length of the column. Therefore, it has a critical buckling load only one-fourth that of the pinned-pinned column.

14.5 ACCURACY OF THE CALCULATED CRITICAL BUCKLING LOAD

The calculated critical buckling load can easily vary from a measured experimental value by a factor of two. Worse yet, the variance is often on the nonconservative side. In other words, the actual buckling strength of a column may be only one-half the calculated critical buckling load.

The reasons for variance are as follows:

1. Slight curvatures in the column due to manufacturing or fabrication inaccuracies.

2. Local plasticity or material failure.

3. Imperfect boundary conditions.

Built-in supports are not perfectly rigid. Pinned supports, on the other hand, are not without friction. Therefore, most actual columns are between cases (a) and (b) of figure 14.6. If calculated as pinned supports, the column test may yield a higher buckling load than the critical. If calculated as fixed, the column will have a lower buckling load than the theoretical critical load. Although exceptions occur, it is usually best to regard the calculated critical buckling load as the upper bound on the strength of a column.

ILLUSTRATIVE PROBLEMS

14.1 A W10 × 100 steel structural shape is used in a column 80 ft long. It is built-in at both ends. What is the theoretical critical load or Euler buckling load for the column?

Solution:
The column has a cross section with a moment of inertia about the two axes of symmetry of 625 and 207 in.^{4}, respectively. The column will buckle in the direction of the lower moment of inertia. The elastic modulus for steel is 29E6 psi.

Due to the built-in ends, the effective length is 40 ft. From relationship 14.5,

$$F_c = \frac{\pi^2 EI}{L^2}$$

$$F_c = \frac{\pi^2 \times 29E6 \text{ lb/in.}^2 \times 207 \text{ in.}^4}{40^2 \text{ ft}^2 \times (12 \text{ in./1 ft})^2}$$

$$F_c = 257 \text{ kips}$$

The theoretical critical load is 257 kips.

14.2 A linkage for a high-speed mechanism contains the member shown in figure 14.7. It has a rectangular cross section and is made of 155A titanium alloy. What is the Euler buckling load for the member?

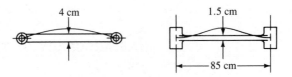

Figure 14.7 Illustrative problem 14.2.

Solution:

In the top view of figure 14.7, the link is a pinned-pinned column. From the side, however, it presumably is attached so that it is fixed-fixed.

In the top view the effective length is the full 85 cm. To buckle to the side, the force would need to overcome a moment of inertia,

$$I = \frac{bh^3}{12} = \frac{1.5 \times 4^3}{12} \text{ cm}^4$$

$$I = 8 \text{ cm}^4$$

The elastic modulus for 155A titanium is 113E9 Pa.

From relationship 14.5,

$$F_c = \frac{\pi^2 \times 113E9 \text{ Pa} \times 8 \text{ cm}^4}{(85 \text{ cm})^2} \times \frac{1 \text{ N/m}^2}{1 \text{ Pa}} \times \frac{1 \text{ m}^2}{(100 \text{ cm})^2}$$

$$F_c = 123.5 \text{ kN}$$

In the side view, the effective length is 42.5 cm. The moment of inertia is

$$I = \frac{4 \times 1.5^3}{12} = 1.125 \text{ cm}^4$$

From relationship 14.5,

$$F_c = \frac{\pi^2 \times 113E9 \text{ Pa} \times 1.125 \text{ cm}^4}{(42.5 \text{ cm})^2} \times \frac{1 \text{ N/m}^2}{1 \text{ Pa}} \times \frac{1 \text{ m}^2}{(100 \text{ cm})^2}$$

$$F_c = 69.5 \text{ kN}$$

The link will buckle in the weakest direction. Therefore, the theoretical critical load is 69.5 kN.

PROBLEM SET 14.1

14.1 A Technovate 9014 is used for a buckling test of tubular specimens of 6061-T6 aluminum. Twelve-in. and 6-in. lengths are used for each of two outside diameters, $\frac{1}{4}$ in. and $\frac{3}{8}$ in. The inside diameters are $\frac{3}{16}$ and $\frac{5}{16}$ in., respectively. Each specimen is tested first with pinned-pinned ends, then with fixed-fixed ends. Calculate the theoretical critical load for each of the eight cases.

  If you have access to a laboratory, perform the laboratory tests and compare the actual buckling loads with each of the calculated theoretical critical loads.

14.2 A buckling test is performed on a rod of dense select douglas fir $\frac{1}{2}$ in. in diameter and 36 in. long. Calculate the Euler buckling load as a benchmark for the test. Assume pinned ends.

14.3 What is the Euler buckling load for an 8-ft-long dense select southern pine 2 × 4?

14.4 A simply supported steel column has a hollow square cross section with an outside dimension 10 cm on a side. The wall thickness is 5 mm. The column is 12 m long. What is the theoretical critical load?

14.5 What is the theoretical critical load of a simply supported column, 40 ft long, made of a 4-in. standard-weight steel pipe?

14.6 Calculate the theoretical critical load for a W6 × 16 wide-flange steel beam 30 ft long that is built-in on the top and pinned at the bottom.

14.6 SLENDERNESS RATIO

The strength of a column is inversely related to its slenderness ratio. The *slenderness ratio* is the effective length divided by the effective width of the cross section. The measure of effective width applicable to buckling is the *radius of gyration, q,* of the cross section. The radius of gyration is defined as follows:

$$q = \sqrt{\frac{I}{A}} \qquad \text{(14.6)}$$

where I is the lesser of the two principal moments of inertia, I_x and I_y, of the cross section as defined by relationship 10.5, and A is the cross-sectional area.

  The radius of gyration is direction-dependent because the moment of inertia, I, has different values for different principal axes of the section. From relationship 10.5,

$$I_x = \sum y^2 \Delta A \quad \text{and} \quad I_y = \sum x^2 \Delta A$$

If the support restraints are identical for both directions, the column will buckle in the weaker (or, more precisely, the more flexible) of the two directions. The smaller value must be used.

  Radius of gyration is a misnomer when applied to areas of structural sections. No radius nor gyration is involved. Nevertheless, like the term *moment of inertia,* it is used because of its computational similarity to a quantity in dynamics where it does have a direct physical meaning. We shall leave unto dynamics those things that are dynamics and take the misnomer for what it actually is. In structural work, the radius of gyration is simply an index useful for comparing the relative slenderness of members.

ILLUSTRATIVE PROBLEM

14.3 What is the slenderness ratio of a wooden column, which is simply supported at one end and built-in at the other, if it is 4 m long and has a rectangular cross section 5 cm by 10 cm on a side?

Solution:
The effective length from case (c) of figure 14.6 is

$$0.707L = 0.707 \times 4 \text{ m} = 2.828 \text{ m}$$

From relationships 10.6 and 10.7,

$$I_x = \frac{bh^3}{12} \quad \text{and} \quad I_y = \frac{hb^3}{12}$$

Taking 5 cm as the b dimension and 10 cm as the h dimension,

$$I_x = 416 \text{ cm}^4 \quad \text{and} \quad I_y = 104.2 \text{ cm}^4$$

Since the column will tend to buckle to the weak direction as shown in figure 14.8, I is the smaller of the two, 104.2 cm^4.

The radius of gyration is given by relationship 14.6,

$$q = \sqrt{\frac{I}{A}} = \sqrt{\frac{104.2 \text{ cm}^4}{5 \text{ cm} \times 10 \text{ cm}}} = 1.444 \text{ cm}$$

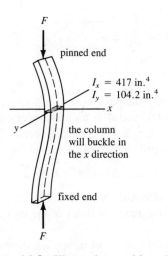

Figure 14.8   Illustrative problem 14.3.

The slenderness ratio is

$$\frac{L}{q} = \frac{2.828 \text{ m}}{1.444 \text{ cm}} \times \frac{100 \text{ cm}}{1 \text{ m}} = 195.9$$

The slenderness ratio is a dimensionless quantity.

14.7 CRITICAL BUCKLING LOAD AS A FUNCTION OF SLENDERNESS RATIO

It is useful to express relationship 14.5 for the critical buckling load in terms of slenderness ratio.

Solving for I from relationship 14.6,

(a) $$I = q^2 A$$

Substituting this result into the relationship 14.5,

$$F_c = \frac{\pi^2 E q^2 A}{L^2}$$

Collecting terms L and q together as the slenderness ratio and dividing through by A gives the critical stress, σ_c, for buckling.

$$\sigma_c = \frac{\pi^2 E}{(L/q)^2} \qquad\qquad \textbf{(14.7)}$$

In other words, the critical stress at which buckling would occur *under ideal conditions* is π^2 times the elastic modulus divided by the slenderness ratio squared.

14.8 THE EFFECT OF PLASTICITY ON BUCKLING

The relationships for critical buckling load or stress, 14.5 and 14.7, are derived for purely elastic conditions. If any local yielding, crushing, or cracking should occur, the analysis is not valid. Relationships 14.5 and 14.7 would then yield results much greater than the actual strength of the column. This case can be envisioned through the diagram shown in figure 14.9, where the yield stress is shown as an upper limit on the critical buckling load.

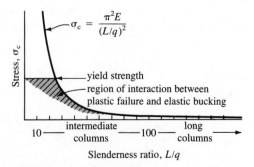

Figure 14.9 Effect of slenderness ratio on buckling stress of columns.

Long columns, those with slenderness ratios greater than approximately 100 depending on the material, have little plastic action associated with buckling. Therefore, the critical loads can be accurately described by the Euler relationships, 14.5 and 14.7. *Short columns,* those with slenderness ratios less than 10, will yield or buckle plastically. Yield stress is an accurate predictor of compression strength of short columns.

Intermediate columns, neither in the short or long range, have a combination of elastic and plastic action. Neither relationships 14.5 or 14.7 nor the yield stress are good predictors in this range. In fact, they indicate greater than actual strength. Unfortunately, most economical columns are in this intermediate range. Therefore, designers must adhere very closely to specific codes for specific materials rather than rely upon theoretical relationships. These codes are supported by extensive testing, more sophisticated analysis, and much experience.

14.9 WOOD COLUMNS

As an example of a column-design code, the *Timber Construction Manual* gives the following formula for *simple solid rectangular columns.*[2]

$$\sigma_{\text{allow}} = \frac{0.30E}{(L/d)^2} \qquad (14.8)$$

where σ_{allow} is the allowable stress in compression parallel to grain, E is the modulus of elasticity, L is the overall unsupported length of the columns, and d is the dimension of the least side of the rectangular column.

Earlier timber codes allow no credit for fixity at the ends. A fixed end in such cases is considered to have the same effective length as a pinned end. An upper limit of 50 is set by the code for the L/d ratio of timber columns. σ_{allow} may in no case be greater than that allowable for direct compression parallel to the grain.

ILLUSTRATIVE PROBLEM

14.4 A truss roof for a church is to be supported on decorative timber columns. The roof is to be designed for 110 lb/ft^2. The columns will be spaced at 40-ft intervals and are 25 ft in length. The square columns are to be laminated dense select douglas fir with all grains in the vertical direction. What should be the cross-sectional dimension of the columns, if they are to be designed according to code?

Solution:
Each column must support a load of 40 ft $\times$ 40 ft $\times$ 110 lb/ft^2 = 176 kips. For dense select douglas fir, $E = 1.76\text{E}6$ psi. Trying a 10-in.-square column, the compression stress is

$$\sigma = \frac{176 \text{ kips}}{100 \text{ in.}^2} = 1\,760 \text{ psi}$$

From relationship 14.8, the allowable stress can be computed.

$$\frac{L}{d} = \frac{25 \text{ ft}}{10 \text{ in.}} \times \frac{12 \text{ in.}}{1 \text{ ft}} = 30$$

$$\sigma_{\text{allow}} = \frac{0.30 \times 1.76E6 \text{ psi}}{(30)^2} = 586 \text{ psi}$$

This column is too small. The compression stress on the column is greater than the allowable stress. Trying a 12-in. column,

$$\sigma = 1\ 222 \text{ psi}$$
$$L/d = 25$$
$$\sigma_{\text{allow}} = 845 \text{ psi}$$

By increasing the column size, the compression stress is reduced and the allowable stress is increased. However, the compression stress is still greater than the allowable.

Trying a 14-in. column,

$$\sigma = 897 \text{ psi}$$
$$L/d = 2.14$$
$$\sigma_{\text{allow}} = 1\ 150 \text{ psi}$$

The 14-in. column has compression stress lower than the allowable stress. However, since this column is likely to be a non-standard item, there is no reason to confine the design to a stock column size since a laminated column can be made to suit.

The required dimension can be calculated by substituting F/d^2 for the stress in 14.8, and solving for d,

$$\frac{F}{d^2} = \frac{0.30E}{(L/d)^2}$$

$$d = \sqrt[4]{\frac{FL^2}{0.30E}}$$

$$d = \sqrt[4]{\frac{180 \text{ kips} \times 25^2 \text{ ft}^2}{0.30 \times 1.76E6 \text{ psi}} \times \frac{144 \text{ in.}^2}{1 \text{ ft}^2}}$$

$$d = 13.23 \text{ in.}$$

The earlier, trial-and-error method is called an *iterative solution*. The latter is called a *closed form* solution.

14.10 STRUCTURAL STEEL COLUMNS

An example of a code for structural steel columns is that of the American Institute of Steel Construction.[3] The Code design of structural steel columns includes formulas for both long and intermediate columns.

Long Columns

Columns are defined as long if they have a slenderness ratio

$$\frac{L}{q} \geq \sqrt{\frac{2\pi^2 E}{\sigma_{\text{yield}}}} \tag{14.9}$$

where σ_{yield} is the yield strength and E is the elastic modulus. Long columns are those with an L/q ratio greater than 126.1 for 36-ksi-yield steel and 107.0 for 50-ksi-yield steel.

The allowable stress for long columns is the critical stress as determined from elastic analysis (relationship 14.7) with a safety factor of 23/12 established by Code.

$$\sigma_{\text{allow}} = \frac{12\pi^2 E}{23(L/q)^2} = \frac{12}{23}\sigma_{\text{c}} \tag{14.10}$$

Short Columns

For very short columns (those with slenderness ratios approaching zero), the allowable stress is 60 percent of the yield stress.

Intermediate Columns

Between very short and long, the allowable stress is given by the following formula, written in terms of the Euler critical stress, σ_{c}, and the yield stress, σ_{yield}.

$$\sigma_{\text{allow}} = \frac{(1 - 1/2B^2)\sigma_{\text{yield}}}{5/3 + 3/8B - 1/8B^3} \tag{14.11}$$

where

$$B = \sqrt{\frac{\sigma_{\text{yield}}}{2\sigma_{\text{c}}}}$$

The above relationship covers slenderness ratios from zero to

$$\frac{L}{q} = \sqrt{\frac{2\sigma_{\text{c}}}{\sigma_{\text{yield}}}}$$

where relationship 14.10 for long columns takes over.[3]

These relationships are shown plotted for steels with minimum specified yield strengths of 36 000 psi and 50 000 psi in figure 14.10.

ILLUSTRATIVE PROBLEM

14.5 What is the safe allowable load for a W12 × 79 column of A36 structural steel with an unbraced length of 30 ft?

Solution:
A36 structural steel has a minimum specified yield strength, σ_{yield}, of 36 ksi. Appendix V(a) shows that the radius of gyration, q, for the W12 × 79 is 3.05 in the

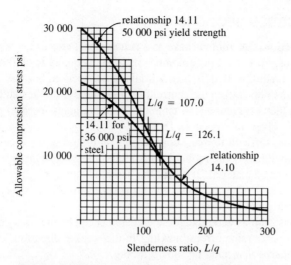

Figure 14.10 Allowable stress in compression for structural steel columns in accordance with AISC formulas.

most flexible direction. Therefore, the slenderness ratio

$$\frac{L}{q} = \frac{30 \text{ ft}}{3.05 \text{ in.}} \times \frac{12 \text{ in.}}{1 \text{ ft}} = 118.03$$

This result is less than the 126.1 that relationship 14.9 establishes as the limit for intermediate columns for A36 steel. Therefore, this is an intermediate column and relationship 14.11 applies.

Using relationship 14.7, the critical buckling stress, σ_c, for an ideal elastic material is computed taking E to be equal to 29 000 ksi.

$$\sigma_c = \frac{\pi^2 E}{(L/q)^2} = \frac{\pi^2 \times 29\text{E}6 \text{ ksi}}{(118.03)^2} = 20.545 \text{ ksi}$$

The factor B is calculated,

$$B = \sqrt{\frac{\sigma_y}{2\sigma_c}} = \sqrt{\frac{36 \text{ ksi}}{2 \times 20.545 \text{ ksi}}} = 0.936$$

Formula 14.11 is used to calculate the allowable stress.

$$\sigma_{\text{allow}} = \frac{(1 - 1/2 \times 0.936^2)36 \text{ ksi}}{(5/3 + 3/8 \times 0.936 - 1/8 \times 0.936^3)}$$

$$\sigma_{\text{allow}} = 10.000 \text{ ksi}$$

The cross-sectional area of the W12 × 79 is 23.2 in.^{2} Therefore, the safe allowable load is

$$23.2 \text{ in.}^2 \times 10.000 \text{ ksi} = 232 \text{ kips}$$

14.11 EFFICIENT COLUMN CROSS SECTIONS

Efficient cross sections for columns have as much material area as far as possible from the centroid of the cross section. If the column is restrained equally in all directions, the most efficient shape is the thin-walled, hollow, circular cross section, because all the material is effectively the same radius from the centroid. Consequently, of the standard shapes available, pipe makes the best columns. A table of allowable loads for steel pipe columns per the AISC Code is given in Appendix IX.

It is sometimes difficult to make connections on pipe. For these cases, square structural tubing is available. Square tubing is only slightly less resistant to buckling than pipe for the same weight of material. It offers the same advantages but is likely to be more expensive. Allowable load data is also given in Appendix IX for selected square tubing shapes of structural steel.

Open sections such as wide flanges and channels can be the most economical choice when the column is restrained from buckling in the weaker direction. Open sections are usually less expensive than pipe or square tubing.

Solid sections are often used in timber construction and sometimes in machinery. In solid sections, round is preferable to square, but both are efficient for laterally unrestrained columns.

ILLUSTRATIVE PROBLEM

14.6 The members of the water tank truss of illustrative problem 7.3, shown again in figure 14.11, are to be made of A36 steel pipe. What sizes should be specified for the members?

Solution:
Using the results of illustrative problem 7.3, the compression force on the leg of the water tank from E to F is the same as the reaction force at F, 37.5 kips. This result is due to the weight of the tank and contents as well as the wind blowing from the left. The effect of the wind force will make the compressive load most critical on that part of the leg EF. Since the wind could blow from any direction, all legs should be designed for 37.5 kips. Although the load can be shown to be less in those parts of the legs in the upper cell, it is probably not worthwhile to make the upper part of the leg of a lighter pipe. Therefore, the legs will be 40-ft lengths of standard-weight pipe. Due to the cross bracing, the effective length is 20 ft. From the table in Appendix IX for standard-weight pipe, 6-in. pipe has an allowable load of 67 kips. Four-in. pipe has an allowable load of only 19 kips, which is insufficient. Six-in. pipe should be specified for the legs.

The diagonal turnbuckle rods should be designed only for tension, since they should buckle and relieve themselves of load when in compression. The horizontal cross-member BE is also a compression member, but its specification is left as an exercise.

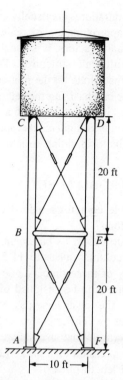

Figure 14.11 Illustrative problem 14.6.

14.12 COLUMNS OF OTHER MATERIALS

This chapter has considered the theoretical Euler buckling analysis. From that perspective, Code approaches used for timber and structural steel were presented and examined. But what about other materials?

Reinforced concrete is a very common and useful column material. There is a well-developed code for its use along with a substantial base of testing and experience. Concrete columns are not included here, because they require more discussion than can be devoted to them in a first comprehensive look at structures.

Much data is available in the literature for columns of other materials such as aluminum and magnesium. However, rational design involves not only the consideration of the material but also the situation in which the material is being applied. Except for standard building and static structures of wood, structural steel, and concrete, the design of a critical column requires specific experience or, lacking that, an extensive program of testing and analysis.

A good experimental program should simulate the extremes of loads, loading rate, vibration, environmental conditions, column sizes, eccentricity, lateral loads, and manufacturing or fabrication inaccuracy that will be encountered in reality. Safety factors must be used that are consistent with the degree of consequences should failure occur. One must be always aware that buckling failures are usually catastrophic, sudden, and without warning. The test is always an approximation of reality, not reality itself. In column testing, due to the influences of many factors, it is usually not valid to extrapolate information from miniature tests to full-scale structures. The simplified Euler relationship 14.5, although it can be grossly inaccurate on the non-conservative side, is nevertheless useful for getting in range and for emphasizing the relative importance of length, elastic modulus, and moment of inertia. In the absence of a design code that systematically includes the experiences of previous designers, the specification of a column or axial compression member for a machine or building structure is a development rather than a design project.

PROBLEM SET 14.2

14.7 What is the Code-allowable compression load on a 4-in. × 4-in. post of dense select southern pine 8 ft long?

14.8 An old house is to be restored. In its basement are redwood support timbers in excellent condition. The columns measure 11.5 by 11.5 in. and are 6 ft long. Estimate the load capacity of these timbers.

14.9 A tower for fire observation is to be designed of dense select douglas fir. Under expected wind conditions, the tower legs may experience their maximum design load of 92 kips in compression. The effective length of the legs is reduced to 24 ft due to cross-diagonal bracing. What size square timbers must be used?

14.10 What is the allowable compression load on a W10 × 39 A36 structural steel column with an unsupported length of 16 ft?

14.11 Using figure 14.10, estimate the maximum unsupported column length for a W6 × 16 structural shape of 50 000-psi-yield-strength steel, in order to have an allowable stress of 15 000 psi.

14.12 What size standard structural steel pipe must be used for a column having an unsupported length of 35 ft that is to carry a concentric compressive load of 102 kips.

14.13 A set of architectural columns are to support a portico in a shopping center. The unsupported column length is 12 ft. Specify a square structural steel tubing for these columns for a concentric compression design load of 120 kips.

14.14 The highway sign truss shown in figure 14.12 was treated earlier in problems 7.1, 7.22, and 7.23 of chapter 7. If the upper horizontal members are to be a continuous standard weight A36 structural steel pipe, specify the pipe size.

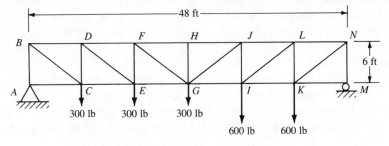

Figure 14.12 Problem 14.14. (Also problems 7.1, 7.22, and 7.23.)

SUMMARY

☐ *Buckling* is bending due to an axial load.

☐ The *critical load,* also called the Euler buckling load, is: ~~Pg 421~~

(14.5) $F_c = \dfrac{\pi^2 EI}{L^2}$

☐ The buckling shape is a sine wave. The *effective length* of a column is the length of one lobe of a sine wave that can be passed through the supports.   423

☐ The *radius of gyration* is a property of the column cross section defined by: 427

(14.6) $q = \sqrt{\dfrac{I}{A}}$ *index for comparing relative slenderness Pg 428*

☐ The *slenderness ratio* is the effective length divided by the radius of gyration. *dimensionless*

☐ The *critical stress* is the critical load divided by the cross-sectional area. It is given in terms of the slenderness ratio by the following:

(14.7) $\sigma_c = \dfrac{\pi^2 E}{(L/q)^2}$

☐ The allowable stress for a rectangular wood column is:

(14.8) $\sigma_{allow} = \dfrac{0.30E}{(L/d)^2}$

☐ The allowable stress for structural steel columns are given by the AISC code formulas and tables such as in Appendix IX.

☐ Columns must be designed using codes developed specifically for the material and the uses intended.

REFERENCES

1. Timoshenko, S. and Gere, J. M. *Theory of Elastic Stability.* New York: McGraw-Hill, 1961.

2. American Institute of Timber Construction. *Timber Construction Manual.* New York: John Wiley & Sons, 1966.

3. American Institute of Steel Construction. *Manual of Steel Construction.* New York: AISC, 1978.

CHAPTER FIFTEEN

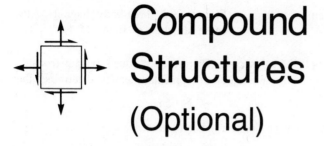

Compound Structures
(Optional)

This chapter ties together the information and techniques developed in all previous chapters for single kinds of loading on members. This chapter applies this information to members having compound loading. Tension, compression, cross-shear, pressure, and bending stresses may be combined on a single member. Mohr's Circle is used to resolve and add the directional components of various kinds of stress. Since technologists are usually accustomed to moving from the specific to the general, Mohr's Circle is derived after the student has become familiar with its use. The Tresca failure criteria are presented so that either the principal stresses or the maximum shear stress can be used to design or rate a structure. Illustrative problems and problems from both civil and mechanical technology are included.

15.1 ECCENTRICALLY LOADED MEMBERS

In other chapters, the problems addressed only structural members experiencing a single loading such as tension, compression, torsion, or bending. In actuality, many structural members are loaded in several ways at the same time. In these cases, the stresses due to each loading combine to produce a net effect upon the member.

As an example, consider the structural member *ABC*, a solid rectangular bar, as shown in figure 15.1. The bar has a jog at *B* that causes it to be eccentrically loaded, because the axis of the force at *C* does not coincide with the axis of the cross section at *A*.

It carries a single load of 80 N at point *C*, which produces both a bending moment, *M*, and a horizontal force, *F*, at the support point, *A*.

The structure is a general planar force system. For equilibrium, the summations of horizontal forces, vertical forces, and moments on the free-body diagram (figure 15.2) must each be zero.

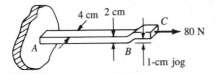

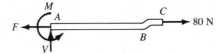

Figure 15.1 A structural member with a combination of loading.

Figure 15.2 Free-body diagram of the entire structure of figure 15.1.

No vertical forces are applied in this case. Therefore, V equals 0. But from the summation of horizontal forces and the summation of moments, it can be determined that

$$F = 80 \text{ N} \quad \text{and} \quad M = 80 \text{ N} \cdot \text{cm}$$

The force, F, is a tension force that produces a tension stress, σ_t.

$$\sigma_t = \frac{F}{A} = \frac{80 \text{ N}}{8 \text{ cm}^2} \times \left(\frac{100 \text{ cm}}{1 \text{ m}}\right)^2 = 0.100 \text{ MPa}$$

The moment is actually a bending moment. From relationship 9.2, the bending stress is

$$\sigma_b = \frac{6M}{bh^2} = \frac{6 \times 80 \text{ N} \cdot \text{cm}}{4 \text{ cm} \times (2 \text{ cm})^2} = 0.300 \text{ MPa}$$

The states of stress for the two conditions are shown in figure 15.3. The stress due to the tension load, F, is uniform over the cross section at the support point, A. The bending stress, on the other hand, is linearly distributed over the cross section from a maximum value of 0.300 MPa tension at the top to 0.300 MPa compression at the bottom. The stress distributions in this case are additive because they are *normal stresses;* that is, tension or compression. The two stresses are in the same horizontal directions. Therefore, the total stress goes linearly from 0.400 MPa tension at the top of the cross section to 0.200 MPa tension at the bottom.

In this example, the single 80-N load was broken down into two different effects. Then the stresses due to the two effects were combined to give the total effect of the load. The recombination could be done by addition in this case because the stresses of both loadings were normal stresses in the same direction. In other cases, the stresses may be in different directions, and shear stress as well as normal stresses may be involved. Since

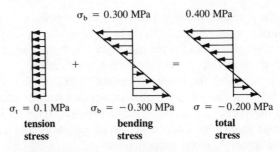

Figure 15.3 Stress at point A due to tension and bending on the structural member shown in figure 15.1.

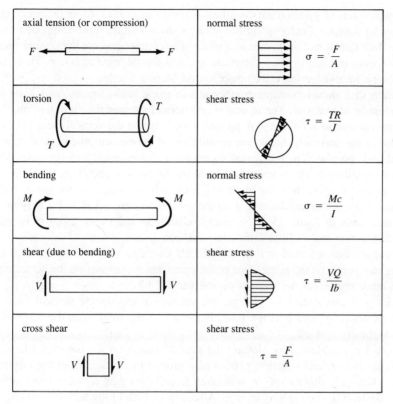

axial tension (or compression)	normal stress
$F \longleftarrow \qquad \longrightarrow F$	$\sigma = \dfrac{F}{A}$
torsion	shear stress
T ... T	$\tau = \dfrac{TR}{J}$
bending	normal stress
M ... M	$\sigma = \dfrac{Mc}{I}$
shear (due to bending)	shear stress
V ... V	$\tau = \dfrac{VQ}{Ib}$
cross shear	shear stress
V ... V	$\tau = \dfrac{F}{A}$

Figure 15.4 Summary of basic loading conditions and resulting stress under elastic action.

handling such combinations is not obvious from previous chapters, this chapter on combined stresses is presented.

To begin, it is useful to review the various kinds of stress patterns produced by different basic loadings. The basic loadings are axial tension or compression, cross shear, torsion, and bending. The conditions and resulting stresses under elastic action are summarized in figure 15.4.

15.2 MOHR'S CIRCLE AND THE DIRECTIONAL COMPONENTS OF STRESS

Such quantities as pressure, temperature, and mass are *scalar* in that they have magnitude only and are independent of direction. Force, displacement, and velocity are *vector quantities*. Vector quantities have directional properties as well as magnitude. In a plane, the directional effects of vector quantities can be resolved using the parallelogram law. A vector such as force can be described in a plane by two components.

Stress has directional properties, but they cannot be resolved using the parallelogram law. Many quantities in structural technology such as stress, strain, curvature, moment of inertia, and radius of gyration are classified as *tensor quantities*. More precisely, they are second-order tensors. Tensor quantities such as stress require three components to fully describe their directional properties in a plane. The components are the normal stresses in the x direction, σ_x, and in the y direction, σ_y, and the shearing stress, τ. These components can be resolved using a technique called Mohr's Circle.

Figure 15.5 shows the three components of plane stress, σ_x, σ_y, and τ, on a small square element of material. The σ_x and σ_y are normal stresses shown in tension. The τ is shear stress shown to act along with its mate clockwise on the vertical faces and counterclockwise on the horizontal faces. For equilibrium of the element, the vertical shears must be equal and opposite. The horizontal shears must also be equal and opposite. Also, for rotational equilibrium, the moment produced by the pairs of shears on the vertical faces must be equal and opposite to the moment produced by the shears on the horizontal faces. All of these conditions can be met if all the face shears are equal and if they are in the directions shown in figure 15.5. Normally, clockwise shears are taken to be positive, although direction makes little difference in shear.

These stresses are used to form the Mohr's Circle as shown in figure 15.6 by constructing two points on the graph. One point represents the stresses on the horizontal faces and the other represents the stresses on the vertical faces.

Point 1 is established by plotting the stresses acting on the vertical faces of the cubical elements, σ_x and positive τ. σ_x is plotted to the right from the origin and τ is plotted vertically upward.

Point 2 is established by plotting the stresses acting on the horizontal faces of the cubical element, σ_y and negative τ. σ_y is *also* plotted to the right from the origin, and τ is plotted vertically downward. σ_x was taken to be larger than σ_y when constructing this Mohr's Circle. The reverse may be true. Also, one or both of the normal stresses, σ, may be negative, in which case they would be plotted to the left rather than to the right on the normal stress axis.

Both points, 1 and 2, lie on the circumference of the Mohr's Circle. The Mohr's Circle is always centered on the normal stress axis, but it may lie anywhere along the axis, even totally or partially negative. The center of the circle is found by drawing a line

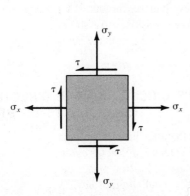

Figure 15.5 The three components, σ_x, σ_y, and τ, of plane stress on an element of material.

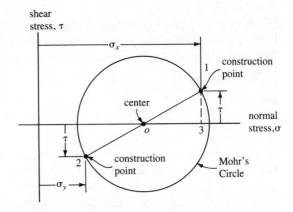

Figure 15.6 The Mohr's Circle for the element shown in figure 15.5.

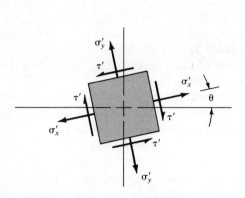

Figure 15.7 The three components of stress, σ'_x, σ'_y, and τ', when the element is rotated by the angle θ.

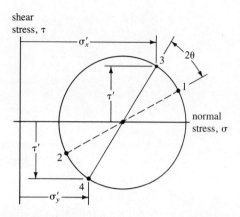

Figure 15.8 Points 3 and 4 on the Mohr's Circle represent the stresses on the faces of the rotated element. Note that the angles on the Mohr's Circle are twice those on the element.

between points 1 and 2. The line will intersect with the normal stress axis at the center of the circle. The center of the Mohr's Circle will always lie on the normal stress axis and have the value of the average of the two normal stresses.

(a)
$$\sigma_{ave} = (\sigma_x + \sigma_y)/2$$

The radius of the Mohr's Circle can be found by using the Pythagorean theorem on triangle $o13$.

(b)
$$R = \sqrt{\left(\frac{\sigma_x - \sigma_y}{2}\right)^2 + \tau^2}$$

The circle of radius R is then constructed from the center at σ_{ave}.

One can use this Mohr's Circle to determine the stresses in any direction. Suppose that the element were to be rotated by the arbitrary angle θ, as shown in figure 15.7. The face stresses in the new direction are called σ'_x, σ'_y, and τ'. Their values are indicated by points 3 and 4 on the Mohr's Circle as shown in figure 15.8. Points 3 and 4 are on a diametral line rotated *twice* the angle θ, in the same direction. Angles on the Mohr's Circle are always twice those on the element. The angle is measured from the diametral line on which lie points 1 and 2. The values of σ'_x, σ'_y, and τ' can be measured from the graph or calculated using geometry. This technique is demonstrated in the following illustrative problem.

ILLUSTRATIVE PROBLEM

15.1 A combination of loadings on a member of a structure produces at a critical location a vertical tension stress of 210 MPa, a horizontal compression stress of 50 MPa, and a shear stress of 80 MPa. From a Mohr's Circle plot, determine graphically the stresses at that point on a coordinate axis rotated 25° clockwise.

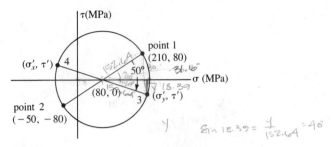

Figure 15.9 Illustrative problem 15.1.

Solution:

No information is given on the direction of τ. Therefore, it is arbitrarily taken to be positive on the horizontal faces. Point 1 is plotted at (210, 80) in figure 15.9, representing the tension stress of 210 MPa and the shear stress of 80 MPa. Point 2 is plotted at (−50, −80), representing a compression stress of 50 MPa and a shear stress of −80 MPa.

A line drawn from point 1 to point 2 intersects the horizontal axis at (80, 0). This intersection is the center of the Mohr's Circle. A circle through points 1 and 2 is drawn about that center. It has a radius of 152.6 MPa. This is the Mohr's Circle for the state of stress at this particular point on the member of the structure.

To find the components of stress at 25° clockwise, a new diameter is drawn rotated $2 \times 25° = 50°$ clockwise. Points 3 and 4, which are the intersections of the new diameter, represent the stresses on the rotated coordinate axis.

The vertical coordinate at point 3 is the shear stress,

$$\tau' = -48 \text{ MPa}$$

The horizontal distance at point 3 is the normal stress,

$$\sigma'_y = 225 \text{ MPa}$$

The vertical distance at point 4 is the shear stress,

$$\tau' = 48 \text{ MPa}$$

The horizontal distance at point 4 is the normal stress,

$$\sigma'_x = -65 \text{ MPa}$$

The plane stress components oriented 25° clockwise from the original set of components are

$$\sigma'_x = 65 \text{ MPa compression}$$
$$\sigma'_y = 225 \text{ MPa tension}$$
$$\tau' = 48 \text{ MPa}$$

15.3 PRINCIPAL STRESSES

Under any combination of stresses, it is possible to rotate the pair of axes to some angle where there is no shear stress. These are called the principal axes. The two normal stresses along these axes are called the *principal stresses*. The principal stresses are indicated by the $\overline{\sigma}_x$ and $\overline{\sigma}_y$. By the geometry of the Mohr's Circle, the following relationship from the principal stresses can be written from relationships (a) and (b).

$$\overline{\sigma}_x = \frac{(\sigma_x + \sigma_y)}{2} + \sqrt{\left(\frac{\sigma_x - \sigma_y}{2}\right)^2 + \tau^2}$$ **(15.1)**

and

$$\overline{\sigma}_y = \frac{(\sigma_x + \sigma_y)}{2} - \sqrt{\left(\frac{\sigma_x - \sigma_y}{2}\right)^2 + \tau^2}$$

The use of Mohr's Circle to determine the principal stresses is demonstrated in the next illustrative problem.

ILLUSTRATIVE PROBLEM

15.2 Due to a combination of bending and torsion on a shaft, a tension stress, σ_x, of 22 ksi is developed parallel to the axis of the shaft along with a shear stress, τ, of 15 ksi in a plane perpendicular to the axis. No stress is developed in the σ_y direction. What are the principal stresses? What is the angle between the original tension stress axis and the principal stress axis?

Solution:

The Mohr's Circle is constructed in figure 15.10. Point 1 has a normal stress of +22 ksi and a shear stress, taken positive, of 15 ksi. Point 2 has a normal stress, σ_y of zero. But, a shear stress must exist on that face, and it must be of the opposite sign. The point is, therefore, plotted at (0, −15 ksi).

A diametral line is constructed between the two points, intersecting the horizontal axis at (11, 0). A circle of radius 18.60 is drawn about that center through the points 1 and 2.

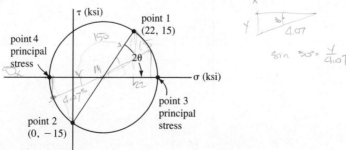

Figure 15.10 Mohr's Circle for illustrative problem 15.2.

The principal stresses are points 3 and 4 on the normal stress axis. This is true because principal stresses are those associated with shear stresses of zero.

The angle 2θ can be measured from the Mohr's Circle to be 54°, or the angle can be computed using trigonometry. The angle of rotation to the principal axis is 27°.

The principal stresses are the normal stresses given by points 3 and 4. They can be measured to be

$$\overline{\sigma}_x = 29.6 \text{ ksi}$$
$$\overline{\sigma}_y = -7.6 \text{ ksi}$$

They can also be calculated from relationships 15.1. Relationships 15.1 represent the horizontal distance to the center, σ_{ave}, plus R to give $\overline{\sigma}_x$ and minus R to give $\overline{\sigma}_y$.

15.4 MAXIMUM SHEAR STRESS

Maximum shear stress occurs when the axes are rotated 45° from the axes of principal stresses. In ductile materials, which fail primarily by shear, maximum shear stress is critical. Figure 15.11 shows the principal stress points as well as the maximum shear stress points on the Mohr's Circle.

It is obvious that the maximum shear stress has a value equal to the radius of the Mohr's Circle. This value can be obtained for any set of σ_x, σ_y, and τ from relationship (b).

Max Shear stress

$$\tau_{max} = \sqrt{\left(\frac{\sigma_x - \sigma_y}{2}\right)^2 + \tau^2} \qquad (15.2)$$

Maximum shear stress is usually accompanied by normal stresses in both directions. The normal stresses may be either positive or negative, but they have the same sign and are

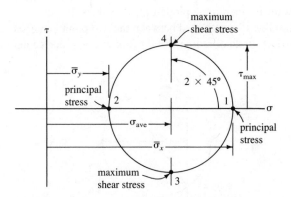

Figure 15.11 Location of the maximum shear stress points and the principal stress points on the Mohr's Circle.

equal in magnitude. They have the average value of the two principal stresses. Repeating relationship **(a)**,

$$\sigma_{ave} = (\sigma_x + \sigma_y)/2 \tag{15.3}$$

ILLUSTRATIVE PROBLEM

15.3 A tensile test of a specimen of 4140 cold-rolled steel shows that the yield strength is 687 MPa. Use Mohr's Circle to determine the shear stress at which yield occurs.

Solution:
A tensile test subjects an element of the metal to a stress in the axial direction, σ_y. No shear stress, τ, or lateral stress, σ_x, is applied. In this case,

$$\sigma_y = 687 \text{ MPa}, \quad \tau = 0, \quad \text{and} \quad \sigma_x = 0$$

Because no τ is applied, σ_y and σ_x in this case are the principal stresses.

Point 1 on the Mohr's Circle is at coordinates (0, 0). Point 2 is at coordinates (687, 0). The Mohr's Circle can be constructed as shown in figure 15.12.

Points 3 and 4 are the points of maximum shear stress on the Mohr's Circle. This circle shows that the maximum shear stress has a value exactly equal to half the tension stress, 45° from the tension axis. This result confirms the discussion of chapters 2 and 3 about the relationship between tension and shear testing.

$$\tau_{max} = 344 \text{ MPa}$$

at 45° from the tension axis.

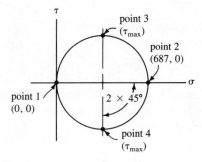

Figure 15.12 Illustrative problem 15.3. The Mohr's Circle for a tension test.

PROBLEM SET 15.1

15.1–15.9
Determine the maximum shearing stress and the principal stresses for the elements stressed as shown in figures 15.13 through 15.21. Also determine the angle clockwise from the y axis to the y' axis of the principal axes. Draw the Mohr's Circle to scale and show the appropriate values on the circle. (Programmable calculators are excellent for relationships 15.1, 15.2, and 15.3.)

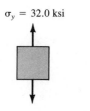

Figure 15.13 Problems 15.1 and 15.17.

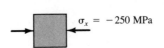

Figure 15.14 Problems 15.2 and 15.16.

Figure 15.15 Problems 15.3 and 15.16.

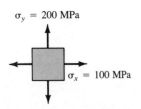

Figure 15.16 Problems 15.4 and 15.16.

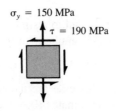

Figure 15.17 Problems 15.5 and 15.16.

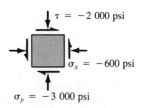

Figure 15.18 Problems 15.6 and 15.18.

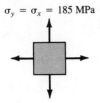

Figure 15.19 Problems 15.7 and 15.16.

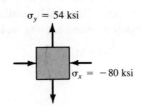

Figure 15.20 Problems 15.8 and 15.19.

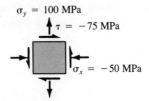

Figure 15.21 Problems 15.9 and 15.16.

15.10–15.15

For the elements stressed as shown in figures 15.22 through 15.27, sketch the Mohr's Circle to scale. Using geometry, determine the three components of stress, σ_x', σ_y', and τ' in the directions shown.

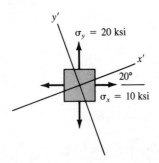

Figure 15.22 Problem 15.10.

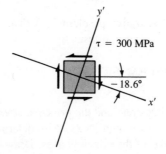

Figure 15.23 Problem 15.11.

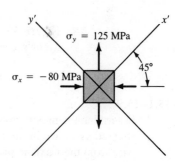

Figure 15.24 Problem 15.12.

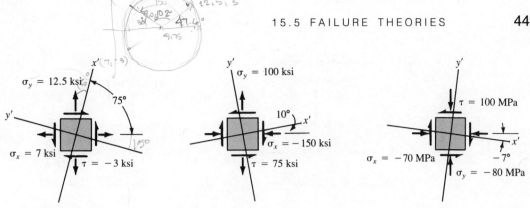

Figure 15.25 Problem 15.13.   **Figure 15.26** Problem 15.14.   **Figure 15.27** Problem 15.15.

15.5 FAILURE THEORIES

The earliest failure theory was formulated by Galileo. It states that material fails when maximum stress reaches failure stress. The present form of that theory is called the *maximum principal stress theory* when stresses occur in two directions. The failure stress could be either yield strength or ultimate strength in accordance with the material and application. The maximum principal stress theory agrees with tests on brittle materials when the ultimate tension or the ultimate compression strengths are used. It also accurately describes ductile materials when the $\overline{\sigma}_x$ and $\overline{\sigma}_y$ are both tension or both compression. It does *not* accurately describe ductile materials when the $\overline{\sigma}_x$ and $\overline{\sigma}_y$ are of opposite signs: one compression, the other tension. When the signs are opposite, the lateral stress tends to intensify the effect of the axial stress.

The *maximum shearing stress theory* states that material fails when the maximum shearing stress as resolved by Mohr's Circle exceeds the maximum shear stress that the material can withstand. This theory more accurately describes ductile materials when $\overline{\sigma}_x$ and $\overline{\sigma}_y$ are of different signs. Ductile materials tend to fail by shear. However, the maximum shearing stress theory predicts far greater strengths than actual when both stresses $\overline{\sigma}_x$ and $\overline{\sigma}_y$ are of the same sign.

It seems that a ductile material under a tension stress in one direction, $\overline{\sigma}_y$, is not reduced in strength by the presence of a lateral tension stress, $\overline{\sigma}_x$. Nor is a material with compression stress, $\overline{\sigma}_y$, reduced in strength by the presence of a lateral compression stress (figure 15.28a). However, a material with a compression stress, $\overline{\sigma}_y$, does seem to be reduced in strength by the presence of a lateral tension stress (figure 15.28b) and vice versa. The lateral stress of the same sign tends to stabilize the material at the slip plane where lateral stresses of opposite sign have a destabilizing effect.

A set of failure criteria, called the *Tresca criteria*, describes the situation accurately. These criteria are the following:

☐ When the stresses, $\overline{\sigma}_x$ and $\overline{\sigma}_y$, have the *same sign*, the maximum principal stress theory (15.4) applies,

$$\boxed{|\overline{\sigma}_x| \leq \sigma_{\text{yield}} \quad \text{and} \quad |\overline{\sigma}_y| \leq \sigma_{\text{yield}}}$$

(15.4)

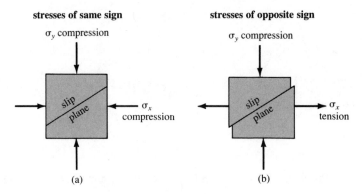

Figure 15.28   Model rationalizing the Tresca failure criteria.

☐   When the stresses, $\bar{\sigma}_x$ and $\bar{\sigma}_y$, have *opposite signs*, the maximum shear stress theory (15.5) applies,

$$|\tau_{max}| \leq (\sigma_{yield})/2 \qquad (15.5)$$

From Mohr's Circle,

$$\tau_{max} = (\bar{\sigma}_x - \bar{\sigma}_y)/2$$

Therefore,

$$\boxed{|\bar{\sigma}_x - \bar{\sigma}_y| \leq \sigma_{yield}} \qquad (15.6)$$

The *Tresca criteria* for yielding can be summarized as shown in figure 15.29. The I and III quadrants are maximum principal stress theory (relationship 15.4). The II and IV quadrants are maximum shear theory (relationships 15.5 and 15.6). A combination of

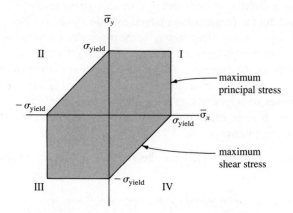

Figure 15.29   The Tresca yield criteria.

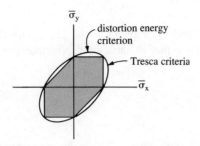

Figure 15.30 Comparison of the Tresca yield criteria and the distortion energy theory.

axial and lateral stress within the boundaries of the hexagon is not likely to produce yield. In design, a safety factor can be used with the Tresca criteria by using the allowable stress rather than the yield stress in relationships 15.4, 15.5, and 15.6.

A more accurate failure theory for ductile materials is the *distortion energy theory*. It states that failure will occur when the shear energy per unit volume is equal to the shear energy at failure in a tension test. The equation describing the failure boundary for the distortion energy theory in terms of the principal stresses is

$$\overline{\sigma}_x{}^2 + \overline{\sigma}_y{}^2 - \overline{\sigma}_x\overline{\sigma}_y \le \sigma_{\text{yield}}{}^2 \qquad \textbf{(15.7)}$$

The distortion energy theory is more refined than Tresca and fits experimental data better. However, the Tresca criteria come very close to the distortion energy theory. In figure 15.30, both are shown on the same graph.

Tresca, the more conservative of the two, is very close indeed, touching the distortion energy ellipse at six points. The Tresca criteria are most often used in preference to the distortion energy criteria.[2]

ILLUSTRATIVE PROBLEM

15.4 A combination of bending, pressure, and torsion on a pipe produces the element stresses shown in figure 15.31. The pressure stresses are 12 000 psi tension in the circumferential direction and zero in the longitudinal direction. The bending stress is 9 000 psi in the longitudinal direction. Bending stress is compressive on the top of the pipe and tensile on the bottom. If the allowable stress is 22 000 psi, what value of shear stress due to torsion is permissible?

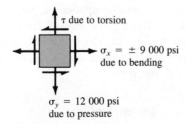

Figure 15.31 Illustrative problem 15.4.

Solution:

The Tresca criteria, relationships 15.4 and 15.5, with the allowable stress rather than the yield, mandate that to satisfy the principal stress criteria:

(a) $|\overline{\sigma}_x| \leq 22\ 000$ psi

and

(b) $|\overline{\sigma}_y| \leq 22\ 000$ psi

on the bottom of the pipe where all stresses are tension. Additionally, the shear stress criterion

(c) $\tau_{max} \leq 11\ 000$

must be satisfied on the top of the pipe, where bending stress is negative and the pressure stress is positive.

Principal Stresses:

Using the first of relationships 15.1 when the principal stress, $\overline{\sigma}_x$, is set equal to 22 000 psi,

$$22\ 000 = \frac{9\ 000 + 12\ 000}{2} + \sqrt{\left(\frac{9\ 000 - 12\ 000}{2}\right)^2 + \tau^2}$$

$$11\ 500^2 = (-1\ 500)^2 + \tau^2$$

$$\tau = \pm 11\ 400 \text{ psi}$$

The second relationship of 15.1, for $\overline{\sigma}_y$, yields the same result. This agreement is as it should be, because there is only one unique answer for the allowable shear stress due to torsion on the tension side of the pipe.

Maximum Shear Stress:

On the compression side of the pipe, condition **(c)** is used with relationship 15.2. Here, σ_x is negative.

$$11\ 000 = \sqrt{\left(\frac{-9\ 000 - 12\ 000}{2}\right)^2 + \tau^2}$$

$$11\ 000^2 = 10\ 500^2 + \tau^2$$

$$\tau = \pm 3\ 280 \text{ psi}$$

The maximum shear stress criterion applied on the compression side governs the torsion-carrying capacity. The significance of the plus or minus sign is that the torsion may be applied in either the clockwise or counterclockwise direction. But, the limiting value is the same. No more than 3 280 psi of shear stress due to torsion may be safely applied to the pipe.

PROBLEM SET 15.2

15.16 Draw a Tresca criteria boundary with the yield strength equal to 200 MPa. Put points on the diagram to represent the stress conditions shown in figures 15.14,

15.15, 15.16, 15.17, 15.19, and 15.21. Which of the points fall outside the boundary, thereby representing a situation in which yield is likely to occur?

In the following problems, if the state of the stress is not within the Tresca criteria, determine the *additional* increase in σ_x, σ_y, or τ stress, whichever is the smallest, that will bring the stress just up to the limit allowed. Draw the Tresca diagram, showing the point representing the original condition and the point representing the condition after the stress is increased.

15.17 Figure 15.13, allowable stress = 40 ksi.
15.18 Figure 15.18, allowable stress = 9 000 psi.
15.19 Figure 15.20, allowable stress = 100 ksi.
15.20 Repeat illustrative problem 15.4 using the distortion energy theory of failure.

15.6 THE DERIVATION OF THE MOHR'S CIRCLE FOR PLANE STRESS

The use of Mohr's Circle up to this point was presented without derivation. To rigorously treat this subject, we will now derive Mohr's Circle from its basic principles. Use was presented before derivation because it is easier to follow the derivation if one is first familiar with the quantities involved and the form of the results.

Consider again the basic element under the action of two normal stresses, σ_y and σ_x, in the axial and lateral directions, respectively, and the shear stress, τ. Suppose a plane is passed through the element at an angle θ from the horizontal as shown in figure 15.32a.

In this case the element is made rectangular so that the plane forms the diagonal of the rectangle. The base width of the element is the distance Δ. Assume that the element has a thickness, t, in the direction perpendicular to the plane of the diagram.

Figure 15.32b shows the free-body diagram of a triangular part of the element cut by the diagonal plane at angle θ. The stresses at the cutting plane are represented by a shearing stress, τ', in the direction parallel to the cutting plane, and by a normal stress, σ', normal to the cutting plane. The horizontal face of the triangular part is $t\Delta$, the vertical is $t\Delta\tan\theta$ and the diagonal is $t\Delta/\cos\theta$.

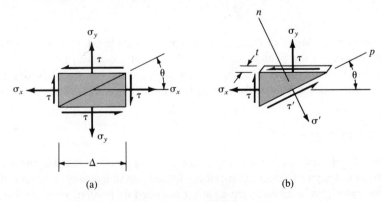

(a)   (b)

Figure 15.32 Free-body diagram of a portion of a stressed element for the purpose of deriving the Mohr's Circle relationship.

Summing the components in the normal direction,

$$\Sigma \, N: \quad \frac{-\sigma' t \Delta}{\cos \theta} + \sigma_y t \Delta \, \cos \theta + \tau t \Delta \, \sin \theta + \sigma_x t \Delta \, \tan \theta \, \sin \theta + \tau t \Delta \, \tan \theta \, \cos \theta = 0$$

Dividing through by $t\Delta$, multiplying through by $\cos \theta$, and moving σ' to the other side of the equal sign,

(a)
$$\sigma' = \sigma_y \cos^2 \theta + \sigma_x \sin^2 \theta + 2\tau \sin \theta \cos \theta$$
$$\sigma' = \sigma_y \cos^2 \theta + \sigma_x - \sigma_x \cos^2 \theta + 2\tau \sin \theta \cos \theta$$

Using the double-angle functions from trigonometry, $\sin (2\theta) = 2 \sin \theta \cos \theta$ and $\cos (2\theta) = 2 \cos^2 \theta - 1$, substituting for $\cos^2 \theta$ and $2 \sin \theta \cos \theta$, and then collecting terms,

(b)
$$\sigma' = \frac{\sigma_x + \sigma_y}{2} - \frac{\sigma_x - \sigma_y}{2} \cos (2\theta) + \tau \sin (2\theta)$$

Summing the components of force in the parallel direction,

$$\Sigma \, P: \quad \frac{\tau' t \Delta}{\cos \theta} - \sigma_x t \Delta \, \tan \theta \, \cos \theta + \sigma_y t \Delta \, \sin \theta + \tau t \Delta \, \tan \theta \, \sin \theta - \tau t \Delta \, \cos \theta = 0$$

Proceeding similarly,

(c)
$$\tau' = \sigma_x \sin \theta \cos \theta - \sigma_y \sin \theta \cos \theta - \tau \sin^2 \theta + \tau \cos^2 \theta$$

Using the double-angle function $\cos (2\theta) = \cos^2 \theta - \sin^2 \theta$,

(d)
$$\tau' = \frac{\sigma_x - \sigma_y}{2} \sin (2\theta) + \tau \cos (2\theta)$$

Relationships **(b)** and **(d)** together describe points on a circle, if σ' is plotted horizontally and τ' is plotted vertically. The center of the circle is on the horizontal axis at $(\sigma_x + \sigma_y)/2$. The circle has a radius of

$$\sqrt{\left(\frac{\sigma_x - \sigma_y}{2} \right)^2 + \tau^2}$$

which is the radius of Mohr's Circle. The angle between the horizontal axis and the radius of the circle going to the plotted point (σ', τ') is exactly equal to 2θ.

It can be concluded, therefore, that relationships **(b)** and **(d)**, derived using the basic principle of static equilibrium on the free-body diagram (figure 15.32b), describe Mohr's Circle.[1]

15.7 PULLING IT ALL TOGETHER

The Mohr's Circle and the Tresca criteria are the tools that permit the understanding, prediction, and design of structural members loaded simultaneously in several different ways. The majority of structures are clearly composed of tension, compression, shear, and bending elements.

Another large group of structures has members in which loading occurs in two different ways. However, the interaction of the two is not important. Examples are beams in which the critical shear stresses are normally not in the same place as the critical normal stresses due to bending. But there are a significant number of structures in which two or more different loading conditions interact in an important way. Usually these structures require a three-dimensional free-body diagram; although except in extreme cases, the critical stress patterns can be described with a two-dimensional Mohr's Circle and judged using plane-stress criteria. Several important cases are presented in the next three illustrative problems.

ILLUSTRATIVE PROBLEMS

15.5 The stainless steel gear drive shaft shown in figure 15.33 is subjected to bending due to a 12.5-kN gear load and a torsion of 0.752 kN · m. The allowable tensile stress for the ductile material is 125 MPa under the alternating loading conditions that occur. Specify the diameter of a solid shaft that will be sufficiently strong under the loads shown.

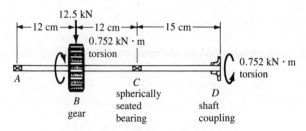

Figure 15.33 Illustrative problem 15.5. A shaft under combined bending and torsion loads.

Solution:
The critical bending moment will occur at point *B* in the shaft directly under the gear load. The gear load of 12.5 kN is developed because of the pressure angle of the gear teeth. As two meshing gears are torqued against one another, a force results which tends to separate the gears.

The spherically seated bearings behave as simple supports. The shaft coupling shows no traverse load. Several coupling styles are designed to transmit torque while offering insignificant traverse loads. Therefore, the shaft overhang (that portion between *C* and *D*) has no effect upon the bending moment in this case. The bending can then be described as a simply supported beam, spanning from bearing *A* to bearing *C*, with a concentrated load caused by the gear at mid-span. It is one of the four common cases of beam loading summarized in table 11.1. The maximum bending moment, which occurs at the gear, is given by

$$M = \frac{FL}{4} = \frac{12.5 \text{ kN} \times 24 \text{ cm}}{4} \times \frac{1 \text{ m}}{100 \text{ cm}}$$

$$M = 0.75 \text{ kN} \cdot \text{m}$$

The bending stress is given by

$$\sigma_B = \frac{M}{Z} = \frac{32M}{\pi d^3}$$

$$\sigma_B = \frac{32 \times 0.75 \text{ kN} \cdot \text{m}}{\pi d^3} = \frac{7.639 \text{ kN} \cdot \text{m}}{d^3}$$

The bending stress is tension at the bottom of the shaft and compression at the top.

Torsion:
The portion of the shaft from B to D transmits the torque of 0.752 kN · m from the gear to the coupling. The shear stress due to the torsion is given by relationship 5.11,

$$\tau = \frac{16T}{\pi d^3} = \frac{16 \times 0.752 \text{ kN} \cdot \text{m}}{\pi d^3}$$

$$\tau = \frac{3.830}{d^3} \text{ kN} \cdot \text{m}$$

This maximum shear stress occurs around the outside surfaces of the shaft from point B to D. At B it interacts with the maximum bending stress. They must be considered together using Mohr's Circle.

Mohr's Circle and the Failure Criteria:
From Mohr's Circle, the maximum shear stress is given by relationship 15.2,

$$\tau_{max} = \sqrt{\left(\frac{\sigma_x + \sigma_y}{2}\right)^2 + \tau^2}$$

Taking the y direction to be in the direction of the axis of the shaft,

$$\sigma_x = 0 \quad \text{and} \quad \sigma_y = \sigma_B = \pm 7.639/d^3 \text{ kN} \cdot \text{m}$$

Since σ_y will be squared, it makes no difference whether the plus or minus value is used.

$$\tau_{max} = \sqrt{\left(\frac{0 + 7.639/d^3}{2}\right)^2 + (3.830/d^3)^2} \text{ kN} \cdot \text{m}$$

$$\tau_{max} = \frac{5.409}{d^3} \text{ kN} \cdot \text{m}$$

The allowable shear stress will be one-half the allowable tensile stress from relationship 15.5.

$$\frac{125}{2} \text{ MPa} = \frac{5.409}{d^3} \text{ kN} \cdot \text{m}$$

$$d = \sqrt[3]{\frac{5.409 \text{ kN} \cdot \text{m}}{62.5 \text{ MPa}} \times \frac{1 \text{ MPa}}{1\,000 \text{ kN/m}^2}}$$

$$d = 4.42 \text{ cm}$$

The shaft must be 4.42 cm in diameter. However, the detail design becomes important because the gear must be fastened to the shaft. This attachment can be done by keys, shoulders, force fits, or even cements. Stress concentrations may be present. It is more accurate to state that the shaft may not have a diameter less than 4.42 cm at point B.

15.6 A high-pressure pipe in an oil refinery is to be supported on rollers so that it is free to expand and contract during temperature changes. The rollers are to be equally spaced. The A53 carbon-steel pipe has a Code-allowable stress of 12 ksi for the temperature range expected. Twelve-in. standard-weight pipe satisfies the design pressure of 600 psi. If, according to Code, other primary bending conditions may raise the stress by no more than 50 percent over the allowable, what is the maximum uniform spacing of the rollers considering that the pipe must support its own weight, contents, and insulation (a total of 110 lb/ft)?

Solution:
The bending conditions are shown in figure 15.34. The bending moment is maximum at the supports. Each span can be idealized as a beam of length L that has built-in ends and a uniformly distributed load over its entire length. The maximum bending moment is given by relationship 13.6,

$$M = \frac{wL^2}{12}$$

The stress due to this bending moment is

$$\sigma_B = \frac{M}{Z} = \frac{wL^2}{12Z}$$

The section modulus for 12-in. standard-weight pipe is 43.8 in.^{3}

$$\sigma_B = \frac{110 \text{ lb/ft} \times L^2}{12 \text{ in./ft} \times 12 \times 43.8 \text{ in.}^3} = 0.017\ 44L^2 \text{ psi}$$

where L is in inches.

The bending stress is tension at the top over the roller support and compression at the bottom.

Pressure:
The circumferential stress in a cylinder as given by relationship 8.13,

$$\sigma_\theta = \frac{PR}{t} = \frac{600 \text{ psi} \times 6 \text{ in.}}{0.375 \text{ in.}}$$

$$\sigma_\theta = 9\ 600 \text{ psi}$$

12-in. SCH40 pipe with insulation and contents (110 lb/ft)

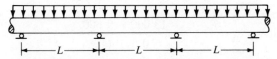

Figure 15.34 Illustrative problem 15.6. A pipeline under combined pressure and bending.

The longitudinal stress is given by relationship 8.12,

$$\sigma_\theta = 4\ 800 \text{ psi}$$

Both pressure stresses are tension.

Mohr's Circle and the Failure Criteria:
The longitudinal pressure stress and the bending stress are both in the y direction. At the top of the pipe,

$$\sigma_y = \sigma_\theta + \sigma_B$$

(a)
$$\sigma_y = 4\ 800 + 0.017\ 44L^2$$

At the bottom, where the bending stress is compressive,

(b)
$$\sigma_y = 4\ 800 - 0.017\ 44L^2$$

The circumferential stress acts around the pipe and can be taken as in the x direction. Therefore,

(c)
$$\sigma_x = \sigma_\theta = 9\ 600 \text{ psi}$$

If stress at the top of the pipe governs, the maximum principal stress must be considered, since both σ_x and σ_y have the same sign. As the length of the span is increased, σ_y increases. Setting relationship **(a)** equal to 1.5 times the allowable, as per Code in the statement of the problem

$$4\ 800 + 0.017\ 44L^2 = 1.5 \times 12\ 000$$

$$L = 870 \text{ in., or } 72.5 \text{ ft}$$

If the stress at the bottom governs, the maximum shear stress must be considered, since σ_x and σ_y will have opposite signs. From relationship 15.2,

$$\tau_{max} = \sqrt{\left(\frac{\sigma_x - \sigma_y}{2}\right)^2 + \tau^2}$$

In this case $\tau = 0$, $\sigma_x = 9\ 600$ psi, and $\sigma_y = -0.017\ 44L^2$. Here, the 4 800 psi of the longitudinal tension stress is advantageous and is ignored, since it would be zero if someone put an expansion bellows in the line. Using the Tresca criteria, relationship 15.5, with the allowable stress,

$$\tau_{max} \leq \frac{1.5 \times 12\ 000}{2} \text{ psi} \leq 9\ 000 \text{ psi}$$

Substituting into relationship 15.2,

$$9\ 000 = \frac{9\ 600 + 0.017\ 44L^2}{2}$$

$$L^2 = \frac{18\ 000 - 9\ 600}{0.017\ 44}$$

$$L = 694 \text{ in., or } 57.8 \text{ ft}$$

The stress on the bottom governs. The rollers may not be spaced more than 57.8 ft apart.

Author's note: The preceding problem is one in which there is a hierarchy of conditions to be satisfied. First, the circumferential pressure stresses must be within the allowable range. These are held to a more stringent safety factor because the bursting of the pipe is thought to be a greater catastrophe than the pipe's sagging under its own weight. Then, primary and secondary bending conditions are considered with lower safety factors, or higher allowables, but including again the pressure stresses. Such hierarchies of conditions are not unusual in design codes.

15.7 A cantilevered billboard sign near a highway is constructed as shown in figure 15.35. The weight of the sign, W, is 10 000 lb. The sign is to be designed for a wind force of 30 psf. The single cantilever leg has a diameter of 40 in. What must be the wall thickness of the leg, if the allowable stress for the rolled and welded A36 structural steel plate is 22 ksi?

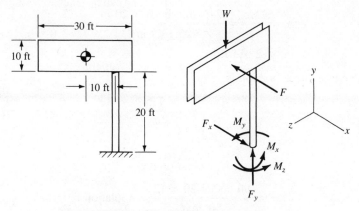

Figure 15.35 Illustrative problem 15.7. A cantilevered billboard under high wind conditions.

Solution:
The total wind force, F, is

$$F = 30 \text{ psf} \times 30 \text{ ft} \times 10 \text{ ft}$$
$$F = 9\,000 \text{ lb}$$

Solving for the static reactions,

$$\sum x = 0: \quad F_x = 9\,000 \text{ lb}$$
$$\sum y = 0: \quad F_y = 10\,000 \text{ lb}$$
$$\sum M_x = 0: \quad M_x + 10\,000 \times 10 = 0$$
$$M_x = -100\,000 \text{ lb} \cdot \text{ft}$$
$$\sum M_y = 0: \quad M_y - 9\,000 \times 10 = 0$$
$$M_y = 90\,000 \text{ lb} \cdot \text{ft}$$
$$\sum M_z = 0: \quad M_z + 9\,000 \times 25 = 0$$
$$M_z = -225\,000 \text{ lb} \cdot \text{ft}$$

For a thin-walled cylinder, where t is the thickness in inches,

$$I = I_x = I_z = \frac{\pi d^3 t}{8} \quad \text{and} \quad J = \frac{\pi d^3 t}{4}$$

$$I = 25\ 132t \quad \text{and} \quad J = 50\ 265t \text{ in.}^4$$

The area equals $\pi dt = 125.66t$ in.^{2}

The stress due to direct compression force,

$$\sigma_c = \frac{10\ 000}{125.66t} = 79.58/t \text{ psi}$$

The bending moment is the resultant of M_x and M_z.

$$M = \sqrt{100\ 000^2 + 225\ 000^2} = 246\ 200 \text{ lb} \cdot \text{ft}$$

The stress due to bending is

$$\sigma_b = \frac{Mc}{I} = \frac{246\ 200 \text{ lb} \cdot \text{ft} \times 20 \text{ in.}}{25\ 132t \text{ in.}^4} \times \frac{12 \text{ in.}}{1 \text{ ft}}$$

$$\sigma_b = 2\ 351/t \text{ psi}$$

The stress due to torsion is

$$\tau = \frac{TR}{J}$$

In this case,

$$T = M_y = 90\ 000 \text{ lb} \cdot \text{ft}$$

$$\tau = \frac{90\ 000 \times 20 \times 12}{50\ 265t} = 430/t \text{ psi}$$

Applying Mohr's Circle and the Tresca Criteria:
Since the worst case occurs on the side of the pipe where the compression stress due to bending adds to that due to the compressive load, σ_y will be taken to be $\sigma_c + \sigma_b$ compression.

$$\sigma_y = 79.58/t + 2\ 351/t = 2\ 431/t \text{ psi compression}$$

$\sigma_x = 0$, since there is no pressure on this pipe
$$\tau = 430t/ \text{ psi}$$

From relationships 15.1 the principal stresses are $\bar{\sigma}_x = 2\ 504.5/t$ and $\bar{\sigma}_y = -73.5/t$. It is in the IV quadrant. Therefore, the maximum shear stress governs. From relationship 15.6, using the allowable rather than the yield, ·

$$|\bar{\sigma}_x - \bar{\sigma}_y| \leq 22\ 000$$

$$|2\ 504.5/t - (-73.5/t)| \leq 22\ 000$$

$$|2\ 578/t| \leq 22\ 000$$

Therefore,

$$t \geq \frac{2\,578}{22\,000}$$

$$t \geq 0.1172 \text{ in.}$$

A wall thickness of $\frac{1}{8}$ in. would be sufficient. However, because of its thinness this would be difficult to weld, transport, and erect. Probably, it would be better to use $\frac{3}{16}$- or $\frac{1}{4}$-in. plate stock.

Author's note: In the aftermath of Hurricane Alicia in 1983, I observed many kinds of demolished sign boards. However, though their superstructures may have been damaged and their advertising totally erased by Alicia's eight hours of driving winds and rain, no cantilevered signs as in the previous illustrative problem were observed to be demolished. Not a great fan of billboards, I have to grudgingly admit to the strength of the design. If we must have billboards, at least they should be constructed so as not to disintegrate into dangerous debris during wind storms.

PROBLEM SET 15.3

15.21 A 1-in.-diameter rod as shown in figure 15.36 is loaded simultaneously with a 10 000-lb tension load and a torque. What level of torque can it be expected to carry without yielding, if the tension yield strength is 32 000 psi?

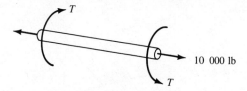

Figure 15.36 Problem 15.21.

15.22 A round stainless steel right-angle bar is to have a stress not exceeding 400 MPa. The bar is cantilevered as shown in figure 15.37. What must be the size of the circular cross section?

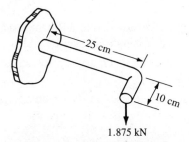

Figure 15.37 Problem 15.22.

15.23 A cylindrical fuel bladder made of reinforced rubber has a diameter of 10 cm and a thickness of 0.1 mm. It is inflated for testing with an internal pressure of 5 bars. The balloon wall is a membrane that will buckle under the slightest bit of compression stress. How much torque can the bladder carry without buckling? In order to visualize the problem, experiment with a toy balloon as shown in figure 15.38.

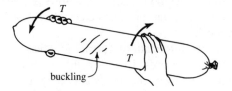

Figure 15.38 Problem 15.23.

15.24 A 6-in. standard-weight steel pipe contains a pressure of 500 psi. In addition, the pipe reactions, F_x and F_y, at end A are calculated to be 1 250 and 780 lb, respectively. What is the worst-case stress that develops at end B where the pipe enters a built-in support? See figure 15.39.

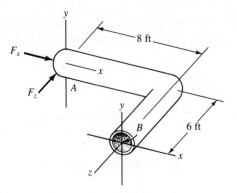

Figure 15.39 Problem 15.24.

SUMMARY

☐ Normal stresses in the same direction due to different conditions, such as tension and bending in *eccentrically loaded members,* may be added.

☐ Stresses have directional properties that must be resolved using *Mohr's Circle.* Stresses are called *tensor quantities.*

☐ The *principal stresses* are the stresses in two mutually perpendicular directions in which there are no shear stresses. They are given by:

(15.1)
$$\bar{\sigma}_x = \frac{(\sigma_x + \sigma_y)}{2} + \sqrt{\left(\frac{\sigma_x - \sigma_y}{2}\right)^2 + \tau^2}$$

and
$$\bar{\sigma}_y = \frac{(\sigma_x + \sigma_y)}{2} - \sqrt{\left(\frac{\sigma_x - \sigma_y}{2}\right)^2 + \tau^2}$$

☐ The *maximum shear stress* occurs when the axes are rotated 45° from the axes of principal stresses.

$$(15.2) \quad \tau_{\max} = \sqrt{\left(\frac{\sigma_x - \sigma_y}{2}\right)^2 + \tau^2}$$

☐ The *Tresca failure criteria* are based on maximum principal stresses when the principal stresses are both of the same sign:

$(15.4) \quad |\overline{\sigma}_x| \le \sigma_{\text{yield}} \quad$ and $\quad |\overline{\sigma}_y| \le \sigma_{\text{yield}}$

and maximum shear stress when the principal stresses are of opposite sign:

$(15.5) \quad |\tau_{\max}| \le (\overline{\sigma}_x - \overline{\sigma}_y)/2 \quad$ or

$(15.6) \quad |\overline{\sigma}_x - \overline{\sigma}_y| \le \sigma_{\text{yield}}$

REFERENCES

1. Juvinall, R. C. *Fundamentals of Machine Component Design*. New York: John Wiley & Sons, 1983.

2. Timoshenko, S. and MacCullough, G. H. *Elements of Strength of Materials*. Princeton, NJ: D. Van Nostrand, 1935.

APPENDIX I

Mechanical Properties of Some Representative Metals, Alloys, and Plastics

Material	Condition	Yield Strength ksi (MPa)	Ultimate Strength ksi (MPa)	Percent Elongation	Modulus of Elasticity psi (Pa)	Specific Gravity
Aluminum 2024 alloy	0	11 (75.8)	27 (186)	22	10E6 (69E9)	2.71
	T4	48 (330)	68 (468)	19	10E6 (69E9)	2.71
Aluminum 6061 alloy	0	8 (55)	18 (124)	30	10E6 (69E9)	2.71
	T6	40 (275)	45 (310)	17	10E6 (69E9)	2.71
Aluminum 7075 alloy	T6	72 (492)	82 (564)	11	10E6 (69E9)	2.71
Aluminum bronze 92% Cu, 8% Al	Annealed	25 (172)	70 (482)	60	17E6 (117E9)	7.78
	Hard	65 (448)	105 (723)	7	17E6 (117E9)	7.78
Brass 70% Cu, 30% Zn	Cold-rolled	63 (434)	76 (523)	8	16E6 (110E9)	8.52
Bronze 90% Cu, 10% Zn	Cold-rolled	54 (372)	61 (420)	5	17E6 (117E9)	8.80
Copper	Annealed	10 (68.9)	32 (220)	45	17E6 (117E9)	8.91
	Cold-rolled	40 (275)	46 (317)	5	17E6 (117E9)	8.91

yield stress

Material	Condition	Yield Strength ksi (MPa)	Ultimate Strength ksi (MPa)	Percent Elongation	Modulus of Elasticity psi (Pa)	Specific Gravity
Dura nickel	Annealed	45 (310)	100 (689)	40	30E6 (207E9)	8.26
	Age-hardened	125 (861)	170 (1171)	25	30E6 (207E9)	8.26
Iron, ductile	Cast	60 (413.4)	80 (551)	10	25E6 (172E9)	7.32
Iron, ingot	Hot-rolled	29 (199.8)	45 (310)	26	30.1E6 (207E9)	7.86
Magnesium A231B alloy	Rolled-sheet	32 (220)	42 (289)	15	6.5E6 (113E9)	1.77
	Rolled-plate	24 (165)	37 (254)	18	6.5E6 (113E9)	1.77
	Extruded	28 (192.9)	38 (261)	14	6.5E6 (113E9)	1.77
Molybdenum	As rolled	75 (517)	100 (689)	30	46E6 (317E9)	10.22
Nickel	Annealed	8.5 (58.6)	46 (317)	30	30E6 (207E9)	8.91
Plastic, delrin	Extruded	4.0	10	19	0.41E6	1.43
Stainless steel Type 316	Annealed	30 (207)	90 (620)	50	28E6 (193E9)	8.02
	Cold-rolled	120 (827)	150 (1033)	8	28E6 (193E9)	8.02
Stainless steel 17-7 PH	Annealed	40 (276)	130 (895)	30	28E6 (193E9)	7.81
	Hardened	185 (1275)	200 (1378)	9	28E6 (193E9)	7.81
Steel AISI 1018	Hot-rolled	32 (220)	58 (399)	25	30E6 (207E9)	7.86
	Cold-rolled	54 (372)	64 (440)	15	30E6 (207E9)	7.86
Steel AISI 1040	Hot-rolled	42 (289)	76 (523)	18	30E6 (207E9)	7.86
	Cold-rolled	71 (489)	85 (585)	12	30E6 (207E9)	7.86
	Heat-treated tempered	86 (592)	113 (778)	23	30E6 (207E9)	7.86
Steel AISI 4140	Cold-rolled	99 (682)	111 (765)	18	30E6 (207E9)	7.86
	Heat-treated tempered	133 (916)	153 (1054)	16	30E6 (207E9)	7.86
Steel, boiler SA 285C	As rolled	30 (207)	55 (379)		30E6 (207E9)	7.86

yield stress & *assummed in tension*

Material	Condition	Yield Strength ksi (MPa)	Ultimate Strength ksi (MPa)	Percent Elongation	Modulus of Elasticity psi (Pa)	Specific Gravity
Steel, pipe A36	As rolled	30 (207)	48 (331)		30E6 (207E9)	7.86
Steel, structural A36	As rolled	36 (248)	58 (400)		30E6 (207E9)	7.86
Titanium	Annealed	70 (482)	90 (620)	23	16.5E6 (113E9)	4.54
Titanium 155A alloy	Annealed	135 (930)	150 (1033)	15	16.5E6 (113E9)	4.59
Tungsten	Hard	360 (2480)	400 (2756)		53E6 (365E9)	19.3
Zirconium	Annealed	16 (110.2)	36 (248)	31	11E6 (75.8E9)	6.5

yield stress in shear = ½ yield strength in tension
pg 128

do 4.27

APPENDIX II

Thread Sizes and Dimensions

THREAD SIZE	THREADS PER INCH	MAJOR DIAMETER (IN.)	MINOR DIAMETER (IN.)	STRESS AREA (IN.^{2})
Coarse-Thread Series (NC)				
$\frac{1}{4}$	20	0.2500	0.1887	0.0317
$\frac{5}{16}$	18	0.3125	0.2443	0.0522
$\frac{3}{8}$	16	0.3750	0.2983	0.0773
$\frac{7}{16}$	14	0.4375	0.3499	0.1060
$\frac{1}{2}$	13	0.5000	0.4056	0.1416
$\frac{1}{2}$	12	0.5000	0.3978	0.1374
$\frac{9}{16}$	12	0.5625	0.4603	0.1816
$\frac{5}{8}$	11	0.6250	0.5135	0.2256
$\frac{3}{4}$	10	0.7500	0.6273	0.3340
$\frac{7}{8}$	9	0.8750	0.7387	0.4612
1	8	1.000	0.8466	0.6051
$1\frac{1}{8}$	7	1.1250	0.9497	0.7627
$1\frac{1}{4}$	7	1.2500	1.0747	0.9684
$1\frac{3}{8}$	6	1.3750	1.1705	1.1538
$1\frac{1}{2}$	6	1.5000	1.2955	1.4041
2	$4\frac{1}{2}$	2.0000	1.7274	2.4971
$2\frac{1}{2}$	4	2.5000	2.1933	3.9976
3	4	3.0000	2.6933	5.9659
$3\frac{1}{2}$	4	3.5000	3.1933	8.3268
4	4	4.0000	3.6935	11.0805

THREAD SIZE	THREADS PER INCH	MAJOR DIAMETER (IN.)	MINOR DIAMETER (IN.)	STRESS AREA (IN.^{2})
		Fine-Thread Series (NF)		
$\frac{1}{4}$	28	0.2500	0.2062	0.0362
$\frac{5}{16}$	24	0.3125	0.2614	0.0579
$\frac{3}{8}$	24	0.3750	0.3239	0.0876
$\frac{7}{16}$	20	0.4375	0.3762	0.1185
$\frac{1}{2}$	20	0.5000	0.4387	0.1597
$\frac{9}{16}$	18	0.5625	0.4943	0.2026
$\frac{5}{8}$	18	0.6250	0.5568	0.2555
$\frac{3}{4}$	16	0.7500	0.6733	0.3724
$\frac{7}{8}$	14	0.8750	0.7874	0.5088
1	12	1.0000	0.8978	0.6624
$1\frac{1}{8}$	12	1.1250	1.0228	0.8549
$1\frac{1}{4}$	12	1.2500	1.1478	1.0721
$1\frac{3}{8}$	12	1.3750	1.2728	1.3137
$1\frac{1}{2}$	12	1.5000	1.3978	1.5799
		8-Thread Series (8N)		
$1\frac{1}{2}$	8	1.5000	1.3466	1.4899
$1\frac{5}{8}$	8	1.6250	1.4716	1.7723
$1\frac{3}{4}$	8	1.7500	1.5966	2.0792
$1\frac{7}{8}$	8	1.8750	1.7216	2.4107
2	8	2.0000	1.8466	2.7665
$2\frac{1}{2}$	8	2.5000	2.3466	4.4352
3	8	3.0000	2.8466	6.4957
$3\frac{1}{2}$	8	3.5000	3.3466	8.9504
4	8	4.0000	3.8466	11.7995
5	8	5.0000	4.8466	18.6694
6	8	6.0000	5.8466	27.1118

SELECTED METRIC THREAD SIZES

THREAD SIZE	PITCH (MM)	MAJOR DIAMETER (MM)	MINOR DIAMETER (MM)	STRESS AREA (MM^2)
M4 × 0.7	0.700	4.000	2.979	8.78
M5 × 0.8	0.800	5.000	3.841	14.2
M6 × 1	1.000	6.000	4.563	20.1
M8 × 1.25	1.250	8.000	6.231	36.6
M10 × 1.5	1.500	10.000	7.879	58.0
M12 × 1.75	1.750	12.000	9.543	84.3
M16 × 2	2.000	16.000	13.204	157
M20 × 2.5	2.500	20.000	16.541	245
M24 × 3	3.000	24.000	19.855	353
M30 × 3.5	3.500	30.000	25.189	561
M36 × 4	4.000	36.000	30.521	817

APPENDIX III

Properties of Common Areas

Circle

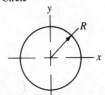

$$A = \pi R^2$$
$$I_{xc} = I_{yc} = \frac{\pi R^4}{4}$$

Ellipse

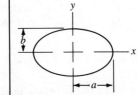

$$A = \pi ab$$
$$I_{xc} = \frac{\pi ab^3}{4}$$
$$I_{yc} = \frac{\pi ba^3}{4}$$

Semicircle

$$A = \frac{\pi R^2}{2}$$
$$\bar{y} = \frac{4R}{3\pi}$$
$$I_{xc} = \frac{\pi R^4}{8}\left(1 - \left(\frac{8}{3\pi}\right)^2\right), \ I_{yc} = \frac{\pi R^4}{8}$$

Parabola

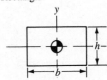

$$A = \frac{4}{3}bh$$
$$\bar{y} = \frac{2}{5}h$$
$$I_{xc} = 0.914bh^3$$
$$I_{yc} = \frac{4}{15}hb^3$$

Rectangle

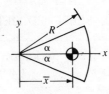

$$A = bh$$
$$I_{xc} = \frac{bh^3}{12}$$
$$I_{yc} = \frac{hb^3}{12}$$

Circular Sector

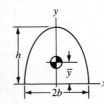

$$A = \alpha R^2$$
$$\bar{x} = \frac{2}{3}\frac{R\sin\alpha}{\alpha}$$
$$I_{xc} = \frac{R^4}{4}\left(\alpha - \frac{1}{2}\sin 2\alpha\right)$$
$$I_{yc} = \frac{R^4}{4}\left(\alpha + \frac{1}{2}\sin 2\alpha\right)$$

Triangle

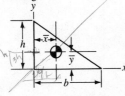

$$A = \frac{bh}{2}$$
$$\bar{y} = \frac{h}{3}, \ \bar{x} = \frac{b}{3}$$
$$I_{xc} = \frac{bh^3}{36}$$
$$I_{yc} = \frac{hb^3}{36}$$

Thin-Wall Hollow Cylinder

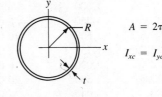

$$A = 2\pi Rt$$
$$I_{xc} = I_{yc} = \pi R^3 t$$

W. Griffel. *Handbook of Formulas for Stress and Strain*. New York: F. Ungar Publishing Co., 1966.

APPENDIX IV

Steel Pipe Dimensions and Properties

$J = 2I$

NOMINAL DIAMETER (IN.)	OUTSIDE DIAMETER (IN.)	WALL THICKNESS (IN.)	WEIGHT PER FT (LB)	AREA, A (IN.^{2})	MOMENT OF INERTIA, I (IN.^{4})	SECTION MODULUS, Z (IN.^{3})	RADIUS OF GYRATION, q (IN.)

Standard Weight (Schedule 40)

NOMINAL DIAMETER	OUTSIDE DIAMETER	WALL THICKNESS	WEIGHT PER FT	AREA, A	MOMENT OF INERTIA, I	SECTION MODULUS, Z	RADIUS OF GYRATION, q
$\frac{1}{2}$	0.840	0.109	0.85	0.250	0.017	0.041	0.261
$\frac{3}{4}$	1.050	0.113	1.13	0.333	0.037	0.071	0.334
1	1.315	0.133	1.68	0.494	0.087	0.133	0.421
$1\frac{1}{4}$	1.660	0.140	2.27	0.669	0.195	0.235	0.540
$1\frac{1}{2}$	1.900	0.145	2.72	0.799	0.310	0.326	0.623
2	2.375	0.154	3.65	1.07	0.666	0.561	0.787
$2\frac{1}{2}$	2.875	0.203	5.79	1.70	1.53	1.06	0.947
3	3.500	0.216	7.58	2.23	3.02	1.72	1.16
$3\frac{1}{2}$	4.000	0.226	9.11	2.68	4.79	2.39	1.34
4	4.500	0.237	10.79	3.17	7.23	3.21	1.51
5	5.563	0.258	14.62	4.30	15.2	5.45	1.88
6	6.625	0.280	18.97	5.58	28.1	8.50	2.25
8	8.625	0.322	28.55	8.40	72.5	16.8	2.94
10	10.750	0.365	40.48	11.9	161	29.9	3.67
12	12.750	0.375	49.56	14.6	279	43.8	4.38

NOMINAL DIAMETER (IN.)	OUTSIDE DIAMETER (IN.)	WALL THICKNESS (IN.)	WEIGHT PER FT (LB)	AREA, A (IN.^{2})	MOMENT OF INERTIA, I (IN.^{4})	SECTION MODULUS, Z (IN.^{3})	RADIUS OF GYRATION, q (IN.)
			Extra Strong (Schedule 80)				
$\frac{1}{2}$	0.840	0.147	1.09	0.320	0.020	0.048	0.250
$\frac{3}{4}$	1.050	0.154	1.47	0.433	0.045	0.085	0.321
1	1.315	0.179	2.17	0.639	0.106	0.161	0.407
$1\frac{1}{4}$	1.660	0.191	3.00	0.881	0.242	0.291	0.524
$1\frac{1}{2}$	1.900	0.200	3.63	1.07	0.391	0.412	0.605
2	2.375	0.218	5.02	1.48	0.868	0.731	0.766
$2\frac{1}{2}$	2.875	0.276	7.66	2.25	1.92	1.34	0.924
3	3.500	0.300	10.25	3.02	3.89	2.23	1.14
$3\frac{1}{2}$	4.000	0.318	12.50	3.68	6.28	3.14	1.31
4	4.500	0.337	14.98	4.41	9.61	4.27	1.48
5	5.563	0.375	20.78	6.11	20.7	7.43	1.84
6	6.625	0.432	28.57	8.40	40.5	12.2	2.19
8	8.625	0.500	43.39	12.8	106	24.5	2.88
10	10.750	0.500	54.74	16.1	212	39.4	3.63
12	12.750	0.500	65.42	19.2	362	56.7	4.33

Design Properties of Selected Wide-Flange Shapes

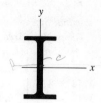

Nominal Size Times Lb per Ft	Area, A (in.²)	Depth, h (in.)	Flange Width (in.)	Flange Thickness (in.)	Web Thickness, t (in.)	x Axis I_x (in.⁴)	x Axis Z (in.³)	x Axis q (in.)	y Axis I_y (in.⁴)	y Axis Z (in.³)	y Axis q (in.)
W36 × 300	88.3	36.72	16.655	1.680	0.945	20300	1110	15.2	1300	156	3.83
W36 × 160	47.1	36.00	12.000	1.020	0.653	9760	542	14.4	295	49.1	2.50
W33 × 200	58.9	33.00	15.750	1.150	0.715	11100	671	13.7	750	95.2	3.57
W30 × 116	34.2	30.00	10.500	0.850	0.564	4930	329	12.0	164	31.3	2.19
W27 × 145	42.7	26.88	13.965	0.975	0.600	5430	404	11.3	443	63.5	3.22
W24 × 130	38.3	24.25	14.000	0.900	0.565	4020	332	10.2	412	58.9	3.28
W24 × 100	29.5	24.00	12.000	0.775	0.468	3000	250	10.1	223	37.2	2.75
W24 × 94	27.7	24.29	9.061	0.872	0.516	2690	221	9.86	108	23.9	1.98
W24 × 76	22.4	23.91	8.985	0.682	0.440	2100	176	9.69	82.6	18.4	1.92
W24 × 55	16.2	23.55	7.000	0.503	0.396	1340	114	9.10	28.9	8.25	1.34
W21 × 112	33.0	21.00	13.000	0.865	0.527	2620	250	8.92	317	48.8	3.10
W21 × 55	16.2	23.55	7.000	0.503	0.396	1340	114	9.10	28.9	8.25	1.34
W21 × 44	13.0	20.66	6.500	0.451	0.348	843	81.6	8.07	20.7	6.38	1.27
W18 × 70	20.6	18.00	8.750	0.751	0.438	1160	129	7.50	84.0	19.2	2.02
W18 × 60	17.7	18.25	7.558	0.695	0.416	986	108	7.47	50.1	13.3	1.68
W18 × 50	14.7	18.00	7.500	0.570	0.358	802	89.1	7.38	40.2	10.7	1.65
W18 × 35	10.3	17.71	6.000	0.429	0.298	513	57.9	7.05	15.5	5.16	1.23
W16 × 64	18.8	16.00	8.500	0.715	0.443	836	104	6.66	73.3	17.3	1.97
W16 × 40	11.8	16.00	7.000	0.503	0.307	517	64.6	6.62	28.8	8.23	1.56
W16 × 36	10.6	15.85	6.992	0.428	0.299	447	56.5	6.50	24.4	6.99	1.52
W16 × 26	7.67	15.65	5.500	0.345	0.250	300	38.3	6.25	9.59	3.49	1.12

10.12
p. 315

Nominal Size Times Lb per Ft	Area, A (in.²)	Depth, h (in.)	Flange		Web Thickness, t (in.)	Elastic Properties					
			Width (in.)	Thickness (in.)		x Axis			y Axis		
						I_x (in.⁴)	Z (in.³)	q (in.)	I_y (in.⁴)	Z (in.³)	q (in.)
W14 × 87	25.6	14.00	14.500	0.688	0.420	967	138	6.15	350	48.2	3.70
W14 × 78	22.9	14.06	12.000	0.718	0.428	851	121	6.09	207	34.5	3.00
W14 × 74	21.8	14.19	10.072	0.783	0.450	797	112	6.05	133	26.5	2.48
W14 × 43	12.6	13.68	8.000	0.528	0.308	429	62.7	5.82	45.1	11.3	1.89
W14 × 38	11.2	14.12	6.776	0.513	0.313	386	54.7	5.88	26.6	7.86	1.54
W14 × 34	10.0	14.00	6.750	0.453	0.287	340	48.6	5.83	23.3	6.89	1.52
W14 × 22	6.49	13.72	5.000	0.335	0.230	198	28.9	5.53	7.00	2.80	1.04
W12 × 79	23.2	12.38	12.080	0.736	0.470	663	107	5.34	216	35.8	3.05
W12 × 65	19.1	12.12	12.000	0.606	0.390	533	88.0	5.28	175	29.1	3.02
W12 × 58	17.1	12.19	10.014	0.641	0.359	476	78.1	5.28	107	21.4	2.51
W12 × 40	11.8	11.94	8.000	0.516	0.294	310	51.9	5.13	44.1	11.0	1.94
W12 × 27	7.95	11.96	6.497	0.400	0.237	204	34.2	5.07	18.3	5.63	1.52
W12 × 16.5	4.87	12.00	4.000	0.269	0.230	105	17.6	4.65	2.88	1.44	0.770
W10 × 100	29.4	11.12	10.345	1.118	0.685	625	112	4.61	207	39.9	2.65
W10 × 49	14.4	10.00	10.000	0.558	0.340	273	54.6	4.35	93.0	18.6	2.54
W10 × 45	13.3	10.10	8.020	0.620	0.350	248	49.1	4.32	53.4	13.3	2.01
W10 × 15	4.41	10.00	4.000	0.269	0.230	68.9	13.8	3.95	2.88	1.44	0.809
W8 × 58	17.1	8.75	8.222	0.808	0.510	227	52.0	3.65	74.9	18.2	2.10
W8 × 31	9.12	8.00	8.000	0.433	0.288	110	27.4	3.47	37.0	9.24	2.01
W8 × 24	7.06	7.93	6.500	0.398	0.245	82.5	20.8	3.42	18.2	5.61	1.61
W8 × 17	5.01	8.00	5.250	0.308	0.230	56.6	14.1	3.36	7.44	2.83	1.22
W8 × 13	3.83	8.00	4.000	0.254	0.230	39.6	9.90	3.21	2.72	1.36	0.842
W6 × 16	4.72	6.25	4.030	0.404	0.260	31.7	10.2	2.59	4.42	2.19	0.967
W6 × 12	3.54	6.00	4.000	0.279	0.230	21.7	7.25	2.48	2.98	1.49	0.918
W5 × 16	4.70	5.00	5.000	0.360	0.240	21.3	8.53	2.13	7.51	3.00	1.26

Refer to American Institute for Steel Construction for more complete table.

APPENDIX V(b)

Design Properties of Selected S Shapes (I Beam)

Nominal Size Times Lb per Ft	Area, A (in.²)	Depth, h (in.)	Flange		Web Thickness (in.)	Elastic Properties					
			Width (in.)	Thickness (in.)		x Axis			y Axis		
						I_x (in.⁴)	Z (in.³) $\frac{I}{c}$	q (in.)	I_y (in.⁴)	Z (in.³) $\frac{I}{c}$	q (in.)
S18 × 70	20.6	18.00	6.251	0.691	0.711	926	103	6.71	24.1	7.72	1.08
S18 × 54.7	16.1	18.00	6.001	0.691	0.461	804	89.4	7.07	20.8	6.94	1.14
S15 × 50	14.7	15.00	5.640	0.622	0.550	486	64.8	5.75	15.7	5.57	1.03
S15 × 42.9	12.6	15.00	5.501	0.622	0.411	447	59.6	5.95	14.4	5.23	1.07
S12 × 50	14.7	12.00	5.477	0.659	0.687	305	50.8	4.55	15.7	5.74	1.03
S12 × 40.8	12.0	12.00	5.252	0.659	0.472	272	45.4	4.77	13.6	5.16	1.06
S12 × 35	10.3	12.00	5.078	0.544	0.428	229	38.2	4.72	9.87	3.89	0.980
S12 × 31.8	9.35	12.00	5.000	0.544	0.350	218	36.4	4.83	9.36	3.74	1.00
S10 × 35	10.3	10.00	4.944	0.491	0.594	147	29.4	3.78	8.36	3.38	0.901
S10 × 25.4	7.46	10.00	4.661	0.491	0.311	124	24.7	4.07	6.79	2.91	0.954
S8 × 23	6.77	8.00	4.171	0.425	0.441	64.9	16.2	3.10	4.31	2.07	0.798
S8 × 18.4	5.41	8.00	4.001	0.425	0.271	57.6	14.4	3.26	3.73	1.86	0.831
S7 × 20	5.88	7.00	3.860	0.392	0.450	42.4	12.1	2.69	3.17	1.64	0.734
S7 × 15.3	4.50	7.00	3.662	0.392	0.252	36.7	10.5	2.86	2.64	1.44	0.766
S6 × 17.25	5.07	6.00	3.565	0.359	0.465	26.3	8.77	2.28	2.31	1.30	0.675
S6 × 12.5	3.67	6.00	3.332	0.359	0.232	22.1	7.37	2.45	1.82	1.09	0.705
S5 × 14.75	4.34	5.00	3.284	0.326	0.494	15.2	6.09	1.87	1.67	1.01	0.620
S5 × 10	2.94	5.00	3.004	0.326	0.214	12.3	4.92	2.05	1.22	0.809	0.643
S4 × 9.5	2.79	4.00	2.796	0.293	0.326	6.79	3.39	1.56	0.903	0.646	0.569
S4 × 7.7	2.26	4.00	2.663	0.293	0.193	6.08	3.04	1.64	0.764	0.574	0.581
S3 × 7.5	2.21	3.00	2.509	0.260	0.349	2.93	1.95	1.15	0.586	0.468	0.516
S3 × 5.7	1.67	3.00	2.330	0.260	0.170	2.52	1.68	1.23	0.455	0.390	0.522

Refer to American Institute for Steel Construction for more complete table.

APPENDIX V(c)

Design Properties of Selected Channels

Nominal Size Times Lb per Ft	Area, A (in.²)	Depth, h (in.)	Flange		Web Thickness (in.)	x Axis			y Axis			
			Width (in.)	Average Thickness (in.)		I_x (in.⁴)	Z (in.³)	q (in.)	I_y (in.⁴)	Z (in.³)	q (in.)	$\bar{x}$ (in.)
C15 × 50	14.7	15.00	3.716	0.650	0.716	404	53.8	5.24	11.0	3.78	0.867	0.799
C15 × 40	11.8	15.00	3.520	0.650	0.520	349	46.5	5.44	9.23	3.36	0.886	0.778
C15 × 33.9	9.96	15.00	3.400	0.650	0.400	315	42.0	5.62	8.13	3.11	0.904	0.787
C12 × 30	8.82	12.00	3.170	0.501	0.510	162	27.0	4.29	5.14	2.06	0.763	0.674
C12 × 25	7.35	12.00	3.047	0.501	0.387	144	24.1	4.43	4.47	1.88	0.780	0.674
C12 × 20.7	6.09	12.00	2.942	0.501	0.282	129	21.5	4.61	3.88	1.73	0.799	0.698
C10 × 30	8.82	10.00	3.033	0.436	0.673	103	20.7	3.42	3.94	1.65	0.669	0.649
C10 × 25	7.35	10.00	2.886	0.436	0.526	91.2	18.2	3.52	3.36	1.48	0.676	0.617
C10 × 20	5.88	10.00	2.739	0.436	0.379	78.9	15.8	3.66	2.81	1.32	0.691	0.606
C10 × 15.3	4.49	10.00	2.600	0.436	0.240	67.4	13.5	3.87	2.28	1.16	0.713	0.634
C9 × 20	5.88	9.00	2.648	0.413	0.448	60.9	13.5	3.22	2.42	1.17	0.642	0.583
C9 × 15	4.41	9.00	2.485	0.413	0.285	51.0	11.3	3.40	1.93	1.01	0.661	0.586
C9 × 13.4	3.94	9.00	2.433	0.413	0.233	47.9	10.6	3.48	1.76	0.962	0.668	0.601
C8 × 18.75	5.51	8.00	2.527	0.390	0.487	44.0	11.0	2.82	1.98	1.01	0.599	0.565
C8 × 13.75	4.04	8.00	2.343	0.390	0.303	36.1	9.03	2.99	1.53	0.853	0.615	0.553
C8 × 11.5	3.38	8.00	2.260	0.390	0.220	32.6	8.14	3.11	1.32	0.781	0.625	0.571
C7 × 14.75	4.33	7.00	2.299	0.366	0.419	27.2	7.78	2.51	1.38	0.779	0.564	0.532
C7 × 12.25	3.60	7.00	2.194	0.366	0.314	24.2	6.93	2.60	1.17	0.702	0.571	0.525
C7 × 9.8	2.87	7.00	2.090	0.366	0.210	21.3	6.08	2.72	0.968	0.625	0.581	0.541
C6 × 13	3.83	6.00	2.157	0.343	0.437	17.4	5.80	2.13	1.05	0.642	0.525	0.514
C6 × 10.5	3.09	6.00	2.034	0.343	0.314	15.2	5.06	2.22	0.865	0.564	0.529	0.500
C6 × 8.2	2.40	6.00	1.920	0.343	0.200	13.1	4.38	2.34	0.692	0.492	0.537	0.512
C5 × 9	2.64	5.00	1.885	0.320	0.325	8.90	3.56	1.83	0.632	0.449	0.489	0.478
C5 × 6.7	1.97	5.00	1.750	0.320	0.190	7.49	3.00	1.95	0.478	0.378	0.493	0.484
C4 × 7.25	2.13	4.00	1.721	0.296	0.321	4.59	2.29	1.47	0.432	0.343	0.450	0.459
C4 × 5.4	1.59	4.00	1.584	0.296	0.184	3.85	1.93	1.56	0.319	0.283	0.449	0.458
C3 × 6	1.76	3.00	1.596	0.273	0.356	2.07	1.38	1.08	0.305	0.268	0.416	0.455
C3 × 5	1.47	3.00	1.498	0.273	0.258	1.85	1.24	1.12	0.247	0.233	0.410	0.438
C3 × 4.1	1.21	3.00	1.410	0.273	0.170	1.66	1.10	1.17	0.197	0.202	0.404	0.437

Refer to American Institute for Steel Construction for more complete table.

APPENDIX V(d)

Design Properties of Selected Structural Tees Cut from W Shapes

Nominal Size Times Lb per Ft	Area, A (in.²)	Depth of Tee, h (in.)	Flange		Stem Thickness (in.)	x Axis				y Axis		
			Width (in.)	Thickness (in.)		I_x (in.⁴)	Z (in.³)	q (in.)	$\bar{y}$ (in.)	I_y (in.⁴)	Z (in.³)	q (in.)
WT9 × 35	10.3	9.00	8.750	0.751	0.438	68.2	9.68	2.57	1.96	42.0	9.60	2.02
WT9 × 25	7.36	9.00	7.500	0.570	0.358	54.0	7.86	2.71	2.13	20.1	5.35	1.65
WT9 × 20	5.88	8.95	6.018	0.524	0.316	44.9	6.75	2.76	2.29	9.54	3.17	1.27
WT9 × 17.5	5.15	8.86	6.000	0.429	0.298	40.1	6.18	2.79	2.38	7.74	2.58	1.23
WT8 × 32	9.41	8.08	8.500	0.715	0.443	48.3	7.72	2.27	1.73	36.7	8.63	1.97
WT8 × 20	5.89	8.00	7.000	0.503	0.307	33.2	5.38	2.37	1.82	14.4	4.11	1.56
WT8 × 18	5.30	7.93	6.992	0.428	0.299	30.8	5.11	2.41	1.89	12.2	3.49	1.52
WT12 × 65	19.2	12.13	14.000	0.900	0.565	223	23.1	3.41	2.46	206	29.4	3.28
WT12 × 55	16.2	12.08	12.042	0.855	0.510	195	20.5	3.47	2.57	125	20.7	2.77
WT12 × 50	14.8	12.00	12.000	0.775	0.468	177	18.7	3.46	2.53	112	18.6	2.75
WT12 × 47	13.8	12.15	9.061	0.872	0.516	186	20.3	3.67	3.00	54.2	12.0	1.98
WT12 × 27.5	8.09	11.78	7.000	0.503	0.396	116	14.1	3.79	3.50	14.4	4.13	1.34
WT10.5 × 56	16.5	10.50	13.000	0.865	0.527	137	16.2	2.88	2.06	159	24.4	3.10
WT10.5 × 22	6.48	10.33	6.500	0.451	0.348	70.9	9.63	3.31	2.97	10.4	3.19	1.27
WT7 × 43.5	12.8	7.00	14.500	0.688	0.420	34.9	5.88	1.65	1.08	175	24.1	3.70
WT7 × 39	11.5	7.03	12.000	0.718	0.428	34.8	5.96	1.74	1.19	103	17.2	3.00
WT7 × 30.5	8.97	6.96	10.000	0.643	0.378	29.2	5.13	1.80	1.25	53.6	10.7	2.45
WT7 × 21.5	6.32	6.84	8.000	0.528	0.308	22.2	4.02	1.87	1.33	22.6	5.64	1.89
WT7 × 17	5.01	7.00	6.750	0.453	0.287	21.1	3.87	2.05	1.54	11.6	3.44	1.52
WT7 × 11	3.24	6.86	5.000	0.335	0.230	14.8	2.90	2.13	1.76	3.50	1.40	1.04
WT6 × 32.5	9.55	6.06	12.000	0.606	0.390	20.6	4.06	1.47	0.985	87.3	14.6	3.02
WT6 × 26.5	7.80	6.03	10.000	0.576	0.345	17.7	3.54	1.51	1.02	48.0	9.61	2.48
WT6 × 25	7.36	6.10	8.077	0.641	0.371	18.7	3.80	1.60	1.17	28.2	6.98	1.96
WT6 × 20	5.89	5.97	8.000	0.516	0.294	14.4	2.94	1.56	1.08	22.0	5.51	1.94

Nominal Size Times Lb per Ft	Area, A (in.^{2})	Depth of Tee, h (in.)	Flange		Stem Thickness (in.)	x Axis				y Axis		
			Width (in.)	Thickness (in.)		I_x (in.^{4})	Z (in.^{3})	q (in.)	$\bar{y}$ (in.)	I_y (in.^{4})	Z (in.^{3})	q (in.)
WT6 × 8.25	2.43	6.00	4.000	0.269	0.230	9.03	2.13	1.93	1.76	1.44	0.721	0.770
WT5 × 24.5	7.20	5.00	10.000	0.558	0.340	10.1	2.40	1.18	0.809	46.5	9.30	2.54
WT5 × 7.5	2.20	5.00	4.000	0.269	0.230	5.46	1.51	1.57	1.37	1.44	0.720	0.809
WT4 × 15.5	4.56	4.00	8.000	0.433	0.288	4.31	1.30	0.973	0.672	18.5	4.62	2.01
WT4 × 12	3.53	3.97	6.500	0.398	0.245	3.53	1.08	1.00	0.695	9.12	2.80	1.61
WT4 × 8.5	2.50	4.00	5.250	0.308	0.230	3.21	1.02	1.13	0.835	3.72	1.42	1.22
WT4 × 6.5	1.92	4.00	4.000	0.254	0.230	2.90	0.976	1.23	1.03	1.36	0.680	0.842
WT3 × 7.75	2.28	3.00	5.995	0.269	0.235	1.44	0.591	0.795	0.559	4.83	1.61	1.46
WT3 × 6	1.77	3.00	4.000	0.279	0.230	1.30	0.558	0.857	0.673	1.49	0.746	0.918
WT2.5 × 8	2.35	2.50	5.000	0.360	0.240	0.840	0.411	0.598	0.457	3.75	1.50	1.26

Refer to American Institute for Steel Construction for more complete table.

APPENDIX V(e)

Design Properties of Selected Structural Tees Cut from S Shapes

Nominal Size Times Lb per Ft	Area, A (in.²)	Depth of Tee, h (in.)	Flange Width (in.)	Flange Thickness (in.)	Stem Thickness (in.)	I_x (in.⁴)	Z (in.³)	q (in.)	$\bar{y}$ (in.)	I_y (in.⁴)	Z (in.³)	q (in.)
						x Axis				y Axis		
ST12 × 60	17.6	12.00	8.048	1.102	0.798	245	28.9	3.72	3.52	42.1	10.5	1.54
ST12 × 52.95	15.6	12.00	7.875	1.102	0.625	205	23.3	3.63	3.19	39.1	9.92	1.58
ST12 × 50	14.7	12.00	7.247	0.871	0.747	215	26.4	3.83	3.84	23.9	6.59	1.27
ST12 × 45	13.2	12.00	7.124	0.871	0.624	190	22.6	3.79	3.60	22.5	6.31	1.30
ST12 × 39.95	11.8	12.00	7.001	0.871	0.501	163	18.7	3.72	3.30	21.1	6.04	1.34
ST10 × 47.5	14.0	10.00	7.200	0.916	0.800	137	19.7	3.13	3.07	24.8	6.90	1.33
ST10 × 42.5	12.5	10.00	7.053	0.916	0.653	118	16.6	3.08	2.85	23.1	6.55	1.36
ST10 × 37.5	11.0	10.00	6.391	0.789	0.641	110	15.9	3.16	3.08	14.8	4.64	1.16
ST10 × 32.7	9.62	10.00	6.250	0.789	0.500	92.3	12.8	3.10	2.80	13.7	4.38	1.19
ST9 × 35	10.3	9.00	6.251	0.691	0.711	84.7	14.0	2.87	2.94	12.1	3.86	1.08
ST9 × 27.35	8.04	9.00	6.001	0.691	0.461	62.4	9.61	2.79	2.50	10.4	3.47	1.14
ST7.5 × 25	7.35	7.50	5.640	0.622	0.550	40.6	7.73	2.35	2.25	7.85	2.78	1.03
ST7.5 × 21.45	6.31	7.50	5.501	0.622	0.411	33.0	6.00	2.29	2.01	7.19	2.61	1.07
ST6 × 25	7.35	6.00	5.477	0.659	0.687	25.2	6.05	1.85	1.84	7.85	2.87	1.03
ST6 × 20.4	6.00	6.00	5.252	0.659	0.462	18.9	4.28	1.78	1.58	6.78	2.58	1.06
ST6 × 17.5	5.14	6.00	5.078	0.544	0.428	17.2	3.95	1.83	1.65	4.93	1.94	0.980
ST6 × 15.9	4.68	6.00	5.000	0.544	0.350	14.9	3.31	1.78	1.51	4.68	1.87	1.00
ST5 × 17.5	5.15	5.00	4.944	0.491	0.594	12.5	3.63	1.56	1.56	4.18	1.69	0.901
ST5 × 12.7	3.73	5.00	4.661	0.491	0.311	7.83	2.06	1.45	1.20	3.39	1.46	0.954
ST4 × 11.5	3.38	4.00	4.171	0.425	0.441	5.03	1.77	1.22	1.15	2.15	1.03	0.798
ST4 × 9.2	2.70	4.00	4.001	0.425	0.271	3.51	1.15	1.14	0.941	1.86	0.932	0.831
ST3.5 × 10	2.94	3.50	3.860	0.392	0.450	3.36	1.36	1.07	1.04	1.59	0.821	0.734
ST3.5 × 7.65	2.25	3.50	3.662	0.392	0.252	2.19	0.816	0.987	0.817	1.32	0.720	0.766

Nominal Size Times Lb per Ft	Area, A (in.^{2})	Depth of Tee, h (in.)	Flange		Stem Thickness (in.)	x Axis				y Axis		
			Width (in.)	Thickness (in.)		I_x (in.^{4})	Z (in.^{3})	q (in.)	$\bar{y}$ (in.)	I_y (in.^{4})	Z (in.^{3})	q (in.)
ST3 × 8.625	2.53	3.00	3.565	0.359	0.465	2.13	1.02	0.917	0.914	1.15	0.648	0.675
ST3 × 6.25	1.83	3.00	3.332	0.359	0.232	1.27	0.552	0.833	0.691	0.911	0.547	0.705
ST2.5 × 7.375	2.17	2.50	3.284	0.326	0.494	1.27	0.740	0.764	0.789	0.833	0.507	0.620
ST2.5 × 5	1.47	2.50	3.004	0.326	0.214	0.681	0.353	0.681	0.569	0.608	0.405	0.643
ST2 × 4.75	1.40	2.00	2.796	0.293	0.326	0.470	0.325	0.580	0.553	0.451	0.323	0.569
ST2 × 3.85	1.13	2.00	2.663	0.293	0.193	0.316	0.203	0.528	0.448	0.382	0.287	0.581
ST1.5 × 3.75	1.10	1.50	2.509	0.260	0.349	0.204	0.191	0.430	0.432	0.293	0.234	0.516
ST1.5 × 2.85	0.835	1.50	2.330	0.260	0.170	0.118	0.101	0.376	0.329	0.227	0.195	0.522

Refer to American Institute for Steel Construction for more complete table.

Design Properties of Selected Equal-Leg Angles

Size and Thickness (in.)	Weight per Ft (lb)	Area (in.²)	x Axis and y Axis				Principal Axis, q (in.)
			I (in.⁴)	Z (in.³)	q (in.)	$\bar{x}$ or $\bar{y}$ (in.)	
L8 × 8 × 1	51.0	15.0	89.0	15.8	2.44	2.37	1.56
$\frac{3}{4}$	38.9	11.4	69.7	12.2	2.47	2.28	1.58
$\frac{5}{8}$	32.7	9.61	59.4	10.3	2.49	2.23	1.58
$\frac{1}{2}$	26.4	7.75	48.6	8.36	2.50	2.19	1.59
L6 × 6 × 1	37.4	11.0	35.5	8.57	1.80	1.86	1.17
$\frac{3}{4}$	28.7	8.44	28.2	6.66	1.83	1.78	1.17
$\frac{5}{8}$	24.2	7.11	24.2	5.66	1.84	1.73	1.18
$\frac{1}{2}$	19.6	5.75	19.9	4.61	1.86	1.68	1.18
$\frac{3}{8}$	14.9	4.36	15.4	3.53	1.88	1.64	1.19
L5 × 5 × $\frac{3}{4}$	23.6	6.94	15.7	4.53	1.51	1.52	0.975
$\frac{5}{8}$	20.0	5.86	13.6	3.86	1.52	1.48	0.978
$\frac{1}{2}$	16.2	4.75	11.3	3.16	1.54	1.43	0.983
$\frac{3}{8}$	12.3	3.61	8.74	2.42	1.56	1.39	0.990
$\frac{5}{16}$	10.3	3.03	7.42	2.04	1.57	1.37	0.994
L4 × 4 × $\frac{3}{4}$	18.5	5.44	7.67	2.81	1.19	1.27	0.778
$\frac{5}{8}$	15.7	4.61	6.66	2.40	1.20	1.23	0.779
$\frac{1}{2}$	12.8	3.75	5.56	1.97	1.22	1.18	0.782
$\frac{3}{8}$	9.8	2.86	4.36	1.52	1.23	1.14	0.788
$\frac{5}{16}$	8.2	2.40	3.71	1.29	1.24	1.12	0.791
$\frac{1}{4}$	6.6	1.94	3.04	1.05	1.25	1.09	0.795

Size and Thickness (in.)	Weight per Ft (lb)	Area (in.²)	x Axis and y Axis				Principal Axis, q (in.)
			I (in.⁴)	Z (in.³)	q (in.)	$\bar{x}$ or $\bar{y}$ (in.)	
L3½ × 3½ × ½	11.1	3.25	3.64	1.49	1.06	1.06	0.683
⅜	8.5	2.48	2.87	1.15	1.07	1.01	0.687
5/16	7.2	2.09	2.45	0.976	1.08	0.990	0.690
¼	5.8	1.69	2.01	0.794	1.09	0.968	0.694
L3 × 3 × ½	9.4	2.75	2.22	1.07	0.898	0.932	0.584
⅜	7.2	2.11	1.76	0.833	0.913	0.888	0.587
5/16	6.1	1.78	1.51	0.707	0.922	0.869	0.589
¼	4.9	1.44	1.24	0.577	0.930	0.842	0.592
3/16	3.71	1.09	0.962	0.441	0.939	0.820	0.596
L2½ × 2½ × ⅜	5.9	1.73	0.984	0.566	0.753	0.762	0.487
¼	4.1	1.19	0.703	0.394	0.769	0.717	0.491
L2 × 2 × ⅜	4.7	1.36	0.479	0.351	0.594	0.636	0.389
¼	3.19	0.938	0.348	0.247	0.609	0.592	0.391
3/16	2.44	0.715	0.272	0.190	0.617	0.569	0.394
L1½ × 1½ × ¼	2.34	0.688	0.139	0.134	0.449	0.466	0.292
⅛	1.23	0.359	0.078	0.072	0.465	0.421	0.296
L1 × 1 × ¼	1.49	0.438	0.037	0.056	0.290	0.339	0.196
⅛	0.80	0.234	0.022	0.031	0.304	0.296	0.196

Refer to American Institute for Steel Construction for more complete table.

APPENDIX V(g)

Design Properties of Selected Unequal-Leg Angles

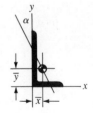

Size and Thickness (in.)	Weight per ft (lb)	Area (in.²)	x Axis				y Axis				Principal Axis	
			I_x (in.⁴)	Z (in.³)	q (in.)	$\bar{y}$ (in.)	I_y (in.⁴)	Z (in.³)	q (in.)	$\bar{x}$ (in.)	q (in.)	$Tan\ \alpha$
L9 × 4 × ½	21.3	6.25	53.2	9.34	2.92	2.17	6.92	3.17	1.05	0.810	0.854	0.220
L8 × 6 × 1	44.2	13.0	80.8	15.1	2.49	2.65	38.8	8.92	1.73	1.65	1.28	0.543
⅞	39.1	11.5	72.3	13.4	2.51	2.61	34.9	7.94	1.74	1.61	1.28	0.547
¾	33.8	9.94	63.4	11.7	2.53	2.56	30.7	6.92	1.76	1.56	1.29	0.551
½	23.0	6.75	44.3	8.02	2.56	2.47	21.7	4.79	1.79	1.47	1.30	0.558
L8 × 4 × ¾	28.7	8.44	54.9	10.9	2.55	2.95	9.36	3.07	1.05	0.953	0.852	0.258
L7 × 4 × ¾	26.2	7.69	37.8	8.42	2.22	2.51	9.05	3.03	1.09	1.01	0.860	0.324
½	17.9	5.25	26.7	5.81	2.25	2.42	6.53	2.12	1.11	0.917	0.872	0.335
⅜	13.6	3.98	20.6	4.44	2.27	2.37	5.10	1.63	1.13	0.870	0.880	0.340
L6 × 4 × ⅞	27.2	7.98	27.7	7.15	1.86	2.12	9.75	3.39	1.11	1.12	0.857	0.421
¾	23.6	6.94	24.5	6.25	1.88	2.08	8.68	2.97	1.12	1.08	0.860	0.428
⅝	20.0	5.86	21.1	5.31	1.90	2.03	7.52	2.54	1.13	1.03	0.864	0.435
½	16.2	4.75	17.4	4.33	1.91	1.99	6.27	2.08	1.15	0.987	0.870	0.440
⁷⁄₁₆	14.3	4.18	15.5	3.83	1.92	1.96	5.60	1.85	1.16	0.964	0.873	0.443
⅜	12.3	3.61	13.5	3.32	1.93	1.94	4.90	1.60	1.17	0.941	0.877	0.446
⁵⁄₁₆	10.3	3.03	11.4	2.79	1.94	1.92	4.18	1.35	1.17	0.918	0.882	0.448
L5 × 3 × ½	12.8	3.75	9.45	2.91	1.59	1.75	2.58	1.15	0.829	0.750	0.648	0.357
⅜	9.8	2.86	7.37	2.24	1.61	1.70	2.04	0.888	0.845	0.704	0.654	0.364
⁵⁄₁₆	8.2	2.40	6.26	1.89	1.61	1.68	1.75	0.753	0.853	0.681	0.658	0.368
¼	6.6	1.94	5.11	1.53	1.62	1.66	1.44	0.614	0.861	0.657	0.663	0.371

Size and Thickness (in.)	Weight per ft (lb)	Area (in.²)	x Axis				y Axis				Principal Axis	
			I_x (in.⁴)	Z (in.³)	q (in.)	$\bar{y}$ (in.)	I_y (in.⁴)	Z (in.³)	q (in.)	$\bar{x}$ (in.)	q (in.)	Tan α
L4 × 3 × $\frac{5}{8}$	13.6	3.98	6.03	2.30	1.23	1.37	2.87	1.35	0.849	0.871	0.637	0.534
$\frac{1}{2}$	11.1	3.25	5.05	1.89	1.25	1.33	2.42	1.12	0.864	0.827	0.639	0.543
$\frac{7}{16}$	9.8	2.87	4.52	1.68	1.25	1.30	2.18	0.992	0.871	0.804	0.641	0.547
$\frac{3}{8}$	8.5	2.48	3.96	1.46	1.26	1.28	1.92	0.866	0.879	0.782	0.644	0.551
$\frac{5}{16}$	7.2	2.09	3.38	1.23	1.27	1.26	1.65	0.734	0.887	0.759	0.647	0.554
$\frac{1}{4}$	5.8	1.69	2.77	1.00	1.28	1.24	1.36	0.599	0.896	0.736	0.651	0.558
L3 × 2 × $\frac{1}{2}$	7.7	2.25	1.92	1.00	0.924	1.08	0.672	0.474	0.546	0.583	0.428	0.414
$\frac{3}{8}$	5.9	1.73	1.53	0.781	0.940	1.04	0.543	0.371	0.559	0.539	0.430	0.428
$\frac{5}{16}$	5.0	1.46	1.32	0.664	0.948	1.02	0.470	0.317	0.567	0.516	0.432	0.435
$\frac{1}{4}$	4.1	1.19	1.09	0.542	0.957	0.993	0.392	0.260	0.574	0.493	0.435	0.440
$\frac{3}{16}$	3.07	0.902	0.842	0.415	0.966	0.970	0.307	0.200	0.583	0.470	0.439	0.446

Refer to American Institute for Steel Construction for more complete table.

APPENDIX VI

Some Softwood Lumber and Timber Sizes

Nominal Size (in.)	Actual Size (in.)	A Area (in.²)	I Moment of Inertia (in.⁴)	Z Section Modulus (in.³)
2 × 2	1.50 × 1.50	2.25	0.42	0.56
2 × 4	1.50 × 3.50	5.25	5.36	3.06
2 × 6	1.50 × 5.50	8.25	20.80	7.56
2 × 8	1.50 × 7.25	10.88	47.63	13.14
2 × 10	1.50 × 9.25	13.88	98.93	21.39
2 × 12	1.50 × 11.25	16.88	177.98	31.64
4 × 4	3.50 × 3.50	12.25	12.51	7.15
4 × 6	3.50 × 5.50	19.25	48.53	17.65
4 × 8	3.50 × 7.25	25.38	111.15	30.66
4 × 10	3.50 × 9.25	32.38	230.84	49.91
4 × 12	3.50 × 11.25	39.38	415.28	73.83
6 × 6	5.50 × 5.50	30.25	76.26	27.73
6 × 8	5.50 × 7.25	39.88	174.66	48.18
6 × 10	5.50 × 9.25	50.88	362.75	78.43
6 × 12	5.50 × 11.25	61.88	652.59	116.02
8 × 8	7.50 × 7.50	56.25	263.67	70.31
8 × 10	7.50 × 9.50	71.25	535.86	112.81
8 × 12	7.50 × 11.50	86.25	950.55	165.31
10 × 10	9.50 × 9.50	90.25	678.76	142.90
10 × 12	9.50 × 11.50	109.25	1 204.03	209.40
12 × 12	11.50 × 11.50	132.25	1 457.51	253.48

APPENDIX VII

Dimensions and Properties of Selected Square Structural Tubing

Nominal Size (in.)	Wall Thickness (in.)		Weight per ft (lb)	Area (in.^{2})	I (in.^{4})	Z (in.^{3})	q (in.)
10 × 10	0.6250	$\frac{5}{8}$	73.98	21.8	304	60.7	3.74
	0.5000	$\frac{1}{2}$	60.95	17.9	260	52.0	3.81
8 × 8	0.6250	$\frac{5}{8}$	56.98	16.8	142	35.5	2.91
	0.3750	$\frac{3}{8}$	36.83	10.8	102	25.4	3.06
7 × 7	0.5000	$\frac{1}{2}$	40.55	11.9	79.2	22.6	2.58
	0.3125	$\frac{5}{16}$	26.99	7.94	57.4	16.4	2.69
6 × 6	0.5000	$\frac{1}{2}$	34.48	10.1	48.6	16.2	2.19
	0.3125	$\frac{5}{16}$	23.02	6.77	35.5	11.8	2.29
5 × 5	0.5000	$\frac{1}{2}$	27.68	8.14	25.7	10.3	1.78
	0.3125	$\frac{5}{16}$	18.77	5.52	19.5	7.81	1.88
4 × 4	0.3750	$\frac{3}{8}$	16.84	4.95	10.2	5.10	1.44
	0.2500	$\frac{1}{4}$	12.02	3.54	8.00	4.00	1.50
3 × 3	0.2500	$\frac{1}{4}$	8.80	2.59	3.16	2.10	1.10
2 × 2	0.1875	$\frac{3}{16}$	4.31	1.27	0.668	0.668	0.726

Outside dimensions across flat sides.
Refer to American Institute for Steel Construction for more complete table.

APPENDIX VIII

Weight, Area, and Perimeter of Standard Individual Reinforcing Bars

| Deformed-Bar Designation | Weight (lb/ft) | Nominal Dim.-Round Sect. | | Perimeter (in.) | Maximum Outside Diameter (in.) |
		Diameter (in.)	Cross-Sectional Area (in.²)		
#3	0.376	0.375	0.11	1.178	$\frac{7}{16}$
#4	0.668	0.500	0.20	1.571	$\frac{9}{16}$
#5	1.043	0.625	0.31	1.963	$\frac{11}{16}$
#6	1.502	0.750	0.44	2.356	$\frac{7}{8}$
#7	2.044	0.875	0.60	2.749	1
#8	2.670	1.000	0.79	3.142	$1\frac{1}{8}$
#9	3.400	1.128	1.00	3.544	$1\frac{1}{4}$
#10	4.303	1.270	1.27	3.990	$1\frac{7}{16}$
#11	5.313	1.410	1.56	4.430	$1\frac{5}{8}$

Allowable Concentric Compression Load in Kips on Standard-Weight Pipe Columns of A36 Steel

Effective Length (ft)	Nominal Pipe Diameter (in.)					
	12	10	8	6	4	3
6	303	246	171	110	59	38
8	299	241	166	106	54	34
10	293	235	161	101	49	28
12	288	229	155	95	43	22
14	282	223	149	89	36	16
16	275	216	142	82	29	12
18	268	209	135	75	23	10
20	261	201	127	67	19	
22	254	193	119	59	15	
24	246	185	111	51	13	
26	238	176	102	43		
28	229	167	93	37		
30	220	158	83	32		
32	211	148	73	29		
34	201	137	65	25		
36	192	127	58	23		
38	181	115	52			
40	171	104	47			

American Institute of Steel Construction
See Appendixes IV and VII for dimensions and weights.

Allowable Concentric Compression Load in Kips on Square Columns of A36 Steel

Effective Length (ft)	Nominal Square Structural Tubing Size (in.)			
	10 × 10	8 × 8	6 × 6	4 × 4
	$\frac{1}{2}$-in. wall	$\frac{3}{8}$-in. wall	$\frac{5}{16}$-in. wall	$\frac{1}{4}$-in. wall
6	370	220	134	66
8	363	214	129	60
10	355	208	123	54
12	347	201	116	48
14	338	193	109	40
16	328	185	101	32
18	318	177	93	25
20	307	168	84	21
22	295	158	74	17
24	284	148	64	14
26	271	137	54	
28	258	126	47	
30	245	115	41	
32	231	102	36	
34	216	91	32	
36	201	81	28	
38	185	73	25	
40	168	66		

American Institute of Steel Construction
See Appendix VII for dimensions and weights.

Answers to Selected Problems

CHAPTER ONE

1.1 29.3 ft/s^2

1.2 -20 m/s^2

1.3 -0.5 m/s^2, 19 050 m/s

1.4 0.500 ft/s^2

1.5 **(a)** 35 900 kN   **(c)** 4.79 ft
(e) 33.8 T   **(g)** 806 ft
(i) 0.000 009 29 N   **(k)** 0.057 3 kg
(m) 0.031 3 lb   **(o)** 0.140 kips

1.6 **(a)** 2 580 N   **(c)** 0.615 N
(e) 0.001 610 m   **(g)** 4.83 m
(i) 25.3 m   **(k)** 5.79E5 N
(m) 8.05 N   **(o)** 1.293 km

1.7 **(a)** 0.401 lb   **(c)** 49.9 miles
(e) 0.000 483 ft   **(g)** 2 470 ft
(i) 0.043 s   **(k)** 55.7 ft
(m) 125 200 ft   **(o)** 6.14 lb

1.8 **(a)** 1.034E4   **(c)** 3.5E6
(e) 1.732E-4 **(g)** 1.432E1
(i) 1.783E2   **(k)** 1.052E2

1.9 **(a)** 700   **(c)** 14.73
(e) 0.001 023   **(g)** 280 000
(i) 0.000 005 55   **(k)** 0.689

1.10 **(a)** 5 030 m   **(c)** 0.000 452 in.
(e) 0.83 m   **(g)** 1 430 m
(i) 9 530 000 m   **(k)** 19 820 N

1.11 **(a)** 7.40E12   **(c)** 3.97E-6
(e) 4.37E2   **(g)** 4.24E1
(i) 6.43E-10

1.12 **(a)** 5.79 N   **(c)** 74.0 kN
(e) 1.610 km   **(g)** 1 208 km

(i) 9.13E-3 N   **(k)** 2 030 kips
(m) 230 T   **(o)** 4.86E3 kg

1.15 $F = 210$ lb

1.16 **(b)** $W = 242$ kg   **(d)** $W = 121$ kg

CHAPTER TWO

2.1 Four sections of drill collar are required. The tension at the drawworks is 167 800 kg.

2.2 61 500 lb

2.3 42.6 ksi

2.5 15.73 MPa

2.7 15.60 ksi

2.9 6 020 psi

2.10 **(a)** 19 970 psi   **(b)** 21 500 psi

2.12 1 278 MPa

2.15 1.213 psi

2.16 134.9 lb per rod

2.18 0.293 in.

2.20 The ultimate strength is 48 MPa.

2.21 The yield strength is 20.6 MPa.

2.24 42 percent elongation

2.26 69.2E9 Pa

2.28 1.982 cm

2.30 0.271 in.

2.32 0.1956 cm

2.34 1.636 factor of safety

2.36 Tungsten

2.38 **(a)** Tungsten   **(b)** Tungsten

2.42 106.0 kips

2.44 20 600 N

2.46 The pipe stretches 103.6 in. over its length due to its own weight.

CHAPTER THREE

3.1 3 780 psi
3.3 207 MPa
3.5 201 kips
3.6 36 000 N. Yielding most likely in I section nearest piston pin.
3.7 192.0 kips
3.9 2.37-cm square
3.11 23 in^2, 18 in., and 79 in.
3.13 2.53E6 psi, from slope of stress-strain diagram
3.15 882 kips
3.17 0.993 mm
3.19 416 kips
3.21 0.958 ksi concrete and 10.7 ksi steel
3.23 0.773 ksi concrete and 7.73 ksi steel
3.26 120.23 ft
3.27 2 cm
3.29 9 900 psi
3.31 34.4 BHN
3.33 13.11 mm^2
3.35 4 680 lb

CHAPTER FOUR

4.1 4 140 lb
4.3 Use 1.000-in.-diameter pin
4.5 2 610 N
4.7 398 MPa
4.9 78.1 ksi
4.11 Use 15.00-mm rivets
4.13 Requires one; use two bolts
4.15 141.4 MPa
4.17 15 in.
4.19 1.300E−2 in.
4.21 0.300
4.23 0.314
4.25 0.328
4.27 11.5E6 psi
4.29 7.84 N
4.31 110 kg

CHAPTER FIVE

5.1 150 lb · in.
5.3 5 540 kip · in.
5.5 1 526 lb · in.
5.7 259 N · m
5.9 6.1 kg · m clockwise
5.11 0.229 N · m clockwise
5.13 7.19 N · m clockwise

5.15 41 lb · in. counterclockwise
5.17 4.45 N · m counterclockwise
5.19 650 kN · m clockwise
5.21 147.0 kg · cm between the 7-cm and the 3-cm pulleys
5.23 34 lb · in. between the 17- and 40-lb · in. torques
5.25 514 N
5.27 127.3 MPa
5.29 3.68 kN · m
5.31 1.670 in.; use 1.750
5.33 17.46 MPa
5.34 6.25 percent reduction in strength, 25 percent reduction in weight
5.36 75 800 mm^4
5.37 1.06 MPa approximately
5.39 1.645 cm
5.41 490 MPa
5.43 0.203 MPa
5.45 0.506 in.; use $\frac{5}{8}$ in.
5.47 13 060 psi

CHAPTER SIX

6.1 (a) 112 N, 27° from horizontal, up and right
    (c) 8.5 lb, 41° from horizontal, down and right
    (e) 1 230 kips, 50° from horizontal, up and left
6.2 (b) 610 N down   (d) +91. kg on y axis
        350 N left   +57. kg on x axis
    (f) −3.8 T on x axis
        −4.6 T on y axis
6.3 (a) 224 lb, 26.6°   (c) 1.939 kips, 68.2°
        from horizontal,   from horizontal, up
        down and right   and left
    (e) 6.72 N, 61.7° from horizontal, down and left
6.4 (b) 3.42 kips right   (d) −0.866 kN left
        9.40 kips up   −0.5 kN down
    (f) 65.4 kg right
        73.9 kg down
6.6 646 kips acting up and to the left at an angle of 41.6° from the horizontal
6.8 1.853 N acting up and to the left at an angle of 86.4° from horizontal
6.10 3.13 lb acting down and to the left at an angle of 12.22°
6.12 28.7 lb down, 40.97 lb horizontally
6.14 6 810 lb down and to the left at an angle of 54.9° from horizontal
6.16 160.3 N up and to the right at an angle of 86.4° from the horizontal
6.18 919 kips cable tension
        650 kips boom compression

6.20 3 530 kN

6.22 A = 1.154E7 kg compression

B = 0.577E7 kg tension

6.23 108.5 T

6.25 $1\frac{1}{4}$-in.-diameter rod

6.27 48.3, 143.4, 716.0, and 7 160 T

6.29 22 800 lb boom compression

22 200 cable tension

6.31 1 537 kg at 49.4° at A

1 167 kg at B

6.33 22.5 kN

6.35 1 944 lb

6.37 281 lb

6.39 It will slide.

CHAPTER SEVEN

7.2 627 lb at A and 1 373 lb at B

7.4 270 kN up at B, 120 kN left and 270 kN down at A, per leg

7.6 10 000 kg down at D and 15 000 kg up at A shared by two legs each, when hook and wind load are removed

7.7 $AB = BC$ = 990 kN compression

$AD = DC$ = 779 kN tension

BD = 900 kN tension

7.9 9/16 rod

7.11 BK = 12.97 compression, BL = 9.17 tension, and JL = 9.17 tension

7.13 AB = 9.50 kips compression, AE = 13.42 tension, BE = 6.00 compression, BC = 17.50 tension, BD = 8.94 tension, DE = 25.5 compression, and EF = 37.5 compression

7.14 It is safe. The maximum stress is 208 psi.

7.15 AC = 500 kN tension, BC = 336 kN compression, and BD = 350 kN compression

7.16 AC = 150 kN compression, AD = 169.7 tension, BC = 0, and BD = 270 compression

7.19 AB = 72 kips compression, CB = 0, and CD = 72 compression

7.21 Maximum tension is in member HJ of 13.34 kN.

7.24 165 psi

7.25 FG = 0.267 MN tension, EG = 1.640 MN compression, and HF = 1.375 MN tension

7.27 BC = 253 lb compression, BD = 1 541 compression, and AD = 1 451 tension

7.28 EG = 1 169 kg tension, FH = 801 compression, and FG = 552 compression

CHAPTER EIGHT

8.1 117 500 strands per cable

8.3 575 lb

8.6 compression = 1.54, 9.6 lb tension at supports

8.8 1 688 kN

8.10 886 lb/ft^2

8.12 compression = 131.1, 220-m cable length

8.16 967 psi

8.18 6 440 psi

8.20 32 000 psi hoop, 16 000 psi girth

8.22 308 psi

8.24 0.46 in.; use $\frac{1}{2}$-in. plate

8.25 0.710 in.; use $\frac{3}{4}$-in. plate

8.27 Order special 3.70-thick plate for lower course.

CHAPTER NINE

9.1 F = 0.2 kg

9.4 A = 37.5 kips, B = 12.5 kips

9.5 A = −7.5 kg, B = 27.5 kg

9.7 A = 519 kips, B = 1 033 kips

9.9 A = 1.14 MN, B = 11.89 MN

9.11 A = 8 690 lb, B = 4 510 lb

9.13 M = 120 kip · ft, A = 10 kips

9.15 F = 121.9 gm, R = 221.9 gm left

9.17 75 kN · m at mid-span

9.19 20 N · m at mid-span

9.21 313 kip · ft, 187.5 kips at support

9.23 −180 kip · ft, 45 kips at support

9.25 84.4 kip · ft and 3.125 at D

9.27 43.8 kN · m and 14.6 kN at C, 58.4 kN · m and 0 at D

9.29 Maximums are 112.5 kip · ft at D and 22.5 kips at B.

9.31 0.867 MPa bending stress and 26 900-Pa shear stress

9.33 16.14 in.; bending governs.

9.35 383 MPa and 230 MPa

9.37 21.4 MPa

9.39 287 lb/ft; shear governs.

9.40 $\frac{1}{2}$-in. planking, 2 × 10 beams 13 ft long on 2-ft spacing, 10 × 14 central girder 20 ft long, 6 × 14 side girders, resting on six columns

CHAPTER TEN

10.1 2.394 in.^{2}, 4.21 in.

10.3 0.987 in., 1.987 in.

10.6 2.77 cm^2, 0.825 cm, and 1.842 cm

10.7 190 in.^{2}, 12.72 in. up on axis

10.10 15.20 cm^2, 4.72 cm up on y axis from bottom of plate

10.11 6.61 in.^{2}, 3.84 in. from flange

10.14 1.917E4 mm^4

10.15 Four

10.17 I_x = 8.95 in.^{4}, I_y = 1.440 in.^{4}

10.19 $I_x = 17.40$ in.^{4}, $I_y = 6.27$ in.^{4}

10.21 $I_x = 1.042$ in.^{4}, $I_y = 0.209$ in.^{4}

10.23 15 070 in.^{4}

10.24 $I_x = 0.00296$ in.^{4} and $I_y = 0.0561$ in.^{4}

10.26 $I_x = 2\,420$ mm^4 and $I_y = 1\,616$ mm^2

10.28 $M_x = 1\,080$ kip $\cdot$ in. and $M_y = 292$ kip $\cdot$ in.

10.30 Top 13.69 ksi, bottom 31.9 ksi

10.31 45.1 MPa

10.33 37.0 N $\cdot$ m

10.35 212 MPa

10.36 18 390 lb

10.39 22.5 kN

10.41 10.5 in. on centers

10.43 549 psi

CHAPTER ELEVEN

11.1 5 kips, 60 kip $\cdot$ ft

11.3 6 000 lb, 300 000 lb $\cdot$ ft

11.5 $h = \sqrt{\dfrac{6F}{Sb}(L - x)}$

11.7 $M = -\dfrac{WL^2}{3}\left(1 - \dfrac{3x}{2L} + \dfrac{x^3}{2L^3}\right)$

11.9 $V = -100x + 500$, $M = -50x^2 + 500x$

11.11 $M = \dfrac{25\,000L^2}{4\pi^2}\left(-\cos\left(\dfrac{2\pi x}{L}\right) + 1\right)$,

$M_{max} = 56.9\text{E}6$ lb $\cdot$ ft

11.13 1d, 2a, 3g, 4c, 5b, 6f, 7a, 8b, 9c, 10e, and 11g

11.14 V (916.6, -83.3, -583), zero at 9.16 in.

M (4 200, 4 170)

11.17 V (4, 3, 2, 1 kN), M (-50, -30, -15,

-5 kN $\cdot$ m)

11.19 V (-100, 0, -200, $+300$), M (0, -500, -500,

$-1\,500$)

11.21 V (396.1, 0 at 5.44 ft, -104.1)

M_{max} 626 kip $\cdot$ ft at 5.44 ft

11.23 V (150 constant to 5 cm, then reduces to 0 at end)

M_{max} $-1\,875$ kg $\cdot$ cm

11.25 V (constant -10 N from left and to mid-span, then straight line from $+35$ N to -15 N at right)

M (straight line from 0 at left to -100 N $\cdot$ cm at mid-span, then parabola from -100 to maximum of $+22.5$ at 17 cm to zero at right)

11.27 V (-500 constant from left to B, then $+1\,875$ constant to right end)

M (straight line from 0 at left to $-7\,500$ at B, then straight line back to zero)

11.29 M (straight line from 0 to 25 000 lb $\cdot$ ft at 5 ft continuing into a parabola reaching a maximum of 43 750 lb $\cdot$ ft at mid-span. Symmetrical)

CHAPTER TWELVE

12.1 1.526E4 in.

12.3 0.518 m

12.5 0.014 in.

12.7 6.74 microinches

12.9 14.98 mm

12.11 $E = 31.4\text{E}6$ psi, $\sigma_y = 72\,800$ psi

12.13 0.230 in. estimated first yield

12.15 0.013 30 in.

12.17 $-0.001\,733$ cm

CHAPTER THIRTEEN

13.1 200 lb/ft

13.2 20 050 N/cm^2

13.4 $-1\,687$ kN $\cdot$ m

13.6 The built-in beam is 50 percent stronger.

13.8 0.33 in.

13.10 $M_{max} = FL/8$

CHAPTER FOURTEEN

14.2 40.2 lb

14.5 9.29 kips

14.7 8 600 lb

14.9 8.22 in.

14.11 8 ft

14.12 10-in. standard steel pipe (Appendix IX)

14.13 6×6 with $\frac{3}{8}$-in. walls

CHAPTER FIFTEEN

15.1 32 and 0 ksi principal stress, 16 ksi maximum shear

15.3 100 and -100 ksi principal stress, 100 ksi maximum shear

15.5 279 and -129.0 ksi principal stress, 204 ksi maximum shear

15.7 185.0 and 185.0 MPa maximum principal stress, zero shear

15.9 131.1 and -81.1 MPa principal stress, 106.1 MPa maximum shear

15.11 181.4 and -181.4 MPa, 239 MPa shear

15.13 13.68 and 5.87 with 1.223 ksi shear

15.15 44.4 and 105.3 with 95.2 MPa shear

15.16 15.14, 15.17, and 15.21 are out of Tresca boundary.

15.17 8 000 psi in y or $-8\,000$ psi in x direction

15.19 Not within Tresca criteria

15.21 2 880 lb $\cdot$ in of torsion stress

15.23 277 N $\cdot$ m

Index